T0182068

Applied and Numerical Harmonic Analysis

More information about this series at http://www.springer.com/series/4968

Alexander I. Saichev · Wojbor A. Woyczyński

Distributions in the Physical and Engineering Sciences, Volume 3

Random and Anomalous Fractional Dynamics in Continuous Media

 Birkhäuser

Alexander I. Saichev
(1946–2013)
ETH Zürich
Zürich, Switzerland

Wojbor A. Woyczyński
Department of Mathematics,
  Applied Mathematics and Statistics
Case Western Reserve University
Cleveland, OH, USA

ISSN 2296-5009   ISSN 2296-5017 (electronic)
Applied and Numerical Harmonic Analysis
ISBN 978-3-030-06467-9   ISBN 978-3-319-92586-8   (eBook)
https://doi.org/10.1007/978-3-319-92586-8

Mathematics Subject Classification (2010): 60E05, 60G22, 60K37, 46F10

Printed on acid-free paper

This book is published under the imprint Birkhäuser, www.birkhauser-science.com by the registered
company Springer Nature Switzerland AG part of Springer Nature
The registered company address is: Gewerbestrasse 11, 6330 Cham, Switzerland

To TANYA and LIZ—
with love and respect

ANHA Series Preface

The *Applied and Numerical Harmonic Analysis (ANHA)* book series aims to provide the engineering, mathematical, and scientific communities with significant developments in harmonic analysis, ranging from abstract harmonic analysis to basic applications. The title of the series reflects the importance of applications and numerical implementation, but richness and relevance of applications and implementation depend fundamentally on the structure and depth of theoretical underpinnings. Thus, from our point of view, the interleaving of theory and applications and their creative symbiotic evolution is axiomatic.

Harmonic analysis is a wellspring of ideas and applicability that has flourished, developed, and deepened over time within many disciplines and by means of creative cross-fertilization with diverse areas. The intricate and fundamental relationship between harmonic analysis and fields such as signal processing, partial differential equations (PDEs), and image processing is reflected in our state-of-the-art *ANHA* series.

Our vision of modern harmonic analysis includes mathematical areas such as wavelet theory, Banach algebras, classical Fourier analysis, time–frequency analysis, and fractal geometry, as well as the diverse topics that impinge on them.

For example, wavelet theory can be considered an appropriate tool to deal with some basic problems in digital signal processing, speech and image processing, geophysics, pattern recognition, biomedical engineering, and turbulence. These areas implement the latest technology from sampling methods on surfaces to fast algorithms and computer vision methods. The underlying mathematics of wavelet theory depends not only on classical Fourier analysis, but also on ideas from abstract harmonic analysis, including von Neumann algebras and the affine group. This leads to a study of the Heisenberg group and its relationship to Gabor systems, and of the metaplectic group for a meaningful interaction of signal decomposition methods. The unifying influence of wavelet theory in the aforementioned topics illustrates the justification for providing a means for centralizing and disseminating information from the broader, but still focused, area of harmonic analyzis. This will be a key role of *ANHA*. We intend to publish with the scope and interaction that such a host of issues demands.

Along with our commitment to publish mathematically significant works at the frontiers of harmonic analysis, we have a comparably strong commitment to publish major advances in the following applicable topics in which harmonic analysis plays a substantial role:

Antenna theory	*Prediction theory*
Biomedical signal processing	*Radar applications*
Digital signal processing	*Sampling theory*
Fast algorithms	*Spectral estimation*
Gabor theory and applications	*Speech processing*
Image processing	*Time–frequency and time-scale*
Numerical partial differential	*analysis*
equations	*Wavelet theory*

The above point of view for the *ANHA* book series is inspired by the history of Fourier analysis itself, whose tentacles reach into so many fields.

In the last two centuries, Fourier analysis has had a major impact on the development of mathematics, on the understanding of many engineering and scientific phenomena, and on the solution of some of the most important problems in mathematics and the sciences. Historically, Fourier series were developed in the analysis of some of the classical PDEs of mathematical physics; these series were used to solve such equations. In order to understand Fourier series and the kinds of solutions they could represent, some of the most basic notions of analysis were defined, e.g., the concept of "function". Since the coefficients of Fourier series are integrals, it is no surprise that Riemann integrals were conceived to deal with uniqueness properties of trigonometric series. Cantor's set theory was also developed because of such uniqueness questions.

A basic problem in Fourier analysis is to show how complicated phenomena, such as sound waves, can be described in terms of elementary harmonics. There are two aspects of this problem: first, to find, or even define properly, the harmonics or spectrum of a given phenomenon, e.g., the spectroscopy problem in optics; second, to determine which phenomena can be constructed from given classes of harmonics, as done, for example, by the mechanical synthesisers in tidal analysis.

Fourier analysis is also the natural setting for many other problems in engineering, mathematics, and the sciences. For example, Wiener's Tauberian theorem in Fourier analysis not only characterises the behaviour of the prime numbers, but also provides the proper notion of spectrum for phenomena such as white light; this latter process leads to the Fourier analysis associated with correlation functions in filtering and prediction problems, and these problems, in turn, deal naturally with Hardy spaces in the theory of complex variables.

Nowadays, some of the theory of PDEs has given way to the study of Fourier integral operators. Problems in antenna theory are studied in terms of unimodular trigonometric polynomials. Applications of Fourier analysis abound in signal processing, whether with the fast Fourier transform (FFT), or filter design, or the adaptive modelling inherent in time–frequency-scale methods such as wavelet theory. The coherent states of mathematical physics are translated and modulated Fourier transforms, and these are used, in conjunction with the uncertainty principle, for dealing with signal reconstruction in communications theory. We are back to the raison d'être of the *ANHA* series!

University of Maryland   John J. Benedetto
College Park, USA   Series Editor

Contents

VI ANOMALOUS FRACTIONAL DYNAMICS   241

Introduction to Volume 3

This book continues our multivolume project which endeavours to show how the theory of distributions, often also called the theory of generalized functions, can be used by a theoretical researcher and graduate student working in the physical/engineering sciences and applied mathematics. But it is also accessible to the population of advanced undergraduate students. Our general goals, the intended audience, and the philosophy we are pursuing here are described in detail in the Introduction to Volume 1, which covers the distributional and fractal (fractional) calculus, the integral transform, and wavelets.

Whereas in the first volume we delved into some foundational topics, and the second volume concentrated on linear and nonlinear dynamics in continuous media, this book concentrates on distributional tools in the theory of generalized stochastic processes and fields, and anomalous fractional random dynamics.

Prerequisites include a typical science or engineering 3–4 semester calculus sequence (including ordinary differential equations, Fourier series, complex variables, and linear algebra) although we review the basic definitions, and facts, especially in probability theory, as needed. Familiarity with the material of Volumes 1 and 2 would help the reader, but is not absolutely necessary for a successful mastering of the material in this book. To make Volume 3 as much self-standing as possible, we have included all the basic facts and definitions concerning the Dirac delta and other distributions in the Appendix following the approach taken in Volume 2.

The book can form a basis of a special one/two semester course on generalized and anomalous stochastic processes. Typically, a course based on this text would be taught in a Mathematics/Applied Mathematics Department or in the Department of Physics. However, in many schools, some other departments (such as Electrical, Systems, Mechanical, and Chemical Engineering) could assume responsibility for it.

The material included in the book is split into two parts:

Part V. *Random dynamics and generalized stochastic processes and fields*, includes chapters on probability theory, random distributions and the white noise, dynamical and statistical characteristics of random fields and waves, the theorey of Burgers turbulence (expanding on the material discussed in

Volume 2), and passive tracer transport in Burgers flows and general randomly moving media.

Part VI. *Anomalous fractional dynamics*, contains an exposition of anomalous fractional diffusions, both linear, nonlinear, as well as multiscale.

As was the case in Volumes 1, and 2, the needs of the applied sciences audience are addressed by a careful and rich selection of examples and problems arising in real-life situations. They form the essential background for our discussions as we proceed through the material. Numerous illustrations help better understanding of the core concepts discussed in the text.

The list of notations is provided following this introduction. To reduce clutter, the formulas are numbered separately in each section (and so are bibliographical references), but, outside the section in which they appear, they are referred to by three numbers. For example, formula (4) in Section 3 of Chapter 9 will be referred to as formula (9.3.4) outside Section 9.3. Sections and chapters can be easily located via the running heads. Bibliographical citations are also numbered separately in each chapter and the references are collected, chapter by chapter, at the end of this volume.

Acknowledgments:

The authors would like to thank David Gurarie of the Mathematics Department, Case Western Reserve University, Cleveland, Ohio; and Valery I. Klyatskin of the Institute for Atmospheric Physics, Russian Academy of Sciences, Moscow, Russia, for their contribution to some of the research discussed in the book. Finally, the helpful anonymous referees' reports are also acknowledged. We are also grateful to the Birkhäuser–Springer editors, Benjamin Levitt, and Christopher Tominich, for encouragement and help in producing the final copy.

The Authors:

Alexander I. SAICHEV received his B.S. in the Radio Physics Faculty at Gorky State University, Gorky, Russia, in 1969, his Ph.D. from the same faculty in 1975 for a thesis on *Kinetic equations of nonlinear random waves*; and his D.Sc. from the Gorky Radiophysical Research Institute in 1983 for a thesis on *Propagation and backscattering of waves in nonlinear and random media*. Since 1980 he has held a number of faculty positions at Gorky State University (now Nizhniy Novgorod University) including the Senior Lecturer in statistical radio physics, Professor of mathematics, and chairman of the mathematics department. Since 1990 he has visited a number of universities in the West including the Case Western Reserve University, University of Minnesota, New York University, and University of California, Los Angeles. He is a co-author of a monograph *Nonlinear Random Waves and Turbulence in Nondispersive Media: Waves, Rays and Particles* and served on editorial boards of *Waves in Random Media* and *Radiophysics and Quantum Electronics*. His research interests included mathematical physics, applied mathematics, waves in random media, nonlinear random waves and the theory of turbulence. In 1997, he was awarded Russian Federation's State Prize and

Gold Medal for research in the area of nonlinear and random fields. Until 2013, when he unexpectedly passed away, he was Professor of Mathematics at the Radio Physics Faculty of the Nizhniy Novgorod University and Professor in the Department of Management, Technology and Economics at the Swiss Federal Institute of Technology (ETH) in Zurich, Switzerland.

Wojbor A. WOYCZYŃSKI received his B.S./M.Sc. in Electrical and Computer Engineering from Wroclaw Polytechnic in 1966, and a Ph. D. in Mathematics in 1968 from Wroclaw University, Poland. He has moved to the U.S. in 1970, and since 1982, he has been Professor of Mathematics and Statistics at Case Western Reserve University in Cleveland, where he served as chairman of the department from 1982 to 1991, and from 2001 to 2002. Before, he has held tenured faculty positions at Wroclaw University, Poland, and at Cleveland State University, and visiting appointments at Carnegie-Mellon University, and Northwestern University. He has also given invited lecture series on short-term research visits at University of North Carolina, University of South Carolina, University of Paris, Gottingen University, Aarhus University, Nagoya University, University of Tokyo, University of Minnesota, the National University of Taiwan in Taipei, and the University of New South Wales in Sydney. He is also (co-)author and/or editor of fourteen books on probability theory, harmonic and functional analysis, and applied mathematics, and currently serves as a member of the editorial board of the *Applicationes Mathematicae*, Springer monograph series UTX, and as a managing editor of the journal *Probability and Mathematical Statistics*. His research interests include probability theory, stochastic models, functional analysis and partial differential equations and their applications in statistics, statistical physics, surface chemistry, hydrodynamics and biomedicine. He is currently Professor of Mathematics, Applied Mathematics and Statistics, and Director of the Case Center for Stochastic and Chaotic Processes in Science and Technology at Case Western Reserve University, Cleveland, Ohio, USA.

Cleveland, March 2018

Notation

$$\lceil \alpha \rceil \quad - \quad \text{least integer greater than or equal to } \alpha$$

$$\lfloor \alpha \rfloor \quad - \quad \text{greatest integer less than or equal to } \alpha$$

$$C \quad - \quad \text{concentration}$$

$$\mathbf{C} \quad - \quad \text{complex numbers}$$

$$C(x) \quad = \quad \int_0^x \cos(\pi t^2/2)\, dt, \text{ Fresnel integral}$$

$$C^\infty \quad - \quad \text{space of smooth (infinitely differentiable) functions}$$

$$\mathcal{D} \quad = \quad C_0^\infty, \text{ space of smooth functions with compact support}$$

$$\mathcal{D}' \quad - \quad \text{dual space to } \mathcal{D}, \text{ space of distributions}$$

$$\bar{D} \quad - \quad \text{the closure of domain } D$$

$$D/Dt \quad = \quad \partial/\partial t + \boldsymbol{v} \cdot \boldsymbol{\nabla}, \text{ substantial derivative}$$

$$\delta(x) \quad - \quad \text{Dirac delta centered at } 0$$

$$\delta(x - a) \quad - \quad \text{Dirac delta centered at } a$$

$$\Delta \quad - \quad \text{Laplace operator}$$

$$\mathcal{E} \quad = \quad C^\infty\text{-space of smooth functions}$$

$$\mathcal{E}' \quad - \quad \text{dual to } \mathcal{E}, \text{ space of distributions with compact support}$$

$$\text{erf}(x) \quad = \quad (2/\sqrt{\pi}) \int_0^x \exp(-s^2)\, ds, \text{ the error function}$$

$$\tilde{f}(\omega) \quad - \quad \text{Fourier transform of } f(t)$$

$$\{f(x)\} \quad - \quad \text{smooth part of function } f$$

$$\lfloor f(x) \rceil \quad - \quad \text{jump of function } f \text{ at } x$$

$$\phi, \psi \quad - \quad \text{test functions}$$

$$\gamma(x) \quad - \quad \text{canonical Gaussian density}$$

$$\gamma_\epsilon(x) \quad - \quad \text{Gaussian density with variance } \epsilon$$

$$\Gamma(s) \quad = \quad \int_0^\infty e^{-t} t^{s-1} dt, \text{ gamma function}$$

$$(h, g) \quad = \quad \int h(x)g(x)dx, \text{ the Hilbert space inner product}$$

$$\chi(x) \quad - \quad \text{canonical Heaviside function, unit step function}$$

$$\hat{H} \quad - \quad \text{the Hilbert transform operator}$$

$$j, J \quad - \quad \text{Jacobians}$$

$$I_A(x) \quad - \quad \text{the indicator function of set } A \ (=1 \text{ on } A, =0 \text{ off } A)$$

$$\text{Im } z \quad - \quad \text{the imaginary part of } z$$

$$\lambda_\epsilon(x) \quad = \quad \pi^{-1}\epsilon(x^2 + \epsilon^2)^{-1}, \text{ Cauchy density}$$

$$L^p(A) \quad - \quad \text{Lebesgue space of functions } f \text{ with } \int_A |f(x)|^p\, dx < \infty$$

$$\mathbf{N} \quad - \quad \text{nonnegative integers}$$

$$\phi = O(\psi) \quad - \quad \phi \text{ is of the order not greater than } \psi$$

$$\phi = o(\psi) \quad - \quad \phi \text{ is of the order smaller than } \psi$$

$\mathcal{PV}$	—	principal value of the integral
$\mathbf{R}$	—	real numbers
$\mathbf{R}^d$	—	d-dimensional Euclidean space
Re z	—	the real part of z
ρ	—	density
sign (x)	$=$	1 if $x > 0$, -1 if $x < 0$, and 0 if $x = 0$
sinc ω	$=$	$\sin \pi\omega / \pi\omega$
$\mathcal{S}$	—	space of rapidly decreasing smooth functions
$\mathcal{S}'$	—	dual to $\mathcal{S}$, space of tempered distributions
$S(x)$	$=$	$\int_0^x \sin(\pi t^2 / 2)\, dt$, Fresnel sine integral
T, S	—	distributions
$T[\phi]$	—	action of T on test function ϕ
T_f	—	distribution generated by function f
$\tilde{T}$	—	generalized Fourier transform of T
z^*	—	complex conjugate of number z
$\mathbf{Z}$	—	integers
∇	—	gradient operator
$\mapsto$	—	Fourier map
$\rightarrow$	—	converges to
$\Rightarrow$	—	uniformly converges to
$*$	—	convolution
$[\![\,.\,]\!]$	—	physical dimensionality of a quantity
$\emptyset$	—	empty set
$\blacksquare$	—	end of proof, example
(x, y)	—	inner product (also called scalar, or dot product) of x, and y
$|\boldsymbol{x}|$	$=$	$\sqrt{(\boldsymbol{x}, \boldsymbol{x})}$, the norm of vector $\boldsymbol{x}$
$\int$	—	the integral over the whole space

Part V

Random Dynamics

Chapter 16

Basic Distributional Tools for Probability Theory

The theory of distributions (generalized functions) is an effective tool in probability theory and the theory of random (stochastic) processes and fields. The aim of this chapter is to introduce the basic definitions and ideas of probability theory, and show how the simple distribution-theoretic concepts of the Dirac delta, its derivatives, and other related distributions can constructively be used to streamline statistical calculations. This approach is seldom used in the standard probability and statistics textbooks but is quite popular and effective in the physical and engineering sciences. In what follows we deliberately shall not resort to the rigorous mathematical measure-theoretic probability theory based on Kolmogorov's axioms; such an approach is beyond the scope of this textbook.[1] In order to develop the distributional tools discussed below, an intuitive "physical" idea of an ensemble of realizations of random quantities, processes, fields, and the concept of statistical averaging over the ensemble of realizations are sufficient.

No prior background in probability theory is needed to study this chapter but the reader familiar with the classical approach will find it easier to immediately see the advantages of the approach taken below.

16.1 Basic concepts

16.1.1. Statistical trials and random events. Until probability theory has been developed, the goal of scientific investigations had been a study of well-determined outcomes given well-determined circumstances. For example, if the initial position and velocity of a bullet were known, the

---

[1]There is, of course, a huge literature on the mathematical probability theory. We just quote here a few of the classics: Feller [1], Loève [2], Billingsley [3], and Kallenberg [4].

© Springer International Publishing AG, part of Springer Nature 2018
A. I. Saichev and W. A. Woyczynski, *Distributions in the Physical
and Engineering Sciences, Volume 3*, Applied and Numerical
Harmonic Analysis, https://doi.org/10.1007/978-3-319-92586-8_16

classical Newtonian mechanics would precisely determine its trajectory. This *deterministic* approach, however, was not always sufficient to describe the real-world situations. If you are a hunter stalking your prey in the woods, you know very well that the bullet's trajectory is also influenced by largely unpredictable local wind gusts.

As long as the environmental fluctuations weakly influence the accuracy of the shot (say, if the bullet's deviation from the perfect Newtonian path is much smaller than the size of the target) then the classical deterministic model of bullet's flight is justified and sufficient. Otherwise, other quantitative methods must be employed. Similar lack of predictability of "experimental" outcomes can be observed in a wide variety of physical phenomena, from quantum mechanics of particles, to turbulent flows of fluids, to the evolution of the large-scale structure of the Universe. The proverbial coin toss or a roll of a die provides other, well exploited in the classroom environment, examples.

Phenomena of this type, where no unique outcome under given conditions is expected, are called *random* or *stochastic*. Random events are often caused by uncontrolled factors,[2] but to study them we would often (but not always) assume that they result from experimenting in "well-controlled" circumstances. Think here about repeated shots taken with a rigidly mounted rifle using bullets of the same type, or repeated tosses of the same fair coin. Each such experiment is often called a *statistical trial*, and its outcome—a *random event*. In what follows we will denote outcomes of random trials by capital letters A, B, C, *etc.* Numerical outcomes associated with random events will be called *random variables* and denoted by capital letters X, Y, Z, *etc.* Think here about the wind velocity or, in the rifle shot experiment, the distance of the bullet from target's center upon impact. In the latter case, a random event could be defined as the bullet hitting any part of the target.

16.1.2. Statistical stability. Before we introduce the key concept of the *probability* of a random event, an important observation must be made: Not all random phenomena encountered in the physical and engineering sciences can be described in probabilistic terms. We will explain what we mean here by recalling the basic principle of any classical physical theory. The latter can only study *deterministically stable* events which can be investigated in repeated experiments under identical conditions. At the first sight, random phenomena do not seem to be fulfilling the stability principle. Indeed, the values of water velocity on the ocean surface at a given point will differ from measurement to measurement. Another example of such a situation is radioactive decay, in which an atom of the radioactive element splits into lighter atoms. Watching different atoms during a fixed time interval it is impossible to predict which ones are going to split.

---

[2]For a more comprehensive discussion of different ways randomness can arise and be formally analyzed, see, M. Denker and W.A. Woyczyński [5], *Introductory Statistics and Random Phenomena: Uncertainty, Complexity and Chaotic Behavior in Engineering and Science*, Birkhauser-Boston 1998.

However, most of the observed random phenomena display a remarkable property of *statistical stability* which we will explain using the following
example.

Example 1. Radioactive decay. Let us denote by A the event of radioactive
decay of an atom, and define a random variable X as follows: X is equal to
1 if the decay occurs, and 0 if it does not. We will call X the *indicator of
the event A*. Although it is impossible to predict exactly when each atom
is going to split, we can watch the decay of an ensemble of a large number
$N \gg 1$ of atoms, consider the corresponding indicator random variables
$X_1, X_2, \ldots, X_N$, and calculate the arithmetic average

$$\bar{X}_N = \frac{1}{N} \sum_{k=1}^{N} X_k, \qquad (1)$$

Then we can observe a different ensemble of N atoms, calculate the analogous average $\bar{X}'_N$, and discover that, the bigger the number N of trials is, the
smaller the difference between $\bar{X}_N$, and $\bar{X}'_N$, is going to be. This phenomenon
justifies the commonly used in atomic physics concept of *half-life* of a radioactive element during which half of all the atoms in a given macroscopic sample
split.

The above-described phenomenon of the vanishing difference between
sample averages obtained for different long series of trials is called *statistical
stability*. It is this stability that permits construction, for a given random
phenomenon, of a theory that will predict, under similar circumstances, the
average outcomes of long series of repeated experiments; the individual outcomes are then *not* the object of the theory. Consequently, like any other
physical theory, probability theory so constructed operates with quantities
that are well-determined by physical experiments.

Remark 1. When is "random," not random.   As we mentioned before
not all phenomena colloquially understood as "random" can be studied via
probability theory. A "random" encounter with a friend in a foreign city, or
a sudden and unpredictable drop of an apple from an apple tree that helped
spark creation of Newtonian mechanics, do not qualify as proper objects of
study for probability theory, and since Adam and Eve no other scientists
we know of tried to repeat their "random" apple tree experiment with any
success.

So, to emphatically restate our basic principle: *probability theory studies
only statistically stable events*. For a concrete phenomenon and associated
with it event, say A, experimenters convince themselves about presence of
this property via empirical tools relying on the concept of *relative frequency*
of the occurrence of the event A. Note that the right-hand side of the equality
(1) is equal to the ratio

$$\mathbf{P}^*(A; N) := \frac{N_A}{N}, \qquad (2)$$

where N is the total number of trials, and N_A is the number of trials that resulted in the event A. The above quantity, $\mathbf{P}^*(A; N)$, is called the *relative frequency* of occurrence of the event A in N trials. For different series of N trials, $\mathbf{P}^*$ can have different values. However, by definition, if the event is statistically stable, the relative frequency converges, as N increases, to a deterministic value

$$\mathbf{P}(A) := \lim_{N \to \infty} \mathbf{P}^*(A; N), \tag{3}$$

which is called the *probability* of event A. Equality (3) gives a *statistical (frequentist) definition* of probability. Notice that since, obviously,

$$0 \leq N_A \leq N,$$

the probability of any event satisfies the following inequalities

$$0 \leq \mathbf{P}(A) \leq 1. \tag{4}$$

Remark 2. Almost sure and almost impossible events. If the probability of an event A is equal to 1, then A is called an *almost sure* event. If the probability is 0, then the event is called *almost impossible*. Note that an almost impossible event can occur occasionally, and that an almost sure event does not always occur. Convince yourself about the validity of these statements experimentally by trying to hit the dead center of a target with a dart.

Remark 3. Infinite number of trials in practice. Strictly speaking, validity of the statistical definition of probability cannot be tested experimentally as nobody can conduct an infinite number of trials and verify the existence of the limit. Therefore, for a physical scientist and an engineer it is necessary (and convenient) to view statistical stability as the law of nature which requires no proof, just an approximate experimental verification. For a mathematician, statistical stability requires a proof based on well-designed fundamental axioms of the abstract axiomatic mathematical probability theory, but it is not the tack we are taking in this book.

Example 2. Coin flip: simulation vs. physics. To illustrate the key concept of statistical stability let us take a look at the classical example of the heads-or-tails game. Assume that a fair coin is being tossed and denote by A the event of heads coming up.

First let us conduct our experiment by using a computer random number generator which produces a random string of 1's and 0's.[3] Two experiments of 4,000 simulated "tosses" have been conducted. Two plots of the evolution of the relative frequencies of 1's, as functions of the number of tosses n, are shown in Fig. 16.1.1. The tendency of both plots to approach the probability of heads $\mathbf{P}(A) = 1/2$, for a fair coin, is clearly visible.

---

[3]For a more detailed discussion of random number generators and what is meant by their "randomness" see, e.g., the above-quoted textbook by Denker and Woyczyński.

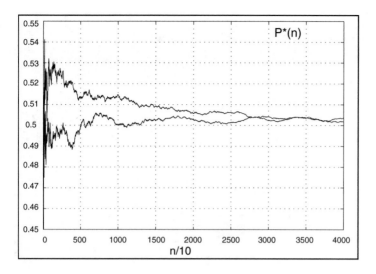

FIGURE 16.1.1.
Plots of the relative frequencies of 1's as functions of the length
n of two substrings of 4,000 1's and 0's produced by a random
number generator simulating the heads-or-tails game. The tendency of both plots to approach the probability of heads P=0.5, is
clearly visible.

But, of course, the coin toss can be considered as a physical phenomenon
governed by elementary mechanical principles. For the sake of simplicity
assume the coin is of radius 1, and in its initial position is flat with its center
and edges at height $h(0) = 1$ over the (soft, sandy) ground which prevents the
coin from bouncing after touching the ground. Imparting the initial velocity
v, and angular velocity ω, gives the following description of the height of the
edges at time t:

$$h(t) = 1 + vt - \frac{1}{2}gt^2 \pm \cos(\omega t)$$

with $g = 9.81 \, \text{m/sec}^2$ being the gravitational constant. At the time t_0, when
the coin touches the ground (and softly settles on it) either the heads side or
the tail side faces up and this determines the outcome of the flip. The result
is completely determined by the values of v and ω, and in the phase space
$\{(v, \omega), v > 0, \omega > 0\}$ the "heads-up" and "tails-up" regions are separated by
the situations when the coin lands vertically on its edge, that is

$$\omega t_0 = 2\pi k \pm \frac{\pi}{2}, \tag{5}$$

for some positive integer k, with $\pm \cos(\omega t) = 0$ and $h(t_0) = 1$. This implies
that

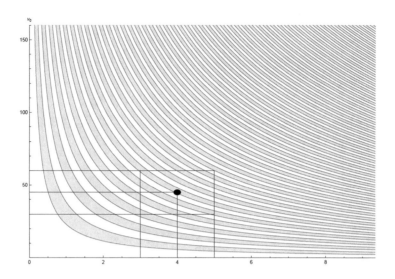

FIGURE 16.1.2.
In the physical coin flip experiment the "heads-up" and "tails-up"
regions in the $\{(v,\omega), v > 0, \omega > 0\}$ phase plane are separated by
hyperbolic curves.

$$vt_0 = \frac{1}{2}gt_0^2,$$

so that $t_0 = 2v/g$. Inserting this value into (5) yields, for $k = 1, 2, \ldots$, the
family of equations,

$$\omega v = \frac{g}{2}\left(2\pi k \pm \frac{\pi}{2}\right), \tag{6}$$

of hyperbolic curves separating different regions in the $\{(v,\omega), v > 0,$
$\omega > 0\}$ phase plane corresponding to "heads-up" (gray) and "tails-up" (white)
outcomes. They are pictured in Figure 16.1.2, above.[4]

Of course, in real life, the person flipping the coin does not have perfect
control over v, and ω, so that if, say, his/her range of values is $v = 45 \pm 10$,
and $\omega = 4 \pm 1$, like indicated in Fig. 16.1.2, then all possible outcomes are
included in the pictured rectangle, and one could "calculate" the likelihood of
the outcome being "heads-up" by computing the ratio of the gray area to the
total area of the rectangle, and see how close it is to $1/2$. A more sophisticated
approach would be to consider an arbitrary smooth test function $\phi(v,\omega)$,
$v > 0, \omega > 0$, such that

$$\int_0^\infty \int_0^\infty \phi(v,\omega)\, dv\, d\omega = 1$$

so that we can think about it as a probability density function of the distribution of all possible values of (v,ω), and see what happens with the tested

---

[4]The picture was produced by my (WAW) student, Andrew Freedman, as part of a
project in one of my stochastic processes classes.

"average" when our vertical and angular velocities increase. More precisely, we want to calculate the limit,

$$p = \lim_{A \to \infty} \int \int_{H} \phi \left(v - A \cos \alpha, \omega - A \sin \alpha \right) dv \, d\omega,$$

where H is the gray area between the hyperbolic curves, and the vector $(A \cos \alpha, A \sin \alpha)$ indicates the magnitude and direction of our "escape" to infinity in the (v, ω)-plane. Taking into account the geometry of the domain H we have, after the change-of-variable $\omega' = \omega - A \sin \alpha$, that

$$p = \lim_{A \to \infty} \int_{A \cos \alpha}^{\infty} \sum_{n=0}^{\infty} \left(\int_{(2n-1/2)\pi g/(2v)}^{(2n+1/2)\pi g/(2v)} \phi \left(v - A \cos \alpha, \omega - A \sin \alpha \right) d\omega \right) dv$$

$$= \lim_{A \to \infty} \int_{A \cos \alpha}^{\infty} \sum_{n=0}^{\infty} \left(\int_{(2n-1/2)\pi g/(2v) - A \sin \alpha}^{(2n+1/2)\pi g/(2v) - A \sin \alpha} \phi \left(v - A \cos \alpha, \omega' \right) d\omega' \right) dv.$$

Asymptotically, in view of the smoothness of ϕ, we can write

$$p = \lim_{A \to \infty} \int_{A \cos \alpha}^{\infty} \sum_{n=0}^{\infty} \phi \left(v - A \cos \alpha, \frac{n \pi g}{v} \right) \frac{\pi g}{2v} (1 + o(1)) dv,$$

which, after changing variables again to $v' = v - A \cos \alpha$, gives

$$p = \frac{1}{2} \int_{0}^{\infty} \int_{0}^{\infty} \phi(v', \omega) \, dv' d\omega \left(1 + o(1) \right) = \frac{1}{2}.$$

Thus the physics says that increasing the vertical velocity and the rate of rotation gets us close and closer to the perfect fifty-fifty outcome of the coin flip.[5]

16.1.3. Random events and probabilities. In this subsection we will introduce certain well-known concepts that are useful in formal computations related to random events and their probabilities, and employ what is known in mathematics as *set-theoretic* notation. Given two random events, A and B, their *union*

$$C = A \cup B.$$

is defined as the event that A or B, or both A and B, occurred. The *intersection*

$$D = A \cap B$$

of events A and B describes the event when both A, and B, occur. Events A and B are said to be *disjoint* if they cannot occur simultaneously, that is, if their intersection is empty, $A \cap B = \emptyset$.

---

[5]This example is discussed in greater detail in Keller [6], and Guttorp [7]. An interesting experiment by Persi Diaconis, to establish the likely practical values of v, and ω, is also described in [6].

Probability of the union of disjoint events A, and B, is equal to the sum of their probabilities, that is,

$$\text{if} \quad A \cap B = \emptyset, \quad \text{then} \quad \mathbf{P}(A \cup B) = \mathbf{P}(A) + \mathbf{P}(B). \quad (7)$$

Indeed, thinking in "frequentist" terms, if the events A and B are disjoint then the related frequencies,

$$N_{A \cup B} = N_A + N_B,$$

and the substitution of this equality into (2), and then into (3), gives the additive formula (7). Similarly, one establishes the fact that if A^c is the event complementary to A, that is, the event that A *did not* occur, then

$$\mathbf{P}(A) + \mathbf{P}(A^c) = 1. \quad (8)$$

If A and B are not mutually exclusive (disjoint) then,

$$\mathbf{P}(A \cup B) = \mathbf{P}(A \cap B^c) + \mathbf{P}(A^c \cap B) + \mathbf{P}(A \cap B)$$

$$= \mathbf{P}(A) + \mathbf{P}(B) - \mathbf{P}(A \cap B),$$

because the events $A \cap B^c$, $A^c \cap B$, and $A \cap B$, are mutually exclusive (pairwise disjoint).

Considering different random events A, and B, we are often interested in quantifying the answer to the following question: How does occurrence of B influence the probability of A happen? For example, if you are into fishing, you may want to know how the fact that the rain is likely tomorrow (event B) will affect the probability of your catching a fish (event A). This is, of course, but a light-hearted example, but more serious situations of this type abound. The question, "Given that one engine of the twin-engine plane fails, what is the probability that the plane crashes on landing?", certainly sounds more compelling.

The answer to these types of questions is given by the concept of *conditional probability*, $\mathbf{P}(A|B)$, of event A given that event B occurred. We will define conditional probability employing the concept of statistical stability, see (3). Assume that in N trials we are tracking occurrence of the two events, A and B. Event A occurred N_A times, event B, N_B times, and both events occurred simultaneously $N_{A \cap B}$ times. Then the relative frequency of the intersection $A \cap B$ of the two events can be written in the form.

$$\mathbf{P}^*(A \cap B; N) = \frac{N_{A \cap B}}{N} = \frac{N_B}{N} \times \frac{N_{A \cap B}}{N_B}. \quad (9)$$

In view of the statistical stability principle, each of the first two fractions above converges, as $N \to \infty$, to the corresponding deterministic probability of events B, and $A \cap B$, respectively, that is

$$\lim_{N \to \infty} \frac{N_{A \cap B}}{N} = \mathbf{P}(A \cap B), \quad \text{and} \quad \lim_{N \to \infty} \frac{N_B}{N} = \mathbf{P}(B).$$

Consequently, the third fraction in (9) also converges to the quantity that is traditionally called the *conditional probability of event A, given that event B occurred* (or, shortly, *probability of A, given B*), and denoted

$$\mathbf{P}(A|B) = \lim_{N \to \infty} \frac{N_{A \cap B}}{N_B}. \tag{10}$$

Taking $N \to \infty$, in (9) we obtain the identity,

$$\mathbf{P}(A \cap B) = \mathbf{P}(A|B) \cdot \mathbf{P}(B). \tag{11}$$

In other words, the probability of the intersection of two events is equal to the product of the probability of one of them, and the conditional probability of the second provided that the first has taken place.

The concept of conditional probability leads us directly to the key notion of *statistical independence* of two random events. It is natural to claim that event A is independent of event B if the conditional probability of A, given B, is independent of the condition B, i.e., if

$$\mathbf{P}(A|B) = \mathbf{P}(A).$$

But substituting this demand into identity (11) we obtain the equality

$$\mathbf{P}(A \cap B) = \mathbf{P}(A) \cdot \mathbf{P}(B), \tag{12}$$

which depends symmetrically on events A and B, and which will be taken as a formal definition of *statistical independence* of events A and B.

We conclude this section by providing an important in many applications extension of formula (11). Consider events $\{H_1, H_2, \ldots, H_n\}$ which are pairwise disjoint, that is,

$$H_k \cap H_l = \emptyset, \qquad \text{for} \qquad k, l = 1, 2, \ldots, n, \quad k \neq l,$$

and such that their union exhausts all possible outcomes. Then an argument similar to that used to derive equation (11) gives the identity

$$\mathbf{P}(A) = \sum_{i=1}^{n} \mathbf{P}(A|H_i) \cdot \mathbf{P}(H_i), \tag{13}$$

which is called the *total probability formula*. Its corollary is the celebrated *Bayes formula for reverse conditional probabilities*:

$$\mathbf{P}(H_i|A) = \frac{\mathbf{P}(A|H_i) \cdot \mathbf{P}(H_i)}{\sum_{i=1}^{n} \mathbf{P}(A|H_i) \cdot \mathbf{P}(H_i).} \tag{14}$$

Indeed,

$$\mathbf{P}(H_i|A) = \frac{\mathbf{P}(H_i \cap A)}{\mathbf{P}(A)} \cdot \frac{\mathbf{P}(H_i)}{\mathbf{P}(H_i)} = \frac{\mathbf{P}(A|H_i) \cdot \mathbf{P}(H_i)}{\mathbf{P}(A)},$$

and an application of the total probability formula immediately gives the Bayes formula.

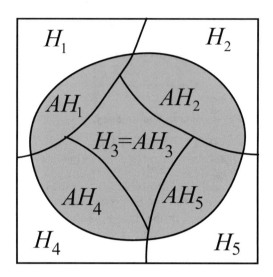

FIGURE 16.1.3.
An illustration of the total probability formula and Bayes formula.

Example 3. Transmission of a binary signal in presence of random errors.
A communication channel transmits binary symbols 0, and 1, with random
errors. The probability that the binary digits 0, and 1, appear at the input of
the channel are, respectively, 0.45 and 0.55. Because of transmission errors,
if the digit 0 appears at the input, then the probability of it being received
as 0 on the receiving end is 0.95. For the digit 1, the analogous conditional
probability is 0.9. Our task is to find the reverse conditional probability that
the digit 1 was on the input, given that 1 was received.

We will define the two random events here as follows: A, corresponds to
the digit 1 being sent, and B corresponds to the digit 1 being received. The
formulation of the problem contains the information that

$$\mathbf{P}(A^c) = 0.45, \qquad \mathbf{P}(A) = 0.55,$$

and

$$\mathbf{P}(B^c|A^c) = 0.95, \qquad \mathbf{P}(B|A) = 0.9,$$

so that

$$\mathbf{P}(B|A^c) = 0.05, \qquad \mathbf{P}(B^c|A) = 0.1.$$

We are seeking $\mathbf{P}(A|B)$, and the Bayes formula gives the answer:

$$\mathbf{P}(A|B) = \frac{\mathbf{P}(B|A) \cdot \mathbf{P}(A)}{\mathbf{P}(B|A^c) \cdot \mathbf{P}(A^c) + \mathbf{P}(B|A) \cdot \mathbf{P}(A)} = \frac{0.9 \cdot 0.55}{0.05 \cdot 0.45 + 0.9 \cdot 0.55} \approx 0.9565.$$

16.2 Random variables and probability distributions

Consider now a random phenomenon in which the measured outcomes are real numbers. These outcomes represent what traditionally are called *random variables* or *random quantities*, and denoted X, Y, etc. The probability of the random event that the random quantity X does not exceed a given real number x will be denoted

$$F_X(x) := \mathbf{P}(X \leq x).$$

Obviously, $F_X(x)$ is a function of x, and it will be called the *cumulative (probability) distribution function* (CDF, in short).[6]

Our next step is to introduce the concept of the *ensemble average* (or, *the mean*) of a random quantity X which will be denoted by the angled brackets, $\langle X \rangle$.[7] In other words, the ensemble average means statistical averaging over an ensemble of all realizations of the random quantity X.

The concept of statistical ensemble average is intuitively natural for a physically minded reader. For operational and computational purposes it is sufficient to define it by requiring that it satisfies the following four properties:

• *Positivity.* If the random quantity X is real-valued, and $X \geq 0$, then $\langle X \rangle \geq 0$;

• *Linearity.* If a, b are real numbers, and X, Y are random quantities, then

$$\langle aX + bY \rangle = a\langle X \rangle + b\langle Y \rangle;$$

• *Normalization.* If a random quantity X is identically equal to number 1, then $\langle X \rangle = 1$.

• *Continuity.* If a sequence $X_1, X_2, \ldots$, of real-valued random quantities increases monotonically to a limit random quantity X, then

$$\langle X \rangle = \lim_{n \to \infty} \langle X_n \rangle.$$

Note that in this context the CDF can be expressed via the formula,

$$F_X(x) = \mathbf{P}(X \leq x) = \langle \chi(x - X) \rangle, \tag{1}$$

---

[6]It is common usage to apply the term "distribution" to both, distributions (generalized functions) like the Dirac delta and the probability distributions. This is particularly unfortunate in a book like ours since both concepts appear simultaneously. So, we will make sure that the two are not confused by always including the adjective "probability" when discussing probability cumulative distribution.

[7]In later chapters, where we delve into more purely mathematical issues of probability theory, we will also adopt the traditional in mathematics terminology: The mean value of X will be called the *expectation* of X and denoted $\mathbf{E}X$.

where $\chi(x)$ denotes the *Heaviside function*[8], which preserves in the averaging process only those values of X for which $X \leq x$, and nullifies those for which $X > x$.

Finally, for a vector-valued random quantity, $\boldsymbol{X} = (X_1, \ldots, X_n)$,

$$\langle \boldsymbol{X} \rangle = (\langle X_1 \rangle, \ldots, \langle X_n \rangle).$$

When calculating the averages of deterministic (test) functions of a random quantity, it is often most convenient to use the derivative of the cumulative distribution function,

$$f_X(x) = F_X{}'(x),$$

called the *probability density function* (or, simply, *probability density*, or PDF) of X. This derivative, in the case of discrete random quantities which take only finitely, or countably many values will be understood as the distributional derivative. Taking into account the above defining equality (1) for $F_X(x)$, and *postulating* that the operation of statistical averaging commutes with linear operators, we get that

$$f(x) = \langle \chi'(x - X) \rangle. \tag{2}$$

Recalling, that the derivative of the Heaviside function is equal to the Dirac delta distribution, we obtain the following formula for the probability density function of a random quantity X:

$$f_X(x) = \langle \delta(x - X) \rangle. \tag{3}$$

In other words, *the probability density function (PDF) of a random quantity X is equal to the statistical average of the Dirac delta evaluated at $x-X$.* Clearly, the probability density satisfies the normalization condition,[9]

$$\int f_X(x)\, dx = 1.$$

Now, consider a random quantity Y which is a deterministic function $g(X)$ of a random quantity X. To find its mean it will be convenient to write $g(X)$ in the distributional spirit as a linear functional

$$Y = g(X) = \int \delta(x - X)g(x)\, dx. \tag{4}$$

---

[8]Recall that the Heaviside function (or, unit step function) $\chi(x)$ is defined as being equal to 0, for $x < 0$, and equal to 1, for $x \geq 0$.

[9]In this chapter we just restrict ourselves to the study of either discrete, or (absolutely) continuous distribution functions, when the formula is correct, and exclude the so-called singular distributions when the derivative of the distribution function need not integrate to 1. We will consider singular probability distributions in Part VI devoted to anomalous fractional dynamics.

Taking the statistical averages of both sides, and taking into account the postulated commutativity of the averaging operation with linear operations (here, integration, and multiplication by deterministic functions), we arrive at the formula

$$\langle Y \rangle = \langle g(X) \rangle = \int \langle \delta(x - X) \rangle g(x) \, dx, \tag{5}$$

which, in view of (3), can be written in the final, convenient for calculations, form

$$\langle Y \rangle = \langle g(X) \rangle = \int f_X(x) g(x) \, dx. \tag{6}$$

It expresses the statistical average of $g(X)$ as a deterministic integral of the product of $g(x)$, and the probability density $f_X(x)$ of the random quantity X.

Example 1. General discrete random quantity. Consider a *discrete random quantity* X taking *finitely* many values, $x_1, x_2, \ldots, x_k$, with probabilities, respectively, $p_1, p_2, \ldots p_k$, which means that for any "test" function $g(x)$,

$$\langle g(X) \rangle = g(x_1)p_1 + g(x_2)p_2 + \cdots + g(x_k)p_k.$$

To satisfy the above-mentioned demands on the ensemble averaging probabilities, the numbers $p_i, i = 1, \ldots, k$, have to be nonnegative, with $p_1 + \ldots + p_k = 1$. In the special case when $p_1 = \cdots = p_k = 1/k$, we say that the random quantity X has a *discrete uniform distribution*. The density of X is the distribution

$$f_X(x) = \langle \delta(x - X) \rangle = \delta(x - x_1)p_1 + \delta(x - x_2)p_2 + \cdots + \delta(x - x_k)p_k.$$

Example 2. Poisson probability distribution. Consider a discrete random quantity x taking *infinitely* many values $x_1, x_2, \ldots$ with probabilities $p_1, p_2, \cdots \geq 0, \sum_i p_i = 1$, which means that, for any test function $g(x)$,

$$\langle g(X) \rangle = \sum_{i=1}^{\infty} g(x_i)p_i.$$

In the special case when $\lambda > 0$, $x_i = i$, $i = 0, 1, 2, \ldots$, and

$$p_i = e^{-\lambda} \frac{\lambda^i}{i!}, \qquad i = 0, 1, 2, \ldots,$$

we say that the random quantity X has a *Poisson probability distribution*. The density of X is the distribution

$$f_X(x) = \langle \delta(x - X) \rangle = \sum_{i=0}^{\infty} \delta(x - i) e^{-\lambda} \frac{\lambda^i}{i!}.$$

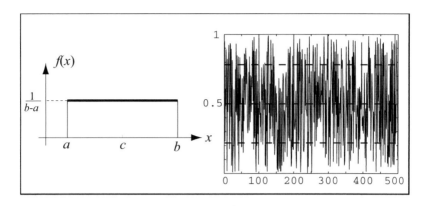

FIGURE 16.2.1.
Left: A graph of the uniform probability density function over the interval $[a, b]$; Right: A sample of 500 random values uniformly distributed on the interval [0,1].

Example 3. Continuous uniform probability distributions. In this case the random quantity X assumes only values in the interval $[a, b]$ with the PDF being constant over this interval, so that

$$f_X(x) = \frac{1}{b-a}\big(\chi(x-a) - \chi(x-b)\big).$$

Fig. 16.2.1 shows the graph of the uniform PDF on the left and a computer-simulated sample of 500 random values uniformly distributed on the interval [0,1].

Example 4. Gaussian (normal) and Cauchy-Lorentz random quantities. (Absolutely) continuous real-valued random quantities are described by densities that are regular functions which integrate to 1. Two particular examples are worth mentioning: the *Gaussian* quantities, with the densities (see, Fig. 16.2.2),

$$\varphi_{\sigma,\mu}(x) = \frac{1}{\sqrt{2\pi\sigma^2}} \exp\left(-\frac{(x-\mu)^2}{2\sigma^2}\right), \text{ with parameters } \sigma > 0, \text{ and } \mu \in \mathbf{R},$$

and the *Cauchy–Lorentz* random quantities, with the densities

$$\lambda_\varepsilon(x) = \frac{1}{\pi}\frac{\varepsilon}{x^2 + \varepsilon^2}, \quad \text{with parameter} \quad \varepsilon > 0.$$

Note that although they both are bell-shaped, their asymptotics at infinity are dramatically different, with the Gaussian density decaying more than exponentially fast, while the Cauchy–Lorentz density decreases at infinity much slower at the inverse parabolic rate.

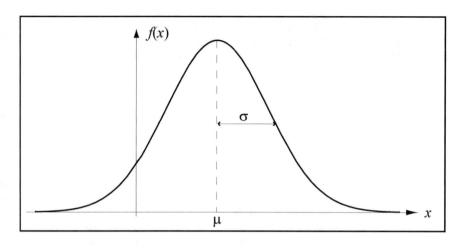

FIGURE 16.2.2.
A graph of a Gaussian probability density function.

16.3 Probability densities of transformations of random quantities

An important question in applied probability theory is how to find the probability density $f_Y(y)$ of a random quantity Y tied through the nonlinear relation,

$$Y = g(X),$$

to another random quantity X with a given density $f_X(x)$. Physical and engineering problems that lead to this type of questions are related, for example, to analysis of statistical properties of signals received via nonlinear filters in the presence of random noises with random amplitudes and phases, or studies of statistical characteristics of experiments with random outcomes which are measured by nonlinear devices. The general algorithm for solving such problems relies on the equality

$$f_Y(y) = \langle \delta(g(X) - y) \rangle. \tag{1}$$

Substituting it into equation (16.2.4), we see that the probability density of the random quantity Y is equal to the singular functional,

$$f_Y(y) = \int \delta(g(X) - y) f_X(x)\, dx. \tag{2}$$

Remark 1. Problems outside the rigorous distribution theory. At this point we should warn the reader that even in very standard problems posed by physical applications one can arrive here at functionals which do not fit neatly in the rigorous distribution theory developed earlier. For example, a radio signal passing through the simplest quadratic detector ($g(x) = x^2$), gives rise to the problem involving a nonmonotone function $g(x)$. Moreover, the density function $f_X(x)$, which in equation (2) plays the role of a distribution-theoretic test function, need not be infinitely differentiable, or have a compact support. Nevertheless, in the examples listed below, and in the numerous other applied problems, one can successfully operate with the functionals of type (2), and define them with the help of equalities of type (2.9.9), p. 62, Volume I, which are outside the rigorous theory. We hasten to add that all the obtained results can be justified mathematically within the framework of rigorous probability theory that the reader is encouraged to study independently of this text.

Let us begin by finding the explicit formula for $f_Y(y)$ in the case when $y = g(x)$ is a smooth, strictly monotone function, well-defined for all possible values of the random quantity X. Utilizing the standard properties of the Dirac delta we obtain that

$$\delta(g(X) - y) = \left| \frac{dg^{-1}(y)}{dy} \right| \delta\big(g^{-1}(y) - X\big), \tag{3}$$

where $x = g^{-1}(y)$ denotes the inverse function to $y = g(x)$. Averaging both sides of the above equation over the statistical ensemble of X gives

$$f_Y(y) = \left| \frac{dg^{-1}(y)}{dy} \right| f_X\big((g^{-1}(y))\big). \tag{4}$$

An obvious observation is that outside the range of the function $g(x)$ the above density is assumed to be equal to 0.

More generally, in the case of nonmonotone smooth function $y = g(x)$, we obtain the formula,

$$\delta(g(X) - y) = \sum_{n=1}^{N} \left| \frac{dg_n^{-1}(y)}{dy} \right| \delta(g_n^{-1}(y) - X), \tag{5}$$

where the summation is extended over all $N(y)$ branches of the inverse function $g^{-1}(y)$, and consequently, in this case,

$$f_Y(y) = \sum_{n=1}^{N(y)} \left| \frac{dg_n^{-1}(y)}{dy} \right| f_X\big(g_n^{-1}(y)\big). \tag{6}$$

Example 1. Linear transformations. In the case of a linear function $y = ax + b$, corresponding, e.g., to the change of measurement units (like moving from the Celsius temperature scale to the Fahrenheit's), with the inverse function $x = (y - b)/a$, equation (4) takes a simple form

$$f_Y(y) = \frac{1}{|a|} f_X\left(\frac{y - b}{a} \right).$$

The multiplier $1/|a|$ guarantees the normalization of the density function of Y and its correct dimensionality.

Example 2. Harmonic signal with a random phase. Assume that Ψ is the random phase of a signal, uniformly distributed in the interval $(-\pi, \pi)$. In other words, its probability density is

$$f_\Psi(\varphi) = \begin{cases} 1/2\pi, & \text{for } |\varphi| < \pi; \\ 0, & \text{for } |\varphi| > \pi. \end{cases}$$

Our task is to find the probability density of the harmonic signal,

$$X = \cos \Psi,$$

with the above-described random phase Ψ. According to the formula (2),

$$f_X(x) = \frac{1}{2\pi} \int_{-\pi}^{\pi} \delta(\cos \varphi - x)\, d\varphi. \tag{3}$$

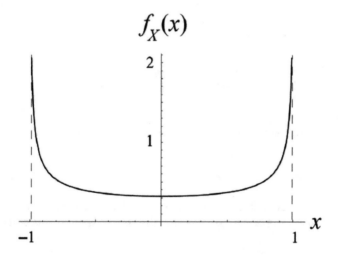

FIGURE 16.3.1.
The probability density of the random quantity $\cos \varphi$, **for** φ **uniformly distributed on the interval** $(-\pi, \pi)$.

The Dirac delta of a composite argument can be transformed utilizing the above-mentioned formula (2.9.9). To accomplish this task notice that, for a fixed $|x| < 1$, equation $x = \cos \varphi$ has two roots, see Fig.16.3.1 (Left),

$$\varphi_{1,2} = \pm \arccos x.$$

Formula (2.9.9) contains derivatives of $g(\varphi) = \cos \varphi$, which, in our case are equal to

$$g'(\varphi_{2,1}) = -\sin \varphi_{2,1} = \pm\sqrt{1 - x^2}.$$

Therefore, the Dirac delta entering (3) can be represented in the form

$$\delta(\cos\varphi - x) = \frac{1}{\sqrt{1-x^2}}\Big[\delta(\varphi - \varphi_1) + \delta(\varphi - \varphi_2)\Big].$$

Substituting it into (3), and making use of the probing property of the Dirac delta, we obtain the final result:

$$f_X(x) = \begin{cases} 1/(\pi\sqrt{1-x^2}), & \text{for } |x| < 1; \\ 0, & \text{for } |x| > 1. \end{cases}$$

This formula is well-known and often used in the signal processing theory. The graph of probability density $f_X(x)$ is plotted on Fig. 16.3.1.

Example 3. Square of a Gaussian random quantity. If the voltage X of the electric current fluctuates randomly with a Gaussian PDF described in Example 16.2.3, then its power is proportional to the square of voltage. Considering thus the PDF of a new random quantity $Y = X^2$ will give us the probability distribution function of the fluctuations' power. An approach similar to the one used in Example 2, remembering that the inverse function in this case has two branches, gives, for each value of the parameter $\sigma > 0$, and $\mu = 0$,

$$f_Y(x) = \frac{1}{\sqrt{2\pi\sigma^2 x}} \exp\left(-\frac{x}{2\sigma^2}\right), \qquad \text{for} \qquad x > 0.$$

The above PDF is called *chi-square probability density function* (with parameter 1). Note that the density has a singularity at $x = 0$.

16.4 Characteristics of random quantities

Among all the ensemble averages $\langle g(X)\rangle$ of the transformed random quantity $g(X)$ there are a few that traditionally have played a prominent role in probability theory. We shall discuss them briefly in this section.

In general, recalling that (see (16.2.4))

$$g(X) = \int g(x)\delta(x - X)\,dx,$$

we obtain the formula for the tested mean of a random quantity X,

$$\langle g(X)\rangle = \int g(x)\langle\delta(x - X)\rangle\,dx = \int g(x)f_X(x)\,dx. \tag{1}$$

16.4.1. Moments, variances, and standard deviations. Selecting $g(x) = x^n, n \in \mathbf{N}$, gives rise to the *n*th *moment*

$$\mu_n = \langle X^n\rangle = \int x^n f_X(x)\,dx, \tag{2}$$

of the random quantity X. Of course, the first moment is just the ensemble average of X. The *variance*, also called the *second central moment*,

$$\text{Var}(X) \equiv \sigma_X^2 = \langle (X - \langle X \rangle)^2 \rangle = \langle X^2 \rangle - \langle X \rangle^2, \tag{3}$$

measures the spread of the values of the random quantity around its mean. It has also a certain optimal property since

$$\sigma_X^2 = \min_c \langle (X - c)^2 \rangle. \tag{4}$$

If Var $(X) = 0$, then X is a constant (in the statistical sense). The disadvantage of the variance as a measure of dispersion is that it scales as the square of X:

$$\text{Var}(cX) = c^2 \text{Var}(X). \tag{5}$$

So, sometimes the *standard deviation* $\sigma(X) = \sqrt{\text{Var}(X)}$, also called the mean root square (MRS), is selected as a better measure of dispersion since

$$\sigma(cX) = |c|\sigma(X). \tag{6}$$

This means that the standard deviation of a random quantity scales like the random quantity itself, and it is measured in the same units as X. Of course, not all the moments need to be well-defined. For a Gaussian random quantity all the moments are finite, but for the Cauchy–Lorentz random quantity, all the absolute moments of orders $n = 1, 2, \ldots$, are infinite.

16.4.2. Characteristic functions. All of the above numerical characteristics contain only partial information about the probability distribution of the random quantity. Of course, the full information about $f_X(x)$ is contained in the means of all compositions,

$$\langle \phi(X) \rangle = \int f_X(x) \phi(x)\, dx, \tag{7}$$

where $\phi(x)$ range over the family of all bounded test functions. Selecting the test function $\phi(x) = \chi(y - x)$ reproduces the cumulative probability distribution function (CDF) of X:

$$\langle \chi(y - X) \rangle = F_X(y). \tag{8}$$

One can select other, more restricted classes of smooth and bounded test functions which also will preserve in their means all the information about the probability distributions of random variables X. One such family is the collection of complex harmonic functions (complex exponentials),

$$\phi(x) = e^{i\omega x}, \qquad -\infty < \omega < \infty, \tag{9}$$

and the corresponding ω-dependent means,

$$\Phi_X(\omega) = \langle e^{i\omega X} \rangle = \int e^{i\omega x} f_X(x)\, dx, \tag{10}$$

which, traditionally, are called the *characteristic function* of the random quantity X. Obviously, up to the change of sign in front of ω, the characteristic function of X is just the Fourier transform of the probability density functions $f_X(x)$. It completely determines the probability density function via the inverse Fourier transform,

$$f_X(x) = \frac{1}{2\pi} \int \Phi_X(\omega) e^{-i\omega x} dx,$$

in view of results of Chapter 3, Volume I.

All the moments of the random quantity X (if they exist) are recoverable from the derivatives (if they exist) of the characteristic function. Indeed, for a given integer n,

$$\frac{d^n \Phi_X(\omega)}{d\omega^n} = i^n \int f_X(x) x^n e^{i\omega x} dx,$$

so that

$$\mu_n = \int f_X(x) x^n e^{i\omega 0} dx = \frac{1}{i^n} \frac{d^n \Phi_X(\omega)}{d\omega^n} \bigg|_{\omega=0}. \tag{11}$$

Example 1. Characteristic functions of uniform densities. For a random quantity X uniformly distributed over the interval $[-1, 1]$, the characteristic function

$$\Phi_X(\omega) = \frac{1}{2} \int_{-1}^{1} e^{i\omega x} dx = \frac{e^{i\omega} - e^{-i\omega}}{2i\omega} = \frac{\sin \omega}{\omega}, \tag{12}$$

and for the uniform discrete probability distribution on the points ± 1, we have

$$\Phi(\omega) = \frac{e^{i\omega} + e^{-i\omega}}{2} = \cos \omega. \tag{13}$$

Example 2. Characteristic functions of Poisson, Gaussian, and Cauchy–Lorentz densities. Via easy calculations one gets characteristic functions for the distributions discussed in Examples 16.2.1-3. Thus, for a discrete random quantity

$$\Phi(\omega) = \sum_{k=1}^{\infty} p_k e^{i\omega x_k},$$

and, in particular, for a Poissonian random quantity we have

$$\Phi(\omega) = \exp[\lambda(e^{i\omega} - 1)]. \tag{14}$$

For a Gaussian density $\gamma(x; \sigma, 0)$ the characteristic function is

$$\Phi(\omega) = \exp[-\sigma^2 \omega^2 / 2], \tag{15}$$

and for the Cauchy–Lorentz density $\lambda_\varepsilon(x)$ it is

$$\Phi(\omega) = \exp[-|\lambda \omega|]. \tag{16}$$

These characteristic functions, following (13), easily yield moments of the above distributions. In particular, both mean and variance of the Poisson distribution are equal to λ. Gaussian distributions have mean zero and variance σ^2, while for the Cauchy–Lorentz distribution neither the mean nor the variance exist, the fact we already mentioned earlier.

Example 3. Characteristic function of a harmonic oscillation with random phase. A similar case has been considered in Example 16.3.2, but this time let us consider the random quantity $X = \sin \Psi$, where Ψ is uniformly distributed on the interval $[-\pi, \pi]$. In this case, the characteristic function is

$$\Phi_X(\omega) = \langle \exp[i\omega \sin \Psi] \rangle = \frac{1}{2\pi} \int_{-\pi}^{\pi} e^{i\omega \sin \psi} d\psi = J_0(-\omega),$$

where $J_0(\omega)$ denotes the Bessel function of the first kind and order zero.

Sometimes, it is more convenient to operate with the logarithm $\log \Phi_X(\omega)$ of the characteristic function (if it is well-defined) than with the characteristic function $\Phi_X(\omega)$ itself. The coefficients κ_k in the Taylor expansion,

$$\log \Phi_X(\omega) = \sum_{k=0}^{\infty} \kappa_k \frac{(i\omega)^k}{k!}, \tag{17}$$

are then called the *cumulants* of X. It is easy to verify (see, Exercise 1) that,

$$\kappa_1 = \langle X \rangle, \qquad \kappa_2 = \text{Var}(X).$$

Finally, it is worthwhile to note an obvious, but important property of the characteristic functions:

$$|\Phi(\omega)| = \left| \int_{-\infty}^{\infty} f(x)e^{i\omega x} dx \right| \leq \int_{-\infty}^{\infty} |f(x)e^{i\omega x}| dx = \int_{-\infty}^{\infty} f(x) dx = 1 = \Phi(0).$$

16.5 Random vectors and their characteristics

16.5.1. Joint densities, conditional densities, and statistical independence. If one considers an n-dimensional random vector,

$$\boldsymbol{X} = (X_1, X_2, \ldots, X_n)$$

then the complete information about its probability distribution is given by the *joint density*,

$$f_{(X_1,\ldots,X_n)}(x_1,\ldots,x_n) = \langle \delta(x_1 - X_1) \cdot \ldots \cdot \delta(x_n - X_n) \rangle. \tag{1}$$

In particular, for any test function $\phi(x_1, \ldots, x_n)$ of n variables,

$$\langle \phi(X_1, \ldots, X_n) \rangle = \int \phi(x_1, \ldots, x_n) f_{(X_1, \ldots, X_n)}(x_1, \ldots, x_n) \, dx_1 \ldots dx_n. \quad (2)$$

Example 1. Rotationally invariant Gaussian random vectors. The rotationally invariant, zero-mean Gaussian random vector with components with identical variances σ^2 is defined by the density,

$$f(x_1, \ldots, x_n) = \left(\frac{1}{\sqrt{2\pi\sigma^2}} \right)^n \exp \left(-\frac{1}{2\sigma^2} \sum_{k=1}^{n} x_k^2 \right).$$

Note the invariance of this Gaussian density under rotations of the n-dimensional space. Indeed, the density can be written in the form

$$f(x_1, \ldots, x_n) = \varphi(r) \equiv \left(\frac{1}{\sqrt{2\pi\sigma^2}} \right)^n \exp \left(-\frac{r^2}{2\sigma^2} \right),$$

where $r = \sqrt{x_1^2 + \cdots + x_n^2}$.

Let us now consider a 2-D random vector, that is a pair of random quantities (X, Y). Then the complete information about its probability distribution is given by the *joint density*,

$$f_{(X,Y)}(x, y) = \langle \delta(x - X) \delta(y - Y) \rangle. \quad (3)$$

In particular, for any test function $\phi(x, y)$,

$$\langle \phi(X, Y) \rangle = \int \phi(x, y) f_{(X,Y)}(x, y) \, dx \, dy. \quad (4)$$

If $g(x, y)$ is a real-valued function of two variables, then $Z = g(X, Y)$ is a one-dimensional random quantity, which in distributional terms can be expressed by the formula

$$g(X, Y) = \int \int g(x, y) \delta(x - X) \delta(y - Y) \, dx dy, \quad (5)$$

which immediately gives the ensemble mean

$$\langle g(X, Y) \rangle = \int \int g(x, y) f_{(X,Y)}(x, y) \, dx dy. \quad (6)$$

The statistical relationship between random components X and Y, of the random vector (X, Y) can be expressed by the conditional probability densities $f_{(X|Y)}(x|y)$, and $f_{(Y|X)}(y|x)$[10], which are determined by the equations,

$$f_{(X,Y)}(x, y) = f_{(X|Y)}(x|y) \cdot f_Y(y) = f_{(Y|X)}(y|x) \cdot f_X(x). \quad (7)$$

---

[10]Pronounced, probability density of X at x, given that $Y = y$, with the analogous phrasing for the second conditional density.

Remark 1. Conditional density's parameter. The conditional density $f_{(x|y)}(x|y)$ depends on two variables x and y, but they have, qualitatively, different meaning. Intuitively, it is perhaps appropriate to call x the *argument*, and the y, the *parameter*. As a function of its argument x, the conditional density satisfies all the properties of the ordinary probability density function, while as a function of parameter y, it does not. The only requirement is that y is restricted to the range of values of the random quantity Y.

Conditional probability density can be used to calculate the *conditional means (conditional expectations)* defined as follows:

$$\langle g(X) \rangle_{Y=y} = \int g(x) f_{(X|Y)}(x|y) \, dx. \tag{8}$$

This time the averaging is conducted only over the subensemble of those realizations that satisfy the above condition $Y = y$. In this language we can rewrite the formula for the conditional density in the form,

$$f_{(X|Y)}(x|y) = \langle \delta(x - X) \rangle_{Y=y}. \tag{9}$$

Now, we can say that the random components of the random vector (X, Y) are *statistically independent* if the conditional density of X given $Y = y$ does not depend on y. In view of equalities (7) we can rephrase the definition as follows:

Definition 1. Statistical independence. Components X and Y, of a random vector (X, Y) are said to be (statistically) independent, if

$$f_{(X,Y)}(x, y) = f_X(x) \cdot f_Y(y), \tag{10}$$

or, equivalently, if

$$\langle \phi(X) \psi(Y) \rangle = \langle \phi(X) \rangle \langle \psi(Y) \rangle, \tag{11}$$

for any (bounded) test functions ϕ, and ψ.

Note that the components of the Gaussian random vector defined in Example 1 are statistically independent. Indeed, the multiplicative formula for the joint densities holds true in this case:

$$f_{(X_1,\ldots,X_n)}(x_1,\ldots,x_n) = \left(\frac{1}{\sqrt{2\pi\sigma^2}} \right)^n \exp\left(-\frac{1}{2\sigma^2} \sum_{k=1}^n x_k^2 \right).$$

$$= \prod_{k=1}^n \left(\frac{1}{\sqrt{2\pi\sigma^2}} \right) \exp\left(-\frac{1}{2\sigma^2} x_k^2 \right) = f_{X_1}(x_1) \cdot \ldots \cdot f_{X_n}(x_n). \tag{12}$$

16.5.2. Densities of sums, products, and ratios of independent random quantities. To calculate the densities of functions of independent random quantities X and Y, we will have to rely on properties of the Dirac delta, such as the formula

$$\delta(kx - a) = \frac{1}{|k|} \delta\left(x - \frac{a}{k}\right), \tag{13}$$

which was derived in Volume 1.

Example 1. Density of the sum of independent random quantities. If $Z = X + Y$, then, by definition, the density

$$f_Z(z) = \langle \delta(z - X - Y) \rangle = \int \int \delta(z - x - y) f_{(X,Y)}(x, y)\, dx dy.$$

In view of the probing property of the Dirac delta,

$$f_Z(z) = \int f_{(X,Y)}(x, z - x)\, dx.$$

If X and Y are independent, then $f_{(X,Y)}(x,y) = f_X(x) f_Y(y)$, and the density of the sum

$$f_Z(z) = \int f_X(x) f_Y(z - x)\, dx = [f_X * f_Y](z), \tag{14}$$

is the simple convolution of the densities of X and Y.

Example 2. Density of the product of independent random quantities. If $Z = X \cdot Y$, then,

$$f_Z(z) = \int \int \delta(z - xy) f_{(X,Y)}(x, y)\, dx dy = \int \int \frac{1}{|x|} f_{(X,Y)}\left(x, \frac{z}{x}\right) dx, \tag{15}$$

so that in the independent case,

$$f_Z(z) = \int \frac{1}{|x|} f_X(x) f_Y(z/x)\, dx. \tag{16}$$

In the special case of independent Gaussian X and Y, with the joint density,

$$f_{(X,Y)}(x, y) = \frac{1}{2\pi\sigma^2} \exp\left(-\frac{x^2 + y^2}{2\sigma^2}\right), \tag{17}$$

we obtain

$$f_Z(z) = \frac{1}{\pi\sigma^2} \int_0^\infty \exp\left(-\frac{1}{2\sigma^2}\left(x^2 + \frac{z^2}{x^2}\right)\right) \frac{dx}{x}. \tag{18}$$

Note, that the above density, pictured in Fig. 16.5.1, can be written in the form,

$$f_Z(z) = \frac{1}{\pi\sigma^2} K_0\left(\frac{|z|}{\sigma^2}\right),$$

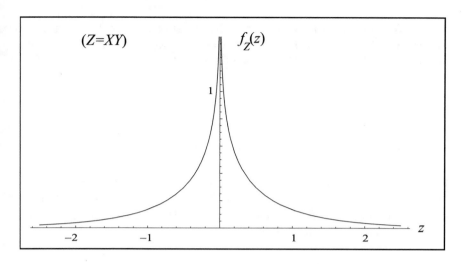

FIGURE 16.5.1.
Probability density function of the product of two independent Gaussian random quantities given by the formula (18), for $\sigma = 1$.

where $K_0(z)$ is the modified Bessel function of the second kind of zeroth order, which gives immediately the following asymptotic properties (see, Watson [8])

$$K_0(z) \sim \ln\left(\frac{2}{z}\right), (z \to 0); \qquad K_0(z) \sim \ln\sqrt{\frac{\pi}{2z}}e^{-z}, (z \to \infty).$$

Intuitively, the product $Z = XY$ can be thought of as a linear function of Y with the random coefficient X. Small values of the coefficient X generate the logarithmic singularity of the density of Z in the neighborhood of 0, and large values of X produce a decay to 0 at $z \to \infty$, of $f_Z(z)$ that is slower than that for the original Gaussian density.

Example 3. Density of the ratio of independent random quantities. Finally, for the ratio, $Z = X/Y$ of two random quantities, we obtain the following formula for the density,

$$f_Z(z) = \left\langle \delta\left(z - \frac{X}{Y}\right) \right\rangle = \langle |Y|\delta(X - Yz)\rangle = \int |y| f_{(X,Y)}(yz, y)\, dy. \quad (19)$$

In the special case of the joint density (17) of independent Gaussian random quantities, the above formula gives the Cauchy–Lorentz density,

$$f_Z(z) = \frac{1}{\pi(1 + z^2)}.$$

16.5.3.   Vector functions of random vectors. In this section we will consider how statistical properties of random vectors in $\mathbf{R}^n$ change under linear and nonlinear transformations of the space. So, let

$$\boldsymbol{\alpha} : \mathbf{R}^n \mapsto \mathbf{R}^n, \tag{21}$$

and the random vector

$$\boldsymbol{Y} = (Y_1, \ldots, Y_n) = \boldsymbol{\alpha}(\boldsymbol{X}) = \Big(\alpha_1(X_1, \ldots, X_n), \ldots, \alpha_n(X_1, \ldots, X_n)\Big). \tag{22}$$

For the sake of simplicity we will assume that the function $\boldsymbol{\alpha}$ is differentiable, and that it is a one-to-one mapping of the range $\Omega_{\boldsymbol{X}}$ of the random vector $\boldsymbol{X}$ onto the range $\Omega_{\boldsymbol{Y}}$ of the random vector $\boldsymbol{Y}$. Under these conditions there exists the inverse differentiable mapping

$$\boldsymbol{X} = \boldsymbol{\beta}(\boldsymbol{Y}).$$

The necessary and sufficient condition here is that the Jacobian of the mapping $\boldsymbol{\beta}$,

$$J(\boldsymbol{y}) = \det \begin{pmatrix} \frac{\partial \beta_1}{\partial y_1}, \ldots, \frac{\partial \beta_1}{\partial y_n} \\ \ldots\ldots\ldots\ldots \\ \frac{\partial \beta_n}{\partial y_1}, \ldots, \frac{\partial \beta_n}{\partial y_n} \end{pmatrix}$$

satisfies the condition

$$0 < |J(\boldsymbol{y})| < \infty.$$

Our next step is to express the joint density $f_{\boldsymbol{Y}}(\boldsymbol{y})$ of the random vector $\boldsymbol{Y}$, in terms of the joint density $f_{\boldsymbol{X}}(\boldsymbol{x})$ of the random vector $\boldsymbol{X}$. For this purpose we'll start with calculating the tested mean

$$\langle g(\boldsymbol{\alpha}(\boldsymbol{X}))\rangle = \int_{\Omega_{\boldsymbol{X}}} g(\boldsymbol{\alpha}(\boldsymbol{x})) f_{\boldsymbol{X}}(\boldsymbol{x})\, d^n x,$$

using an arbitrary real-valued test function g of n variables. Passing to the new variables of integration,

$$\boldsymbol{y} = \boldsymbol{\alpha}(\boldsymbol{x}) \iff \boldsymbol{x} = \boldsymbol{\beta}(\boldsymbol{y}),$$

and remembering that the infinitesimal volume changes according to the formula,

$$d^n x = |J(\boldsymbol{y})| d^n y,$$

we obtain the formula

$$\langle g(\boldsymbol{Y})\rangle = \int_{\Omega_{\boldsymbol{Y}}} g(\boldsymbol{y}) f_{\boldsymbol{X}}(\boldsymbol{\beta}(\boldsymbol{y})) |J(\boldsymbol{y})|\, d^n y.$$

On the other hand, the same ensemble mean can be expressed through the joint density $f_Y(y)$ of the random vector Y,

$$\langle g(Y) \rangle = \int_{\Omega_Y} g(y) f_Y(y) \, d^n y.$$

Since the above two equalities hold true for any test function g, we obtain the final expression of the joint density of Y in terms of the density of X:

$$f_Y(y) = f_X(\beta(y)) |J(y)|. \tag{23}$$

Example 1. Modulus and argument of a random complex-valued quantity. Consider the following problem which often appears in engineering applications. Let

$$Z = a + X + iY, \tag{24}$$

be a complex-valued random quantity with a being a deterministic real constant (signal), and X and Y, being zero-mean, independent Gaussian random quantities (noise), with variance σ^2. In polar coordinates, we have the representation

$$Z = Re^{i\Theta}$$

where R is the modulus of Z, and Θ is the argument (angular phase) of Z, and the question is how to determine the joint density of (R, Θ) in terms of the joint density of (X, Y).

In our case, the mapping under consideration transforms the (X, Y) plane into the domain

$$\{R \in [0, \infty), \Theta \in [-\pi, \pi)\} \tag{25}$$

via the pair of functions

$$x = r \cos \theta - a, \qquad y = r \sin \theta, \tag{26}$$

with the Jacobian

$$J = \det \begin{pmatrix} \frac{\partial x}{\partial r}, & \frac{\partial x}{\partial \theta} \\ \frac{\partial y}{\partial r}, & \frac{\partial y}{\partial \theta} \end{pmatrix} = \det \begin{pmatrix} \cos \theta, & -r \sin \theta \\ \sin \theta, & r \cos \theta \end{pmatrix} = r. \tag{27}$$

Hence, in view of (23), the joint density of (R, Θ) is expressed in terms of the joint density $f_{(X,Y)}$ as follows:

$$w_{(R,\Theta)}(r, \theta) = r f_{(X,Y)}(r \cos \theta - a, r \sin \theta). \tag{28}$$

Substituting the explicit expression for the Gaussian joint density f in the above formula we finally obtain that, on the domain (25),

$$w_{(R,\Theta)}(r, \theta) = \frac{r}{2\pi\sigma^2} \exp\left(-\frac{r^2 + a^2 - 2ar \cos \theta}{2\sigma^2} \right). \tag{29}$$

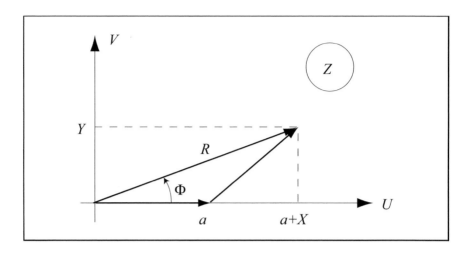

FIGURE 16.5.2.
Illustration of the setup for Example 1.

Observe that in the case $a = 0$, the above joint density splits into the product of the Rayleigh probability density for the modulus R, and uniform density for the argument (phase) Θ,

$$w_R(r) = \frac{r}{\sigma^2} \exp\left(-\frac{r^2}{2\sigma^2}\right),\ r \geq 0, \qquad w_\Theta(\theta) = \frac{1}{2\pi},\ \theta \in [-\pi, \pi). \qquad (30)$$

Thus, in this case, the modulus and the argument are statistically independent random quantities.

If $a \neq 0$, then R and Θ are no longer statistically independent. Calculating, in this general case, the probability density of R by integrating (29) over the interval $[-\pi, \pi)$ in θ, we obtain

$$w_R(r) = \frac{r}{\sigma^2} \exp\left(-\frac{r^2 + a^2}{2\sigma^2}\right) I_0\left(\frac{ar}{\sigma^2}\right), \qquad (31)$$

where

$$I_0(z) = \frac{1}{2\pi} \int_{-\pi}^{\pi} e^{z \cos \theta} d\theta$$

is the familiar modified Bessel function of first kind and order zero. The density $w_R(r)$ is usually called the *generalized Rayleigh density*. Note that it depends on the dimensionless parameter $s = a/\sigma$ which in electrical engineering applications is called the *signal-to-noise ratio*. For different values of s, the generalized Rayleigh densities are shown in Fig. 16.5.3.

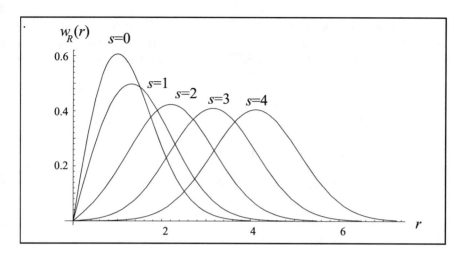

FIGURE 16.5.3.
Generalized Rayleigh densities for different values of the signal-to-noise parameter s.

Example 2. Densities of linear transformations. In the case of a linear transformation of a random vector,

$$Y = AX,$$

where $A = (a_{ij})$ is an invertible matrix with nonvanishing determinant, the corresponding Jacobian is simply the determinant of the inverse matrix

$$B = A^{-1}, \quad \text{and} \quad X = BY.$$

So, in this case, the joint density of the transformed random vector Y is given by the formula,

$$f_Y(y) = f_X(By)|\det B|. \tag{32}$$

16.5.4. Mixed moments, covariances, and correlations. For real-valued random quantities, moments of different orders, $\langle X^k \rangle, k = 1, 2, \ldots$, provided numerical characteristics of their statistical properties. For random vectors, $X = (X_1, X_2, \ldots, X_n)$, the similar role is played by the collection of mixed moments

$$\left\langle X_1^{k_1} \cdot X_2^{k_2} \cdot \ldots \cdot X_n^{k_n} \right\rangle,$$

indexed by a set of integers, $\{k_1, k_2, \ldots, k_n\}$. In this subsection we will focus our attention on the special case when $n = 2$, and $k_1 = k_2 = 1$.

So, let X and Y, be a pair of random quantities. In many applications one needs to answer the following question: Given measurements of the values of X, how well one can predict the values of Y? Of course, the prediction cannot be perfect as both X and Y, are random, and we have to make use of their *statistical dependence* as fully expressed by their joint probability density $f_{(X,Y)}(x,y)$ in the case when this joint density does not factor into the product of two marginal densities $f_X(x)$, and $f_Y(y)$, thus producing statistical independence of X and Y.

The problem can be formulated as a search for a function $\varphi(x)$ such that the random quantity $Y' = \varphi(X)$ approximates well (in a certain sense to be defined) the random quantity Y. The simplest φ we can select is linear, and the simplest error to consider is the mean-square error, which leads to the following *linear prediction problem*: Find

$$Y' = aX + b, \tag{32}$$

which minimizes the error

$$\epsilon^2 = \langle (Y - Y')^2 \rangle. \tag{33}$$

Substituting (32) into (33) we obtain the expression,

$$\epsilon^2 = \langle (Y - (aX+b))^2 \rangle = \langle Y^2 \rangle + a^2 \langle X^2 \rangle + b^2 - 2a\langle XY \rangle - 2b\langle Y \rangle + 2ab\langle X \rangle. \tag{34}$$

The problem requires that we assume that the second moments of X and Y, are finite, but it also shows the need to consider the mixed second moment

$$\langle XY \rangle = \int \int xy f_{(X,Y)}(x,y)\,dxdy$$

which is sometimes called the *correlation* of X and Y.

The optimal values of a and b minimizing the error ϵ^2 in (34) can now be found by solving the pair of linear equations,

$$\left\{ \frac{\partial \epsilon^2}{\partial a} = 0, \quad \frac{\partial \epsilon^2}{\partial b} = 0 \right\} \iff \left\{ \begin{array}{c} a\langle X^2 \rangle + b\langle X \rangle = \langle XY \rangle, \\ a\langle X \rangle + b = \langle Y \rangle, \end{array} \right.$$

which yield

$$a = \frac{\langle XY \rangle - \langle X \rangle \langle Y \rangle}{\sigma_X^2}, \qquad b = \langle Y \rangle - a\langle X \rangle, \tag{35}$$

with

$$\sigma_X^2 = \langle X^2 \rangle - \langle X \rangle^2 = \langle (X - \langle X \rangle)^2 \rangle,$$

denoting the usual variance of X. In analogy with this terminology, the expression in the numerator of (35) will be called the *covariance* of X and Y:

$$\mathrm{Cov}\,(X,Y) = \langle XY \rangle - \langle X \rangle \langle Y \rangle = \left\langle (X - \langle X \rangle)(Y - \langle Y \rangle) \right\rangle. \tag{36}$$

Substituting (35) into (34) we find the minimal mean-square error of our linear prediction,

$$\min \epsilon^2 = \sigma_Y^2 (1 - \rho_{(X,Y)}^2), \tag{37}$$

where the parameter

$$\rho_{(X,Y)} = \frac{\text{Cov}\,(X,Y)}{\sigma_X \sigma_Y} = \frac{\langle (X - \langle X \rangle)(Y - \langle Y \rangle) \rangle}{\sqrt{\langle (X - \langle X \rangle)^2 \rangle} \sqrt{\langle (Y - \langle Y \rangle)^2 \rangle}} \tag{38}$$

is called the *correlation coefficient* of X and Y. Because $\epsilon^2 \geq 0$, the correlation coefficient necessarily satisfies the inequality

$$-1 \leq \rho_{(X,Y)} \leq 1. \tag{39}$$

The relationship (37) implies that the closer is the correlation coefficient $\rho_{(X,Y)}$ to ± 1 (for fixed standard deviations of X and Y), the smaller the error of approximation ϵ^2. So, in a sense, the correlation coefficient describes the degree of linear dependency between X and Y. If $\rho_{(X,Y)} = 0$, then there is no such dependence, and we say that X and Y are *uncorrelated*. If $\rho_{(X,Y)} = \pm 1$, then there is a deterministic linear dependence between them. If X and Y are statistically independent then they are always uncorrelated. But the opposite implication is not true, just check the random vector (X,Y) with uniform probability density on the unit circle. The random quantities X and Y, are uncorrelated but not independent.

For a general n-dimensional vector-valued random quantity $\boldsymbol{X} = (X_1, X_2, \ldots, X_n)$, the natural numerical characteristics quantifying the degree of linear dependence between different components are the *covariances*,

$$\text{Cov}\,(X_i, X_j) = \langle (X_i - \langle X_i \rangle)(X_j - \langle X_j \rangle) \rangle = \langle X_i X_j \rangle - \langle X_i \rangle \langle X_j \rangle, \tag{40}$$

$i, j = 1, 2, \ldots, n$, that together form the (always symmetric) *covariance matrix* Σ, the diagonal terms thereof being, of course, variances of components of the random vector $\boldsymbol{X}$.

Note that the covariance matrix plays the critical role in calculation of the variances of scalar products of random vectors with fixed deterministic vectors or, equivalently, of any linear combination of the components of $\boldsymbol{X}$. Indeed, if

$$\boldsymbol{Y} = (\boldsymbol{\alpha}, \boldsymbol{X}) = \sum_{k=1}^{n} \alpha_k X_k, \tag{41}$$

then the variance of $\boldsymbol{Y}$,

$$\sigma_Y^2 = \left\langle \left(\sum_{k=1}^{n} \alpha_k (X_k - \langle X_k \rangle) \right)^2 \right\rangle = \sum_{k=1}^{n} \sum_{l=1}^{n} \text{Cov}\,(X_k, X_l) \alpha_k \alpha_l. \tag{42}$$

The above quadratic form in $\alpha_1, \ldots, \alpha_n$ is positive-definite because the left-hand side of (42) is always nonnegative.

Finally, observe that if the components of the random vector $\boldsymbol{X}$ are uncorrelated then, for any $\boldsymbol{\alpha} = (\alpha_1, \ldots, \alpha_n) \in \mathbf{R}^n$,

$$\mathrm{Var}\left(\sum_{k=1}^{n} \alpha_k X_k\right) = \sum_{k=1}^{n} \alpha_k^2 \sigma_{X_k}^2,$$

and in particular,

$$\mathrm{Var}\left(\sum_{k=1}^{n} X_k\right) = \sum_{k=1}^{n} \sigma_{X_k}^2,$$

that is, the variance of the sum is equal to the sum of variances.

16.5.5. Characteristic functions of random vectors.

The characteristic function of a random vector $\boldsymbol{X}$ is defined by the formula,

$$\Phi(\boldsymbol{\omega}) \equiv \Phi(\omega_1, \ldots, \omega_n) = \left\langle e^{i(\boldsymbol{\omega}, \boldsymbol{X})} \right\rangle = \left\langle \exp\left(i \sum_{k=1}^{n} \omega_k X_k\right) \right\rangle. \qquad (42)$$

It is a function of n real variables, and it obviously depends on the joint density of the random vector $\boldsymbol{X}$. Also, in the very special case of

$$\boldsymbol{\omega} = (\omega, \ldots, \omega)$$

the characteristic function of $\boldsymbol{X}$ gives the characteristic function on the sum of random quantities. Indeed, if

$$Y = \sum_{k=1}^{n} X_k, \qquad (43)$$

then

$$\Phi_Y(\omega) = \Phi_X(\omega, \ldots, \omega) = \left\langle \exp\left(i\omega \sum_{k=1}^{n} X_k\right) \right\rangle. \qquad (44)$$

If the components of $\boldsymbol{X}$ are statistically independent then, in view of the multiplicative property of expectations in such a case,

$$\Phi_X(\boldsymbol{\omega}) = \prod_{k=1}^{n} \Phi_{X_k}(\omega_k), \qquad (45)$$

where $\Phi_{X_k}(\omega_k)$ is the characteristic function of the random quantity X_k. In particular, in the case of a random vector with components that are independent and identically distributed, the characteristic function of the sum (43) is of the form,

$$\Phi_Y(\omega) = \Phi_{X_1}^n(\omega). \qquad (46)$$

The additive property of variances of sums of independent random quantities discussed at the end of Subsection 16.5.4 can be obtained also for other characteristics which can be derived by considering the logarithm of the characteristic function of X,

$$\Psi_X(\omega) = \log \Phi_X(\omega).$$

Expanding Ψ_X into the power series gives

$$\Psi_X(u) = \sum_{m=1}^{\infty} \frac{\kappa_m}{m!}(i\omega)^m,$$

which gives the following representation for the characteristic function

$$\Phi_X(\omega) = \exp\left(\sum_{m=1}^{\infty} \frac{\kappa_m}{m!}(i\omega)^m\right), \tag{47}$$

with the coefficients κ_m usually called the *cumulants of order m* of the random quantity under consideration.

In view of the multiplicative property (45) of the characteristic function of the sum of independent random quantities, we immediately see that the cumulant of the sum of independent random quantities is equal to the sum of their cumulants. In particular, if $Y = X_1 + X_2 + \cdots + X_n$, and X_k are independent and identically distributed, then

$$\Phi_Y(\omega) = \exp\left(\sum_{m=1}^{\infty} \frac{n\kappa_m}{m!}(i\omega)^m\right). \tag{48}$$

Performing the Taylor expansion on (47) and taking into account formula (11) give the following relationships between cumulants κ_n and moments μ_n, for $n = 1, 2, 3, 4$:

$$\begin{aligned}
\kappa_1 &= \mu_1, \\
\kappa_2 &= \mu_2 - \mu_1^2 = \sigma_X^2, \\
\kappa_3 &= \mu_3 - 3\mu_1\mu_2 + 2\mu_1^3 \\
\kappa_4 &= \mu_4 - 3\mu_2^2 - 4\mu_1\mu_3 + 12\mu_1^2\mu_2 - 6\mu_1^4.
\end{aligned} \tag{49}$$

The first two cumulants are equal, respectively, to the mean and the variance of X.

Remark 1. Cumulants and the shape of the probability density function. Cumulants contain a wealth of information about the nature of the probability density function itself. In particular, if all the cumulants of order ≥ 2 vanish then the random quantity is a constant, since

$$\Phi_X(\omega) = e^{i\kappa_1\omega} \iff f_X(x) = \delta(X - \kappa_1). \tag{50}$$

In general, the first cumulant just affects the "center of mass" of the probability density but not its shape.

Next assume that all the cumulants of order three or higher vanish, i.e., $\kappa_n = 0, n = 3, 4, \ldots$, while $\kappa_1 \neq 0, \kappa_2 > 0$. Then, the density is necessarily Gaussian,

$$\Phi_X(\omega) = \exp\left(i\kappa_1\omega - \frac{\kappa_2}{2}\omega^2\right) \iff f_X(\omega) = \frac{1}{\sqrt{2\pi\kappa_2}}\exp\left(-\frac{(x-\kappa_1)^2}{2\kappa_2}\right).$$

Often, to study the shape properties of the density one considers the centered and normalized dimensionless random quantity,

$$\hat{X} = \frac{X - \kappa_1}{\sqrt{\kappa_2}} = \frac{X - \langle X \rangle}{\sigma}. \tag{51}$$

It is normalized in the sense that its standard deviation is equal to 1. In a similar fashion one can introduce dimensionless *cumulant coefficients* of order m:

$$\varkappa_m = \frac{\kappa_m}{\kappa_2^{m/2}} = \frac{\kappa_m}{\sigma^m}, \qquad m = 3, 4, \ldots. \tag{52}$$

As observed before, vanishing of all the cumulants of order higher than 2 implies that the density is Gaussian. So, one may be curious about what happens if we impose similar restriction of cumulants vanishing only for orders greater than 3, 4, etc. Here, we run into a surprise. The Marcinkiewicz Theorem (see, [9]) says that in this situation the function (47) is no longer positive definite, and thus does not represent the characteristic function of any random quantity. In other words, the Gaussian distribution is the only distribution whose cumulant generating function is a polynomial, that is the only distribution having a finite number of nonzero cumulants. Thus the only three possibilities are:

(a) Only the first cumulant is nonzero, and the random quantity is a constant;

(b) Only the first two cumulants are nonzero, and the random quantity is Gaussian;

(c) The random quantity is neither constant nor Gaussian, and in this case all the cumulants are nonzero.

So in a sense, the behavior of higher cumulant coefficients can be used to describe how different is a given density from the Gaussian density. The smaller the higher cumulant coefficients, the closer the density is to the Gaussian density. Commonly, the third- and fourth-order cumulant coefficients are utilized in practice:

$$\varkappa_3 = \frac{\kappa_3}{\sigma^3}, \quad \text{and} \quad \varkappa_4 = \frac{\kappa_4}{\sigma^4}. \tag{53}$$

The first, called *skewness*, measures the asymmetry of the density, and the second, called *kurtosis*, measures the deviation of the density from the thin tails characterizing the Gaussian density.

16.6 Fundamental probability laws

There are several fundamental laws of probability which sooner or later appear on the horizon of any researcher studying random phenomena. In this section we will discuss four of them: the law of binomial probabilities, the law of rare limit events, the law of large numbers, and the central limit theorem also called the stability of fluctuations law.

16.6.1. Bernoulli trials One of the fundamental requirements for scientific experiments is their reproducibility. For a given voltage on the terminals of a given resistor, the current intensity flowing through it will always be the same. At face value this principle does not seem to be satisfied for random events. Consecutive experiments, conducted independently in identical conditions, need not lead to the same outcomes.

A series of statistically independent experiments in which, with the same probability p, a random event A occurs are called *Bernoulli trials*. The cointossing experiment discussed at the beginning of the Chapter is an example of such an experiment with, ideally, $p = 1/2$.

The question in this subsection is: What is the probability distribution of the random number Y of times the even A occurred in n trials? Formally, we are interested in the random quantity

$$Y = X_1 + X_2 + \cdots + X_n, \tag{1}$$

where X_k are the indicators of the random event A, that is, $X_k = 1$, if the event A occurs in the kth trial, $X_k = 0$, if it does not.

Under the assumption of statistical independence of consecutive trials, the characteristic function $\Phi_Y(\omega)$ of the random quantity Y is easy to find:

$$\Phi_Y(\omega) = \Phi_Y(\omega|n) = (q + pe^{i\omega})^n = \sum_{k=0}^{n} \binom{n}{k} p^k q^{n-k} e^{ik\omega}, \tag{2}$$

where $q = 1 - p$ is the probability of the event complementary to A. Remembering that the characteristic function of the density $\delta(x - k)$ is $e^{ik\omega}$, we immediately obtain the singular probability distribution function for Y,

$$f_Y(x|n) = \Phi_Y(\omega|n) = (q + pe^{i\omega})^n = \sum_{k=0}^{n} \binom{n}{k} p^k q^{n-k} \delta(x - k). \tag{3}$$

Thus the probability of k successes in n Bernoulli trials is

$$P_n(k) = \binom{n}{k} p^k q^{n-k}, \qquad k = 0, 1, 2, \ldots, n. \tag{4}$$

Thus above probability distribution is usually called the *binomial distribution*.

16.6.2.   Poisson distribution: the law of rare events.   In the physical and economical applications, one often encounters the situation when the number of Bernoulli trials is very large, but the probability of the random event of interest occurring is very small. Think here about a large number of gas atoms, the very small fraction thereof are subject to radioactive decomposition, or the large number of cars produced annually by the Ford Corporation, with a negligible number displaying manufacturing defects. In such situations it make sense to consider not the binomial probability distribution but its limit when

$$n \to \infty, \qquad np_n = \lambda, \tag{5}$$

where λ is a positive and finite constant. The random event with the probabilities obtained in the above-described limit process will be called here a *rare event*, and the corresponding probability law, the *Poisson distribution*.

To find the probability distribution of the number Z of rare events let us consider the binomial characteristic function (2) in the limit (5), which gives

$$\Phi_Z(\omega) = \lim_{n \to \infty} \Phi_Z(\omega|n) = \lim_{n \to \infty} \left(1 + \frac{\lambda(e^{i\omega} - 1)}{n}\right)^n = \exp\left(\lambda(e^{i\omega} - 1)\right). \tag{6}$$

Noting the Taylor expansion,

$$\Phi_Z(\omega) = e^{-\lambda} \sum_{k=0}^{\infty} \frac{\lambda^k}{k!} e^{ik\omega},$$

and replacing the complex exponential by the Dirac deltas, we finally arrive at the explicit formula for the singular probability density (discrete probability distribution) of Z:

$$f_Z(x) = \sum_{k=0}^{\infty} P_\lambda(k)\delta(x - k), \tag{7}$$

where

$$P_\lambda(k) = e^{-\lambda}\frac{\lambda^k}{k!}, \qquad k = 0, 1, 2, \ldots, \tag{8}$$

is called the Poisson probability distribution with parameter λ. An easy calculation (see, Exercises) shows that the mean and variance are identical here,

$$\mu_1 = \langle Z \rangle = \lambda, \qquad \text{and} \qquad \sigma_Z^2 = \lambda.$$

Remark 1. A bit of history.   The law of rare events was discovered by Siméon Poisson in 1837 in a publication that was analyzing judicial practice in probabilistic terms. For a long time it remained unnoticed. But in 1898 it suddenly came to the general attention when it was discovered that the law perfectly described the death rate from horse kicks in fourteen German cavalry regiments in the years 1874–1894. With 196 soldiers killed in those 20 years, the average number of death per year, per regiment, is $\lambda = 196/280 = 0.7$, so that the probability of no deaths in the regiment in a particular year was

$$P_{0.7}(0) = e^{-0.7} = 0.497,$$

and the probabilities of death of one or two soldiers were

$$P_{0.7}(1) = 0.7 \cdot e^{-0.7} = 0.348, \qquad \text{and} \qquad P_{0.7}(2) = \frac{1}{2}(0.7)^2 \cdot e^{-0.7} = 0.122.$$

Thus the predicted Poisson model number of regiment·years with zero, one, and two death would be, correspondingly,

$$N_0 \approx 0.497 \cdot 280 \approx 139, \quad N_1 \approx 0.348 \cdot 280 \approx 97, \quad N_2 \approx 0.122 \cdot 280 \approx 34.$$

The actual numbers were

$$N_0 \approx 140, \qquad N_1 \approx 91, \qquad N_2 \approx 32 \qquad !!!$$

Remark 2. Applications of the Poisson distribution to physical problems. Let us return to the problem of radioactive decay in more detail. Consider a particle detector counting particles emitted during the process which provides a number alpha-particles emitted in the time interval $[0, t]$. Let's find the probability $P_t(k)$ that number was equal to k.

To accomplish this task let us split the interval $[0, t]$ into n equal subintervals of length $\Delta t = t/n$, such that the probability of registering one particle in each subinterval is very small, the probability of registering more than one particle is negligibly small. Additionally, assume that the probability of registration of exactly one particle is proportional to the length of the subinterval,

$$p \sim \nu \Delta t = \nu \frac{t}{n} \ll 1.$$

The physics dictates that the events of emission of alpha-particles in non-intersecting subintervals are statistically independent. So, the situation can be described probabilistically as a series of Bernoulli trials where the "success" means registration of a particle in a given subinterval Δt, and "failure" means that no particle was registered. If n is very large, then the registration probability $p = \nu \Delta t$ is vanishingly small, and the parameter,

$$\lambda = pn = \nu \frac{t}{n} n = \nu t, \tag{9}$$

describes the intensity of emissions from the radioactive source. Substituting this λ into the formula describing the Poisson probability distribution (8), we find the desired probability of registering k particles by the time t:

$$P_t(k) = \frac{(\nu t)^k}{k!} e^{-\nu t}. \tag{10}$$

The physical meaning of parameter ν is clear. It describes the mean number of registered particles per unit time. The resulting random function, that is continuous time-dependent family of random quantities, $Z(t), t > 0$, with independent and stationary increments, is usually called the *Poisson stochastic process*.

Another example of application of the Poisson distribution to physical problems arises when one considers gas molecules with concentration ρ, meaning that the mean number of molecules in a ball of radius r is

$$\lambda = \frac{4}{3}\pi r^3 \rho.$$

Then, using (10), we obtain the probability of finding no particles in that volume to be

$$P_r(0) = \exp\left(-\frac{4}{3}\pi\rho r^3\right). \tag{11}$$

16.6.3. The exponential distribution. Let us take a look at the law of rare events and distribution (10) of the Poisson process, $Z(t)$, from another perspective: the random time, T, elapsed from the time the counter was turned on to the arrival of the first particle. Obviously, if $T > t$, then there were no particles registered by time t, so

$$\mathbf{P}(T > t) = P_t(0) = e^{-\nu t}, \qquad t > 0. \tag{12}$$

Hence, the cumulative probability distribution of T is given by the formula,

$$F_T(t) = \mathbf{P}(T \leq t) = 1 - P_t(0) = 1 - e^{-\nu t}, \qquad t > 0, \tag{13}$$

with the probability density function,

$$f_T(t) = \frac{dF_T(t)}{dt} = \nu e^{-\nu t}, \qquad t > 0. \tag{14}$$

The above probability distribution is called the *exponential distribution* with parameter ν; the mean value of T is equal to $1/\nu$. The cumulative distribution and the density of the exponential distribution are shown in Fig. 16.6.1.

The characteristic function of the exponential distribution is easy to calculate:

$$\Phi_T(\omega) = \nu \int_0^\infty e^{i\omega t - \nu t}\, dt = \frac{\nu}{\nu - i\omega}. \tag{15}$$

Remark 1. Interarrival times are also exponential. If we denote by T_k the random time of arrival at the counter of the kth particle (so that $T = T_1, T_0 = 0$), then the assumption of the statistical independence of arrivals of particles in nonoverlapping time intervals implies that the consecutive *interarrival times* $T_k - T_{k-1}, k = 1, 2, \ldots$, also have the exponential probability distribution with parameter ν (see, Exercises).

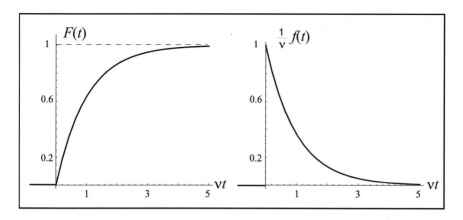

FIGURE 16.6.1.
The cumulative distribution (13) and the density (14) of the exponential probability distribution.

Also, recall that the formula (11) provided us with the probability that there are no randomly distributed molecules within the ball of radius r. Utilizing the assumption of statistical independence of positions of different molecules, and placing the ball center at each molecule, we conclude, by a reasoning similar to (12), that the probability that the random distance R from the molecule to its closest neighbor exceeds r is given by the right-hand side of (11). This, in turn, gives us the probability density distribution of the shortest distance between two molecules:

$$f_R(r) = \frac{dP_r(0)}{dr} = 4\pi\rho r^2 \exp\left(-\frac{4}{3}\pi r^3 \rho\right). \tag{16}$$

16.6.4. Stability of averages: Law of Large Numbers. The stability of averages of random quantities, discussed briefly at the beginning of this chapter, is a fundamental problem in probability theory and statistical physics. Recall that if a random vector (X, Y) with finite-variance components satisfies the following condition, Cor $(X, Y) = 0$, then the vanishing of the mixed term in the quadratic form defining the variance of their sum implies that

$$\text{Var}\,(X + Y) = \text{Var}\,(X) + \text{Var}\,(Y). \tag{17}$$

Thus, if we consider a sequence $X, X_1, X_2, \ldots$, of pairwise uncorrelated (or statistically independent, but here this stronger assumption is not needed) and identically distributed random quantities, then

$$\text{Var}\left(\frac{X_1 + \cdots + X_n}{n} - \langle X \rangle\right) = \frac{\text{Var}\,(X)}{n} \to 0, \tag{18}$$

as $n \to \infty$. This statement is called the *law of large numbers in the mean square*. It says that the arithmetic means converge to the ensemble average,

$$\bar{X}_n = \frac{X_1 + \cdots + X_n}{n} \to \langle X \rangle, \qquad n \to \infty,$$

or, in other words, that long sequences of uncorrelated identically distributed random quantities become almost deterministic, with their distributions narrowly concentrated about their common ensemble average $\langle X \rangle$.

Another way to think about it is to observe that, adopting the notation,

$$S_n = X_1 + \cdots + X_n,$$

the dimensionless ratio,

$$\frac{\sigma_{S_n}}{\langle S_n \rangle} = \frac{1}{\sqrt{n}} \frac{\sigma_X}{\langle X \rangle}, \tag{19}$$

can be thought of as a measure of randomness of the random sum S_n. It clearly decays to 0 as $n \to \infty$.

The meaning of the above law of large numbers can be further elucidated by noticing that not only the mean square deviation of the averages from the statistical mean converge to 0, but also the probability of large deviations of the averages from the mean also decay to zero.

To prove this phenomenon, we shall employ the *Chebyshev's Inequality*,

$$\mathbf{P}(|X - \langle X \rangle| \geq \epsilon) = \langle \chi(|X - \langle X \rangle| - \epsilon) \rangle \leq \frac{\mathrm{Var}(X)}{\epsilon^2}, \tag{20}$$

which immediately follows from the positivity and linearity property of the ensemble averages, and by applying the observation that, for any $\epsilon > 0$,

$$|\chi(x - \epsilon)| \leq \frac{x^2}{\epsilon^2}, \qquad x \in \mathbf{R},$$

see Fig. 16.6.2, to $x = |X - \langle X \rangle|$.

Applied to the arithmetic average $\bar{X}_n$, in combination with the equality in (18) which computed the variance of $\bar{X}_n$), it immediately gives another version of the law of large numbers: For any $\epsilon > 0$,

$$P(|\bar{X}_n - \langle X \rangle| \geq \epsilon) \leq \frac{\mathrm{Var}(X)}{n\epsilon^2} \to 0, \qquad \text{as} \qquad n \to \infty. \tag{21}$$

In other words, the probability of large deviations of the average $\bar{X}_n$ from the statistical mean $\langle X \rangle$ becomes smaller, and smaller, as $n \to \infty$.

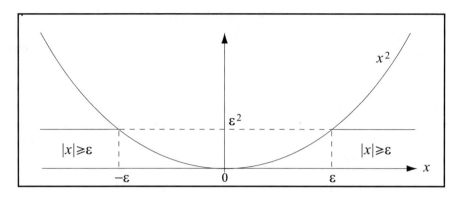

FIGURE 16.6.2.
Geometric interpretation of the Chebyshev's Inequality(20)

Given that, for any sequence of random quantities (Y_n), the following two types of convergence (as $n \to \infty$) are equivalent:

$$P(|Y_n| \geq \epsilon) \to 0, \quad \text{for any} \quad \epsilon > 0,$$

and

$$\langle g(Y_n) \rangle \to 0, \quad \text{for any bounded, continuous } g(x), \text{ with } g(0) = 0,$$

the latter being, of course, the weak convergence of the probability densities of Y_n's to the Dirac delta, $\delta(x)$, the above statement (21) is often called the *Weak law of large numbers*.

Remark 1. Transition from microscopic to macroscopic principles in physics. The law of large numbers discussed above represents one of the most fundamental principles in nature, explaining all kinds of phenomena. In particular, it explains the seeming contradiction between the chaotic behavior of particles at the microscopic molecular level, and the smooth hydrodynamic behavior in continuous media at the macroscopic level. In spite of the broad random distribution of velocities of gas molecules we can reliably measure the stable characteristics of gas, like pressure, temperature.

16.6.5. Stability of fluctuations in the Law of Large Numbers: Gaussian probability distributions and the Central Limit Theorem. In general, we say that probability distributions of random quantities $Y_n, n = 1, 2, \ldots$, *weakly converge* to the probability distribution of Y if, for every bounded continuous function $g(x)$,

$$\langle g(Y_n) \rangle = \int f_{Y_n}(x) g(x) \, dx \longrightarrow \int f_Y(x) g(x) \, dx = \langle g(Y) \rangle, \quad \text{as} \quad n \to \infty.$$
$$(22)$$

The above condition is equivalent to the convergence of characteristic functions,

$$\langle e^{i\omega Y_n} \rangle \to \langle e^{i\omega Y} \rangle, \quad n \to \infty, \tag{23}$$

for every $\omega \in \mathbf{R}$. Introduction of the concept of weak convergence, especially in the context of characteristic functions, immediately creates new opportunities for a more subtle study of the arithmetic averages, $\bar{X}_n = (X_1 + \cdots + X_n)/n$, of independent and identically distributes random quantities. In particular, this technique provides answers to questions about fluctuations of averages $\bar{X}_n$ about their limit, that is, their common mean $\langle X \rangle$.

But, first, note that control of the distributions of the arithmetic averages, $\bar{X}_n$, requires more than just uncorrelatedness of random quantities $X_1, X_2, \ldots$ which permitted cancellations of mixed terms in the quadratic forms which appeared in the proof of the weak law of large numbers. Here, we will need a stronger assumption that $X_1, X_2, \ldots$, are statistically independent, in which case the characteristic functions,

$$\Phi_{\bar{X}_n}(\omega) = \left\langle \exp\left(i\omega \frac{X_1 + \cdots + X_n}{n} \right) \right\rangle = \left[\Phi_X(\omega/n) \right]^n. \tag{24}$$

Thus, just the assumption of finiteness of the ensemble average $\langle X \rangle$ (and not necessarily of the variance, as was the case in the law of large numbers in the mean square) yields the asymptotic relation

$$\Phi_X(\omega) = 1 + i\langle X \rangle(\omega) + o(\omega) = \exp\left[i\langle X \rangle(\omega) + o(\omega) \right], \quad (\omega \to 0),$$

which immediately implies that, as $n \to \infty$,

$$\Phi_{\bar{X}_n}(\omega) = \exp\left[i\langle X \rangle(\omega) + no(\omega/n) \right] \to e^{i\langle X \rangle \omega},$$

which gives the following general stronger version of the Weak Law of Large numbers:

General Weak Law of Large Numbers: *If $X, X_1, X_2, \ldots$, are independent and identically distributed random quantities with finite first moments, then the probability distributions of the averages $\bar{X}_n$ weakly converge to the Dirac delta $\delta(x - \langle X \rangle)$, as $n \to \infty$.*

To take a closer look at fluctuations of arithmetic averages $\bar{X}_n$ about their mean $\langle X \rangle$ observe that formula (24) explicitly calculates their characteristic function in terms of the characteristic function of the independent summands X_n.

Then, note that although the probability distributions of the centered averages, $\bar{X}_n - \langle X \rangle$, collapse to the Dirac delta at zero, their renormalizations,

$$Y_n := \frac{\sqrt{n}}{\sigma_X} (\bar{X}_n - \langle X \rangle), \tag{25}$$

by the magnifying constant factors, $\sqrt{n}/\sigma(X)$, do not, since their variances (their existence must be assumed here !),

$$\text{Var}(Y_n) \equiv \text{Var}\left(\frac{\sqrt{n}}{\sigma_X}\left(\bar{X}_n - \langle X \rangle\right)\right) = 1, \qquad n = 1, 2, \ldots.$$

So, now, a study of the limit probability distribution of this renormalization becomes a sensible proposition. If the variance (second moment) of X is finite, the characteristic function $\Phi_X(\omega)$ enjoys the asymptotics

$$\Phi_X(\omega) = \exp\left[i\langle X \rangle \omega - \frac{\sigma_X^2}{2}\omega^2 + o(\omega^2)\right], \qquad (\omega \to 0).$$

Therefore, arguing as in the case of the general weak law of large numbers discussed earlier in this subsection, we see (always remembering the statistical independence of $X_1, X_2, \ldots$), that the characteristic functions of the renormalized arithmetic averages Y_n have the following asymptotic behavior:

$$\Phi_{Y_n}(\omega) = \left\langle \exp\left[i\omega\frac{\sqrt{n}}{\sigma}(\bar{X}_n - \langle X_1 \rangle)\right] \right\rangle = \left[\Phi_X\left(\frac{\omega}{\sigma_X\sqrt{n}}\right)\right]^n \exp\left[-i\omega\langle X \rangle\frac{\sqrt{n}}{\sigma_X}\right]$$

$$= \exp\left[-\frac{1}{2}\omega^2 + no\left(\frac{\omega^2}{n}\right)\right] \longrightarrow \exp\left(-\frac{1}{2}\omega^2\right), \qquad \text{as} \qquad n \to \infty,$$

with the limit equal to the characteristic function of the standard Gaussian probability distribution. Hence, we have proved the celebrated *Central Limit Theorem*, which is sometimes also called the *Stability of Fluctuations Law*.

Central Limit Theorem *If real-valued, independent, identically distributed random quantities $X, X_1, X_2, \ldots$, possess finite variance σ_X^2, then the probability distributions of the centered and renormalized arithmetic averages $Y_1, Y_2, \ldots$, (25), weakly converge, as $n \to \infty$, to the standard Gaussian probability distribution with the density,*

$$f_{Y_\infty}(x) = \frac{1}{\sqrt{2\pi}}\exp\left(-\frac{x^2}{2}\right).$$

Remark 1. The case of finite cumulants. In the case when all the cumulants $\kappa_n, n = 1, 2, \ldots$, of X are finite, we can rewrite the characteristic function of the renormalized centered averages,

$$Z_n = \frac{1}{\sqrt{n}}\sum_{k=1}^{n}(X_k - \langle X_k \rangle), \qquad (26)$$

as follows:

$$\Phi_{Z_n}(\omega) = \exp\left[n\sum_{m=1}^{\infty}\frac{\kappa_m}{m!}\left(\frac{i\omega}{\sqrt{n}}\right)^m\right]$$

$$= \exp\left(-\frac{\kappa_2}{2}\omega^2\right) \exp\left(-i\frac{\kappa_3}{6\sqrt{n}}\omega^3 + \frac{\kappa_4}{24n}\omega^4 + \dots\right). \qquad (27)$$

So, for any $\omega \in \mathbf{R}$,

$$\lim_{n\to\infty} \Phi_{Z_n}(\omega) = \Phi_{Z_\infty}(\omega) = \exp\left(-\frac{\kappa_2}{2}\omega^2\right),$$

and consequently, remembering that $\kappa^2 = \sigma_X^2$, probability density of the limit random quantity Z_∞ is Gaussian:

$$f_{Z_\infty}(x) = \frac{1}{\sqrt{2\pi}\sigma_X} \exp\left(-\frac{x^2}{2\sigma_X^2}\right). \qquad (28)$$

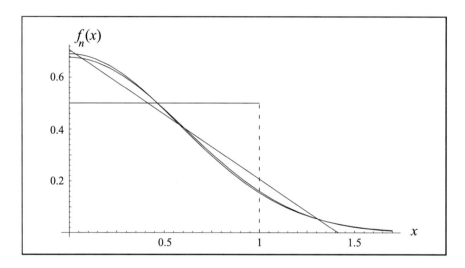

FIGURE 16.6.3.
Figure 16.6.3 shows (only on the positive half-line since the densities are symmetric) the plots of the corresponding probability density functions for Z_1 (rectangle), Z_2 (triangle), Z_8, and also the limit Gaussian density.

Example 1. From uniformly distributed X_n, to Gaussian Z_∞. As an explicit illustration of the above Central Limit Theorem behavior, let us consider the sequence of independent random quantities, $X, X_1, X_2, \dots$, with the identical uniform probability distribution on the interval $[-1, 1]$. Here, the mean $\langle X \rangle = 0$, and the characteristic function of the normalized sums (26) is

$$\Phi_{Z_n}(\omega) = \left(\frac{\sqrt{n}\sin(\omega/\sqrt{n})}{\omega}\right)^n.$$

Figure 16.6.3 shows (only on the positive half-line since the densities are symmetric) the plots of the corresponding probability density functions for Z_1, Z_2 and Z_8, and also the limit Gaussian density, which in this case is equal to

$$f_{Z_\infty}(x) = \frac{\sqrt{3}}{\sqrt{2\pi}} \exp\left(-\frac{3x^2}{2}\right). \tag{29}$$

Note that, for just eight independent, uniformly distributed on $[-1,1]$, random summands, the density of their normalized average is already very close to the limiting Gaussian density.

Remark 3. Accuracy of Gaussian approximation. Let us try, at the physical level of rigorousness, to estimate the number n of summands for which the density of the normalized sums (26) become practically Gaussian. Note that the deviation from the limit is determined by the second factor in (27) which is significantly different from 0 for ω approximately within the interval $[-1/\sqrt{\kappa_2}, 1/\sqrt{\kappa_2}]$. So, if at the boundaries of this interval the above factor

$$\exp\left(-i\frac{\kappa_3}{6\sqrt{n}}\omega^3 + \frac{\kappa_4}{24n}\omega^4 + \ldots\right)\Big|_{\omega=1/\sqrt{\kappa_2}} \approx 1, \tag{30}$$

we can expect the density of Z_n to approximate the Gaussian density well. It is clear that to achieve (30), the skewness and kurtosis characteristics of X, and n, have to satisfy the following conditions:

$$|\kappa_3| \ll 6\sqrt{n}, \qquad |\kappa_4| \ll 24n. \tag{31}$$

So, often, in this context, a good approximation is obtained already for $n = 10$; for $n = 20$, the approximation is excellent.

16.7 Gaussian random vectors

The Central Limit Theorem discussed in the previous section guarantees the prominent role of the Gaussian probability density in studies of real-valued random quantities. In physical and engineering problems, multidimensional random quantities appear naturally, and now we will turn to a brief study of the corresponding Gaussian random vectors in the n-dimensional Euclidean space $\mathbf{R}^n$.

Before formally defining the concept let us begin with an observation that if $(X_1, \ldots, X_n)$ and $(Y_1, \ldots, Y_n)$ are two independent random vectors with independent components such that, as $n \to \infty$, the renormalized sums

$$n^{-1/2} \sum_{k=1}^{n} X_k \longrightarrow X, \quad n^{-1/2} \sum_{k=1}^{n} Y_k \longrightarrow Y,$$

with the Gaussian limits X and Y, then, for linear combinations

$$Z_k = aX_k + bY_k, \qquad k = 1, \ldots, n,$$

the renormalized sums converge, as $n \to \infty$, to a limit Z, which also has a Gaussian probability distribution. Indeed,

$$n^{-1/2} \sum_{k=1}^{n} Z_k \longrightarrow Z = aX + bY.$$

The proof is easily accomplished using the characteristic functions technique.

The above observation justifies the following definition of a Gaussian random vector:

Definition 1. Gaussian random vectors: An n-dimensional random vector $\zeta = (Z_1, \ldots, Z_n)$ is said to be *Gaussian* if, for any $a_1, \ldots, a_n \in \mathbf{R}$, the linear combination

$$a_1 Z_1 + \cdots + a_n Z_n, \tag{1}$$

has a Gaussian probability distribution on $\mathbf{R}$.

The theory of random Gaussian vectors requires use of linear algebra and properties of matrices which we will summarize in the next subsection.

16.7.1.   A brief review of fundamentals of matrix theory.   Let us start with introduction of the basic concepts and notations of matrix theory. For an $n \times n$ matrix,

$$\mathbf{A} = (a_{ij}) = \begin{pmatrix} a_{11}, \ldots, a_{1j}, \ldots, a_{1n} \\ \cdots\cdots\cdots\cdots \\ a_{i1}, \ldots, a_{ij}, \ldots, a_{in} \\ \cdots\cdots\cdots\cdots \\ a_{n1}, \ldots, a_{nj}, \ldots, a_{nn} \end{pmatrix},$$

the *transposed matrix* $\mathbf{A}^T$ is defined by interchanging rows and columns in matrix $\mathbf{A}$

$$\mathbf{A}^T := (a_{ji}), \tag{2}$$

Matrix $\mathbf{A}$ is called *symmetric* if $\mathbf{A} = \mathbf{A}^T$, that is, if $a_{ij} = a_{ji}, i, j, = 1, \ldots, n$. Recall that the covariance matrices which appeared earlier in this chapter were symmetric matrices.

The (noncommutative !!) product of two matrices,

$$\mathbf{C} = \mathbf{AB} \iff c_{ij} = \sum_{m=1}^{n} a_{im} b_{mj}, \tag{3}$$

can be thought of as just a generalization of the operation of multiplication of a matrix by a vector,

$$y = Ax \iff y_i = \sum_{j=1}^{n} a_{ij} x_j, \tag{4}$$

and the scalar product of two vectors,

$$(x, y) = \sum_{j=1}^{n} x_j y_j. \tag{5}$$

Matrix multiplication is *associative*, that is,

$$A(BC) = (AB)C = ABC, \tag{6}$$

so that the parentheses in the above identities are not needed. The *inverse* A^{-1} of matrix A is defined by the condition

$$AA^{-1} = I = (\delta_{ij}),$$

where I is the identity matrix with the main diagonal entries $\delta_{ii} = 1$, and off diagonal entries $\delta_{ij} = 0, i \neq j$. The above definition immediately implies that

$$B = A^{-1} \iff A = B^{-1}, \quad \text{and} \quad (AB)^{-1} = B^{-1}A^{-1}, \tag{7}$$

so that, in particular,

$$(ABC)^{-1} = C^{-1}B^{-1}A^{-1}. \tag{8}$$

The determinant of a quadratic matrix is multiplicative, that is

$$\det(AB) = \det(A)\det(B), \qquad \det(A^T) = \det(A). \tag{9}$$

The covariance matrices Σ introduced earlier in this chapter are positive-definite, which implies that they possess a set of orthonormal *eigenvectors*,

$$\{\gamma_1, \ldots, \gamma_n\}, \qquad (\gamma_m, \gamma_l) = \delta_{ml}, \tag{10}$$

satisfying the condition,

$$\Sigma\gamma_m = \rho_m^2 \gamma_m, \tag{11}$$

where $\rho_m^2 \geq 0$ are the *eigenvalues* of the matrix Σ. In what follows we will assume that $\det(\Sigma) > 0$, so that the eigenvalues are strictly positive, $\rho_m^2 > 0$.
   Now consider the matrix

$$G = (\gamma_1, \ldots, \gamma_n) \tag{12}$$

the columns thereof are the eigenvectors of $\boldsymbol{\Sigma}$. The orthonormality implies immediately that in this case the transpose is identical with the inverse,

$$\boldsymbol{G}^T = (\delta_{ij}) \implies \boldsymbol{G}^T = \boldsymbol{G}^{-1}, \tag{13}$$

and the transformation,

$$\boldsymbol{G}^T \boldsymbol{\Sigma} \boldsymbol{G} = \boldsymbol{D} = (\rho_i^2 \delta_{ij}), \tag{14}$$

produces the diagonal matrix with eigenvalues of the matrix $\boldsymbol{\Sigma}$ on the main diagonals, and zeros off the diagonal. Thus $\boldsymbol{\Sigma}$ itself can be written in the form

$$\boldsymbol{\Sigma} = \boldsymbol{G} \boldsymbol{D} \boldsymbol{G}^T. \tag{15}$$

Now, the multiplicative property of the determinants, and the orthogonality of $\boldsymbol{G}$, and $\boldsymbol{G}^T$, gives us the formula

$$\det \boldsymbol{\Sigma} = \det \boldsymbol{D} = \prod_{k=1}^{n} \rho_k^2. \tag{16}$$

Moreover,

$$\boldsymbol{D}^{-1} = \left(\frac{\delta_{ij}}{\rho_i^2}\right) = \boldsymbol{G}^T \boldsymbol{\Sigma}^{-1} \boldsymbol{G} \qquad \Longleftrightarrow \qquad \boldsymbol{\Sigma}^{-1} = \boldsymbol{G} \boldsymbol{D}^{-1} \boldsymbol{G}^T. \tag{17}$$

Finally, observe that an arbitrary quadratic function in n variables $x_1, \ldots, x_n$, can be written in the matrix/scalar product form,

$$\sum_{k=1}^{n} q_{ij} x_i x_j = (\boldsymbol{x}, \boldsymbol{Q} \boldsymbol{x}) = (\boldsymbol{Q}^T \boldsymbol{x}, \boldsymbol{x}). \tag{18}$$

16.7.2. Densities of Gaussian random vectors. Let us begin by finding the probability density function of a Gaussian random vector

$$\boldsymbol{Y} = \boldsymbol{A} \boldsymbol{X}, \tag{19}$$

obtained by a linear transformation, via matrix $\boldsymbol{A}$, of a Gaussian random vector $\boldsymbol{X} = (X_1, \ldots, X_n)$, with statistically independent components with zero mean and the diagonal covariance matrix,

$$\boldsymbol{D} = \left(\langle X_m X_l \rangle\right) = \left(\rho_m^2 \delta_{ml}\right). \tag{19}$$

The covariances of the Gaussian components $Y_1, \ldots, Y_n$ of $\boldsymbol{Y}$,

$$\sigma_{ij} = \langle Y_i Y_j \rangle = \sum_{m=1}^{n} \sum_{l=1}^{n} a_{im} a_{jl} \langle X_m X_l \rangle = \sum_{m=1}^{n} a_{im} a_{jm} \rho_m^2,$$

so that the covariance matrix $\boldsymbol{\Sigma}$ of $\boldsymbol{Y}$ can be written in the form

$$\boldsymbol{\Sigma} = \boldsymbol{A} \boldsymbol{D} \boldsymbol{A}^T. \tag{20}$$

Thus, in view of the change-of-variables formula in Subsection 16.5.3, the probability density of $\boldsymbol{Y}$,

$$f_Y(\boldsymbol{y}) = f_X(\boldsymbol{A}^{-1}\boldsymbol{y})|\det \boldsymbol{A}^{-1}|, \tag{21}$$

where

$$f_X(\boldsymbol{x}) = \frac{1}{(\sqrt{2\pi})^n \sqrt{\det \boldsymbol{D}}} \exp\left(-\frac{1}{2}\sum_{k=1}^{n} \frac{x_k^2}{\rho_k^2}\right)$$

$$= \frac{1}{(\sqrt{2\pi})^n \sqrt{\det \boldsymbol{D}}} \exp\left(-\frac{1}{2}(\boldsymbol{x}, \boldsymbol{D}^{-1}\boldsymbol{x})\right), \tag{22}$$

since the density of the random vector with statistically independent components is the product of individual marginal densities of the components.

Now, using (22), (21) can be rewritten as follows:

$$f_Y(\boldsymbol{y}) = \frac{1}{(\sqrt{2\pi})^n \sqrt{\det \boldsymbol{D} (\det \boldsymbol{A}|)^2}} \exp\left(-\frac{1}{2}(\boldsymbol{A}^{-1}\boldsymbol{y}, \boldsymbol{D}^{-1}\boldsymbol{A}^{-1}\boldsymbol{y})\right),$$

$$= \frac{1}{(\sqrt{2\pi})^n \sqrt{\det \boldsymbol{D} (\det \boldsymbol{A}|)^2}} \exp\left(-\frac{1}{2}(\boldsymbol{y}, (\boldsymbol{A}^{-1})^T \boldsymbol{D}^{-1} \boldsymbol{A}^{-1}\boldsymbol{y})\right),$$

so that, finally, in view of (20),

$$f_Y(\boldsymbol{y}) = \frac{1}{(\sqrt{2\pi})^n \sqrt{\det \boldsymbol{\Sigma}}} \exp\left(-\frac{1}{2}(\boldsymbol{y}, \boldsymbol{\Sigma}^{-1}\boldsymbol{y})\right), \tag{23}$$

where $\boldsymbol{\Sigma}$ is the covariance matrix of the Gaussian random vector $\boldsymbol{Y}$.

In the case of a general Gaussian random vector $\boldsymbol{Y}$, with the nonzero mean $\boldsymbol{\mu}$, the density (23) just needs to be shifted, so that

$$f_Y(\boldsymbol{y}) = \frac{1}{(\sqrt{2\pi})^n \sqrt{\det \boldsymbol{\Sigma}}} \exp\left(-\frac{1}{2}\big((\boldsymbol{y} - \boldsymbol{\mu}), \boldsymbol{\Sigma}^{-1}(\boldsymbol{y} - \boldsymbol{\mu})\big)\right). \tag{24}$$

16.7.3.   Explicit formulas for densities of 2-D Gaussian random vectors.   In the two-dimensional case we can write the densities of Gaussian random vectors, (X, Y), more explicitly. The density is determined by its moments of first and second order, including the mixed second moments,

$$\langle X \rangle = \mu_X, \ \langle Y \rangle = \mu_Y, \ \text{Var}\, X = \sigma_X^2, \ \text{Var}\, Y = \sigma_Y^2, \ \text{Cov}\,(X, Y) = \sigma_X \sigma_Y \rho,$$

where $\rho = \rho_{(X,Y)}$ stands for the correlation coefficient of the components X, and Y. The covariance matrix and its determinant are

$$\mathbf{\Sigma} = \begin{pmatrix} \sigma_X^2, & \sigma_X\sigma_Y\rho, \\ \sigma_X\sigma_Y\rho, & \sigma_Y^2 \end{pmatrix}, \qquad \det \mathbf{\Sigma} = \sigma_X^2\sigma_Y^2(1-\rho^2), \qquad (25)$$

with the inverse,

$$\mathbf{\Sigma} = \begin{pmatrix} \frac{1}{\sigma_X^2(1-\rho^2)}, & \frac{-\rho}{\sigma_X\sigma_Y(1-\rho^2)} \\ \frac{-\rho}{\sigma_X\sigma_Y(1-\rho^2)}, & \frac{1}{\sigma_Y^2(1-\rho^2)} \end{pmatrix}.$$

Substituting this expression into the two-dimensional version of (24) we obtain the general formula for the probability density of a two-dimensional Gaussian random vector in terms of its moments,

$$f_{(X,Y)}(x,y) = \frac{1}{2\pi\sigma_X\sigma_Y\sqrt{1-\rho^2}} \times$$

$$\exp\left[-\frac{1}{2(1-\rho^2)}\left(\frac{(x-\mu_X)^2}{\sigma_X^2} - \frac{2\rho(x-\mu_X)(y-\mu_Y)}{\sigma_X\sigma_Y} + \frac{(y-\mu_Y)^2}{\sigma_Y^2}\right)\right].$$

In the case of uncorrelated components, the correlation coefficient $\rho = 0$, and the above join probability density splits into two factors,

$$f_{(X,Y)}(x,y) = \frac{1}{\sqrt{2\pi}\sigma_X}\exp\left[-\frac{(x-\mu_X)^2}{2\sigma_X^2}\right] \times \frac{1}{\sqrt{2\pi}\sigma_Y}\exp\left[-\frac{(x-\mu_Y^2}{2\sigma_Y^2}\right],$$

so *uncorrelated Gaussian random quantities are necessarily independent.*

In the special case of components with zero means the joint density of (X,Y), is, simply,

$$f_{(X,Y)}(x,y) = \frac{1}{2\pi\sigma_X\sigma_Y\sqrt{1-\rho^2}}\exp\left[-\frac{1}{2(1-\rho^2)}\left(\frac{x^2}{\sigma_X^2} - \frac{2\rho xy}{\sigma_X\sigma_Y} + \frac{y^2}{\sigma_Y^2}\right)\right].$$

$$(26)$$

In numerous applications, in addition to knowledge of joint densities, $f(x,y)$, it is essential to obtain the conditional probability densities, $f(x|y)$ of random quantity X, given that the value of $Y = y$. In the case of Gaussian random vectors, we can obtain these conditional densities with the help of the fact that statistical independence and uncorrelatedness are equivalent here. We shall demonstrate how to accomplish it by considering the following example.

Example 1. Conditional Gaussian densities. Consider two random quantities

$$X = aX_1 + bX_2, \qquad Y = cX_2, \qquad (27)$$

where X_1 and X_2 are independent, zero-mean Gaussian random quantities with variance 1. Given the coefficients a, b, c, in (27), we can calculate the entries of the covariance matrix (25) of the random vector (X,Y):

$$\sigma_X^2 = a^2 + b^2, \quad \sigma_Y^2 = c^2, \quad \sigma_X\sigma_Y\rho = bc.$$

Thus,

$$c = \sigma_Y, \quad b = \sigma_X \rho, \quad \text{and} \quad a = \sigma_X \sqrt{1 - \rho^2}. \tag{28}$$

Let's find the statistical characteristics of X, given $Y = y$. From the second equality in (27), and (28), we see that the conditioning is equivalent to the statement that $X_2 = y/\sigma_Y$, which gives

$$X_{|y} = \sigma_X \sqrt{1 - \rho^2} X_1 + \frac{\sigma_X}{\sigma_Y} \rho y.$$

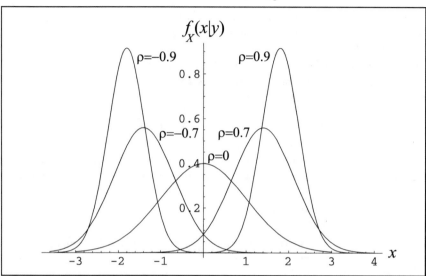

FIGURE 16.7.1.
Plots of conditional probability density functions (30) for $\sigma_X = \sigma_Y = 1$ and $y = 2$, and several values of the correlation coefficient ρ.

Hence, $X_{|y}$ is also a Gaussian random quantity with the *conditional mean*, and *conditional variance*,

$$\langle X | Y = y \rangle = \mu_{X|y} = \frac{\sigma_X}{\sigma_Y} \rho y, \qquad \sigma^2_{X|y} = \sigma^2_X (1 - \rho^2). \tag{29}$$

Substituting the above parameters into the general expression for Gaussian probability density function, we obtain the formula for the *conditional density function*,

$$f_X(x | Y = y) = \frac{1}{\sqrt{2\pi(1 - \rho^2)} \sigma_X} \exp\left[-\frac{1}{2(1 - \rho^2)} \left(x - \frac{\sigma_X}{\sigma_Y} \rho y \right)^2 \right]. \tag{30}$$

In the case of (X, Y) with nonzero means, μ_X and μ_Y, conditional density takes the form.

$$f_X(x | Y = y) = \frac{1}{\sqrt{2\pi(1 - \rho^2)} \sigma_X} \exp\left[-\frac{1}{2(1 - \rho^2)} \left((x - \mu_X) - \frac{\sigma_X}{\sigma_Y} \rho (y - \mu_Y) \right)^2 \right]. \tag{31}$$

For $\rho = 0$, the above conditional densities become just the usual Gaussian densities of X. However, as $|\rho|$ increases, the conditional densities concentrate more, and more, around the conditional means. The phenomenon is illustrated in Fig. 16.7.1. In the limit $|\rho| \to 1$, they weakly converge to the Dirac delta,

$$\lim_{\rho \to 1} f_X(x|Y = y) = \delta\left((x - \mu_X) - \frac{\sigma_X}{\sigma_Y}(y - \mu_Y)\right).$$

16.7.4. Characteristic functions of Gaussian random vectors.
To calculate the characteristic function of the n-dimensional Gaussian random vector $\boldsymbol{Y}$ with mean $\boldsymbol{\mu}$, and covariance matrix $\boldsymbol{\Sigma}$, let us first consider an auxiliary Gaussian real-valued random quantity $Z = (\boldsymbol{\omega}, \boldsymbol{Y})$, where $\boldsymbol{\omega}$ is an arbitrary n-dimensional deterministic vector. Then the characteristic function of $\boldsymbol{Y}$ is given by the formula,

$$\Phi_Y(\boldsymbol{\omega}) = \langle e^{i(\boldsymbol{\omega}, \boldsymbol{Y})}\rangle = \exp\left[i(\boldsymbol{\mu}, \boldsymbol{\omega}) - \frac{1}{2}(\boldsymbol{\omega}, \boldsymbol{\Sigma}\boldsymbol{\omega})\right]. \tag{32}$$

In particular, the two-dimensional characteristic function of a Gaussian random vector (X, Y), with parameters described in Subsection 16.7.3, is of the form,

$$\Phi_{(X,Y)}(\omega, \psi) = \langle e^{i(\omega X + \psi Y)}\rangle$$
$$= \exp\left[i\mu_X \omega + i\mu_Y \psi - \frac{1}{2}\left(\sigma_X^2 \omega^2 + \sigma_Y^2 \psi^2 + 2\sigma_X \sigma_Y \rho \omega \psi\right)\right]. \tag{33}$$

In the often studied in applications case of zero means and equal variances σ^2 of the components, the above formula becomes particularly simple,

$$\Phi_{(X,Y)}(\omega, \psi) = \exp\left[-\frac{\sigma^2}{2}\left(\omega^2 + \psi^2 + 2\rho\omega\psi\right)\right]. \tag{34}$$

Since characteristic functions contain full information about the probability distributions of the corresponding random vectors, they can be employed to calculate other statistical characteristics of the latter. Often, this approach is easier to implement than performing calculations involving the original probability density functions. We will illustrate the power of this tool on several examples.

Example 1. Conditional characteristic functions. Earlier in this chapter we have found the probability density functions of a product $Z = XY$ of two Gaussian random quantities. Now, let us take a look at a more complicated problem involving the random quantity,

$$Z = X_1 Y_1 + X_2 Y_2, \tag{35}$$

where (X_1, X_2, Y_1, Y_2) is a Gaussian random vector with zero means and equal variances of σ^2 of the independent components. Let us begin by finding the conditional probability density of Z given the values of X_1, and X_2. In this case, Z is simply a linear combination of the components of (Y_1, Y_2), with the characteristic function

$$\Phi_{Z|X_1 X_2}(\omega) = \langle e^{i\omega Z|X_1 X_2} \rangle = \exp\left[-\frac{\sigma^2\omega^2}{2}(X_1^2 + X_2^2)\right].$$

Averaging it over (X_1, X_2) we observe that, in view of the statistical independence of X_1, and X_2, the mean splits into the product of two identical factors,

$$\left\langle \exp\left(-\frac{\sigma^2\omega^2}{2}X_1^2\right)\right\rangle = \frac{1}{\sqrt{2\pi}\sigma}\int \exp\left[-\frac{1}{2}\left(\sigma^2\omega^2 + \frac{1}{\sigma^2}\right)x^2\right]dx = \frac{1}{\sqrt{1 + \sigma^4\omega^2}}.$$

Thus the characteristic function of Z is the square of the above expression,

$$\Phi_Z(\omega) = \frac{1}{1 + \sigma^4\omega^2}.$$

Taking the inverse Fourier transform gives us the probability density function of Z,

$$f_Z(z) = \frac{1}{2\sigma^2}\exp\left(-\frac{|z|}{\sigma^2}\right) \tag{36}$$

which we recognize as the well-known *Laplace probability distribution*.

Calculation of the covariance (as a function of ρ), of the transformations, $g(X)$, and $g(Y)$, of random quantities, X and Y,

$$B(\rho) = \langle g(X)g(Y)\rangle = \int\int g(x)g(y)f(x, y)\,dxdy, \tag{37}$$

can also be easily accomplished via the characteristic function. Let us first derive a useful relationship between $B(\rho)$ and $g(x)$. Differentiating (34) n times we get the expression,

$$\frac{\partial^n}{\partial\rho^n}\Phi(\omega, \psi) = (-1)^n\sigma^{2n}(\omega\psi)^n\Phi(\omega, \psi).$$

Taking the inverse Fourier transform yields the equation,

$$\frac{\partial^n}{\partial\rho^n}f = \sigma^{2n}\frac{\partial^{2n}}{\partial x^n\partial y^n}f, \tag{38}$$

expressed in the language of the probability density function. Now, differentiating (37) n times in ρ, and taking into account (39), we obtain that

$$\frac{d^n B(\rho)}{d\rho^n} = \sigma^{2n}\int\int g(x)g(y)\frac{\partial^{2n}f}{\partial x^n\partial y^n}dxdy,$$

which, after integration by parts, leads to the following equation for $B(\rho)$:

$$\frac{d^n B(\rho)}{d\rho^n} = \sigma^{2n}\left\langle g^{(n)}(X)g^{(n)}(Y)\right\rangle, \tag{39}$$

which we will exploit in the next four examples.

Example 2. Linear transformations. Take $g(x) = x$. Equation (38), for $n = 1$, gives

$$\frac{dB(\rho)}{d\rho} = \sigma^2.$$

Observing that the initial condition

$$B(\rho)|_{\rho=0} = \langle X \rangle \langle Y \rangle = 0,$$

the unique solution of the above equation is

$$B(\rho) = \sigma^2 \rho = \langle XY \rangle.$$

Example 3. Quadratic transformations. Here, $g(x) = x^2$, and $n = 2$ in (38), so that

$$\frac{d^2 B(\rho)}{d\rho^2} = 4\sigma^4.$$

The obvious initial conditions are

$$B(\rho)|_{\rho=0} = \langle X^2 \rangle \langle Y^2 \rangle = \sigma^4, \qquad \frac{dB(\rho)}{d\rho}\bigg|_{\rho=0} = 4\sigma^2 \langle X \rangle \langle Y \rangle = 0,$$

so that

$$B(\rho) = \sigma^2(1 + 2\rho^2) = \langle X^2 Y^2 \rangle. \tag{40}$$

Example 4. Digitizing transformations. Many useful applications transformations convert continuous data into digital data. An example of the relevant function is

$$g(x) = \text{sgn}(x),$$

with the derivative $g'(x) = 2\delta(x)$. Consequently, in view of (39),

$$\frac{dB(\rho)}{d\rho} = 4\sigma^2 \langle \delta(X)\delta(Y) \rangle = 4\sigma^2 f(x = 0, y = 0).$$

The value $f(0,0)$ can be found by substituting in (26), $x = y = 0, \sigma_X = \sigma_Y = \sigma$, which yields the equation,

$$\frac{dB(\rho)}{d\rho} = \frac{2}{\pi\sqrt{1 - \rho^2}}, \qquad B(\rho)|_{\rho=0} = 0,$$

the unique solution thereof is

$$B(\rho) = \frac{2}{\pi} \int_0^\rho \frac{d\rho}{\pi\sqrt{1 - \rho^2}} = \frac{2}{\pi} \arcsin(\rho). \tag{41}$$

Example 4. Product of absolute values. Finally, let us consider the correlation of the "amplitudes" of two Gaussian random quantities,

$$B(\rho) = \langle |X||Y| \rangle.$$

From (38), and Example 3, it follows that

$$\frac{dB(\rho)}{d\rho} = \sigma^2 \langle \text{sgn}(X)\text{sgn}(Y) \rangle = \sigma^2 \frac{2}{\pi} \arcsin(\rho).$$

In this case it is more convenient to consider the initial condition at $\rho = 1$,

$$B(\rho)|_{\rho=1} = \langle X^2 \rangle = \sigma^2.$$

As a result, we, finally, obtain

$$B(\rho) = \sigma^2 \left(1 - \frac{2}{\pi} \int_\rho^1 \arcsin(\rho) d\rho \right) = \sigma^2 \frac{2}{\pi} \left(\sqrt{1 - \rho^2} + \rho \arcsin(\rho) \right). \quad (42)$$

16.8   Level crossing and fixed points problems for random functions

In this section we describe two examples of physical problems that can be solved using a combination of distributional and statistical techniques developed earlier in this chapter.

16.8.1.   Level crossing problem.   In our youth we all enjoyed reading Jules Verne's novel about the brave Phileas Fogg who wagered that he could travel around the globe in 80 days. On his way, he would run into various obstacles which nearly prevented him from winning the wager. Following with excitement the development of events, we paid no attention to another aspect of his adventures: What should be fair odds in the wager between the supporters and opponents of Phileas Fogg. Before we proceed to resolve this dilemma we need to formulate it in rigorous scientific terms.

Let $\xi(t)$ be the distance covered by Ph. Fogg during time t. Because of random breaks in his progress, $\xi(t)$ is a random (stochastic) process whose statistical properties are prescribed by an ensemble of trips undertaken by several Ph. Foggs, when each Ph. Fogg overcomes his own random difficulties. If one is familiar with statistics of train catastrophes, and other (in insurance companies' parlance) "acts of God," one can calculate statistical properties of the process $\xi(t)$. However, in order to find correct odds in our wager, one needs to know the statistics of the random time needed to complete the trip around the globe. The corresponding mathematical problem can be formulated as follows. Let τ be a solution of the equation,

$$\xi(\tau) = L$$

where L is the total distance to be covered by the traveler. So, using the known statistics of the random process $\xi(t)$, we need to find probabilistic properties of the random quantity τ and, in particular, the probability that $\tau < t^*$, where t^* is a given time (in our case, 80 days). It is exactly the probability that Ph. Fogg's supporters win the wager. Different modifications of the above problem emerge in various physical and engineering applications such as statistical analysis of reliability problems involving the mean failure times of electronic devices. They are collectively known under the term *level crossing problems* for random processes. The tools of distribution theory offer an elegant solution to some of these problems. The following example illustrates our approach.

Example 1. Level crossing for monotone random function.   Let $\eta = \alpha(x)$ be a monotonically increasing random function whose probability density $f_\alpha(y; x) = \langle \delta(\alpha(x) - y) \rangle$ is known. Here, and in what follows, the arguments and parameters of probability densities will be separated by semicolons, with the arguments appearing to the left of it, while parameters appear to the right. In this case, y is the argument of the probability density reflecting statistical properties of the process $\alpha(x)$ for the given value of parameter x. Our job is to find the probability density, $f_\beta(y; x) = \langle \delta(\beta(y) - x) \rangle$, of the random function $\xi = \beta(y)$, inverse to function $\eta = \alpha(x)$.

The relation between these two probability densities will be found through averaging the differential equality,

$$\frac{\partial}{\partial x}\delta(\alpha(x) - y) + \frac{\partial}{\partial y}\delta(\beta(y) - x) = 0,$$

over the statistical ensemble. As a result, taking into account the above definitions of the probability densities involved, we shall get that

$$\frac{\partial}{\partial x}f_\alpha(y; x) + \frac{\partial}{\partial y}f_\beta(x; y) = 0, \tag{1}$$

Solving this equation with respect to the unknown density function $f_\beta(x; y)$ we finally obtain that

$$f_\beta(x; y) = -\int_{-\infty}^{y} \frac{\partial}{\partial x}f_\alpha(z; x)\, dz. \tag{2}$$

The above equality can also be readily obtained from purely probabilistic considerations, noting that, in view of the monotonic nature of functions $\alpha(x)$ and $\beta(y)$, we have the equality of probabilities,

$$P(\alpha(x) > y) = P(\beta(y) < x).$$

Rewriting it in terms of the cumulative probability distribution functions gives the equality,

$$1 - F_\alpha(y; x) = F_\beta(x; y),$$

which is clearly equivalent to (2). However, in more complex, and more interesting cases, where a nonmonotone function of a random vector field may be of interest, the purely probabilistic arguments are not that obvious, and formalized manipulations with the help of distribution theory provide a quicker answer.

16.8.2.   Average number of level crossings by a random process. Singular generalized functions of nonmonotonic composite argument, prohibited in the rigorous theory, are used widely when solving problems related to calculation of statistics of the number of level crossings by a random function. In particular, the average number of crossings can be obtained averaging the functional (2.9.11) in Volume 1. Interchanging the averaging operation with the integral yields the formula,

$$\langle N(y, z)\rangle = \int_0^z \langle|\alpha'(x)|\delta(\alpha(x) - y)\rangle\, dx, \tag{3}$$

for the average number of crossings.

If we denote by $f(\alpha, \gamma; x)$ the joint probability density of the process $\alpha(x)$ and its derivative $\alpha'(x)$, then the average inside the integral in (3) can be rewritten as follows:

$$\langle N(y, x)\rangle = \langle|\alpha'(x)|\delta(\alpha(x) - y)\rangle = \int |\gamma| f(y, \gamma; x)\, d\gamma.$$

This average is obviously equal to the average frequency of crossings, i.e., to the number of level crossings per unit length of the x-axis.

Example 1. Frequency of level crossing for Gaussian processes. If $\alpha(x)$ is a stationary Gaussian random process, then the joint probability density of $\alpha(x)$ and its derivative does not depend on x, and it factors into the product of probability densities of α and α':

$$f(\alpha, \gamma; x) = \frac{1}{2\pi\sigma\sigma'} \exp\left[-\frac{(\alpha - \langle\alpha\rangle)^2}{2\sigma^2} - \frac{\gamma^2}{2\sigma'^2}\right]. \tag{4}$$

Here, σ^2 and σ'^2 are variances of random functions $\alpha(x)$ and $\alpha'(x)$, and $\langle\alpha\rangle$ is the ensemble average of the process $\alpha(x)$. Inserting the density function (4) in formula (3) yields the average frequency of crossings

$$\langle N(y, x)\rangle = \frac{\sigma'}{\pi\sigma} \exp\left[-\frac{y - \langle\alpha\rangle)^2}{2\sigma^2}\right].$$

Note that the frequency of crossings is maximal at the "median line" $y = \langle\alpha\rangle$. This is not surprising since the process has a tendency to return to its "median line". Moreover, the frequency of crossings rapidly decays with growth of the distance from the "median line" (i.e., for $|y - \langle\alpha\rangle| > \sigma$).

16.8.3.   The fixed point problem.   The fixed point problem for random mappings is closely related to the level crossing problem considered above. Notice that the random function $\eta = \alpha(x)$ represents a random point mapping from $\mathbf{R}$ into $\mathbf{R}$. Assume, for the sake of definiteness, that the mapping is differentiable, monotonically increasing, and is a contraction, that is it satisfies the condition,

$$0 < \alpha'(x) < 1,$$

for every $x \in \mathbf{R}$. In view of the Mean Value Theorem of calculus, this implies that

$$|\alpha(x_1) - \alpha(x_2)| < |x_1 - x_2|,$$

for any $x_1, x_2 \in \mathbf{R}$—hence the name contraction. Such a mapping has a unique fixed point ξ^* satisfying equation

$$\alpha(\xi^*) = \xi^*.$$

To find the fixed point start with an arbitrary initial number x_0 and consider a sequence of random numbers defined recursively by the formula,

$$\xi_{n+1} = \alpha(\xi_n).$$

It is not hard to prove that the sequence converges to the unique fixed point. The general class of results of this type is known in mathematics as the *Banach Fixed Point Theorem* (see, e.g., Granas and Dugundji [10]).

A question arises then as to how the probability density of a fixed point ξ^* is expressed through the known statistics of the mapping $\alpha(x)$ itself. The answer can be obtained by considering the Dirac delta $\delta(\alpha(x) - x)$. In view of the properties of the Dirac delta,

$$(1 - \alpha'(x))\delta(\alpha(x) - x) = \delta(x - \xi^*).$$

Furthermore, since

$$\alpha'(x)\delta(\alpha(x) - x)\Big|_{y=x} = \delta(\beta(y) - x)\Big|_{y=x},$$

where $\xi = \beta(y)$ is the inverse point mapping, we can rewrite the preceding equality in the form,

$$\delta(x - \xi^*) = \delta(\alpha(x) - x)\delta(\beta(y) - x)\Big|_{y=x}.$$

Finally, averaging it, and recalling the definition of the density function with the help of the average and the Dirac delta, we arrive at the equality,

$$f_*(x) = f_\alpha(x; x) - f_\beta(x; x), \tag{5}$$

where by $f_*(x)$ we denoted the desired probability density of the fixed point of the random point mapping. Using equality (16.8.2) we can exclude $f_\beta(x; x)$ from (5) and obtain that

$$f_*(x) = \frac{\partial}{\partial x} \int_{-\infty}^{x} f_\alpha(z; x)\, dz. \tag{6}$$

Then, via a purely probabilistic argument, we can show that

$$F_*(x) = F_\alpha(x; x),$$

which gives a relationship between the cumulative distribution functions of a random fixed point ξ^*, and the CDF of a point mapping $\alpha(x)$, which is equivalent to formula (6).

16.9 Exercises

1. In modeling and simulation of random phenomena it is often necessary to produce random numbers with the desired cumulative distribution function $F_Y(y)$, by superposing a function $g(x)$ on the outcomes produced by a random number generator producing uniformly distributed random numbers on the interval $(0, 1)$. Find $g(x)$, for the exponential CDF,

$$F_Y(y) = 1 - e^{-\mu y}, \qquad \text{for} \qquad y \geq 0,$$

and equal to zero on the negative half-line. *Hint:* Let $y = g(x)$ be a monotonically increasing function defined for all $x \in \mathbf{R}$. In this case, to find the probability distribution of the random quantity $Y = g(X)$, given that the probability distribution of X is known, it is most convenient to observe that the corresponding cumulative distribution functions are related by the equation,

$$F_Y(y) = F_X(g^{-1}(y)), \tag{1}$$

2. Let X be a real-valued random quantity, with a continuously differentiable CDF, $f_X(x)$, transmitted through a linear detector with the transfer function

$$y = g(x) = x\chi(x).$$

Find the CDF and PDF of the random quantity $Y = g(X)$.

3. Find the CDF and PDF of the random quantity Y on the output of the exponential detector

$$Y = \exp(sX), \tag{2}$$

where $s > 0$, and $m \in \mathbf{R}$ are given constants, if the input is a random quantity of Gaussian probability distribution with mean zero and variance one.

4. Assume that in (2), $m = -s^2/2$, and denote the corresponding random quantity by $Y(s)$. Calculate its mean, variance, and the probability that it exceeds its mean value. Investigate the behavior of these characteristics as s increases.

5. Taking advantage of the technique of characteristic functions, calculate all the moments of order $n = 1, 2, \ldots$, of the Gaussian random quantity X with mean zero and variance σ^2. Then, also calculate the following means,

$$\langle e^{aX} \cos(bX) \rangle, \qquad \langle X \sin(bX) \rangle.$$

6. Let $Z = X_1 + X_2 + X_3 + X_4$, be the sum of four independent random quantities with uniform probability density function on the interval $[-L, L]$. Find all the moments of Z.

7. Taking advantage of the additivity property of the cumulants for sums of independent random quantities, calculate the first four cumulants, and central moments, of the binomial probability distribution.

8. Find the kurtosis, κ_4, of the random quantity with the PDF,

$$f(x) = p\delta(x) + q\frac{1}{\sqrt{2\pi}\sigma} \exp\left(-\frac{x^2}{2\sigma^2}\right), \qquad p + q = 1.$$

9. Derive the formula (20) in Subsection 16.5.2 from (19) and (17).

10. Prove that the 2-D random vector with uniform PDF on the unit disk has uncorrelated, but not independent components.

11. Prove that not only the first, but all the interarrival times in the Poisson process have exponential PDFs.

12. Calculate the mean and variance of the Poisson probability distribution.

Chapter 17

Random Distributions: Generalized Stochastic Processes

In this chapter we will study a formal mathematical theory of random distributions or, in other words, distributions with values that are not real or complex numbers, but random quantities. Such random distributions can be thought of as generalized stochastic processes, or generalized stochastic fields, depending on whether we are discussing distributions on one-dimensional, or multi-dimensional spaces. The origins of the theory can be traced back to the 1953 works of Itô [1], and the monograph by Gelfand and Vilenkin [2], Chapter III. We start with the classical functional approach, but later also explain how the theory can be developed via sequential approximations by regular random functions, the approach which may be more appealing to the practitioners in the physical and engineering sciences.

17.1 Random distributions: basic concepts

Recall (see Volume 1, p.8, and Appendix A), that, by definition, a deterministic distribution (generalized function) on $\mathbf{R}$ is a real-valued linear continuous functional on the space $\mathcal{D}$ of infinitely differentiable functions ϕ with compact support on $\mathbf{R}$, equipped with the topology of uniform convergence, and also for all its derivatives, on each compact set. The concept of random distribution extends the idea of a deterministic distribution by allowing the linear functional (operator in this case) to take values in a certain space of random quantities.[1]

---

[1] We will stick here to the terminology of "random quantities" rather than the standard in mathematical literature term "random variables" as it seems to be more intuitive for our intended audience.

© Springer International Publishing AG, part of Springer Nature 2018
A. I. Saichev and W. A. Woyczynski, *Distributions in the Physical and Engineering Sciences, Volume 3*, Applied and Numerical Harmonic Analysis, https://doi.org/10.1007/978-3-319-92586-8_17

Formally, a *random distribution*, also commonly called a *generalized random function*, or *generalized stochastic process*, or *field*, depending on whether we are discussing distributions on $\mathbf{R}^1$, or $\mathbf{R}^n$, $n > 1$, is a continuous linear mapping

$$\mathcal{D} \ni \phi \longmapsto X[\phi] \in L^2 \tag{1}$$

where

$$L^2 = \{X : \langle X^2 \rangle < \infty\} \tag{2}$$

denotes the space of random quantities X with finite second moment, or equivalently, finite variance. One can also consider other random distributions with values in more general spaces of random quantities, and we will discuss them later in this chapter. Since condition (2) assures that the ensemble means, $\langle X[\phi] \rangle$, are well defined, without loss of generality we can subtract this deterministic distribution from X, and consider only the "fluctuating" part of the random distributions which has the mean equal to zero. Thus our standing assumption in this section (unless explicitly stated otherwise) is that

$$\langle X[\phi] \rangle = 0, \qquad \phi \in \mathcal{D}. \tag{3}$$

The continuity of the random functional $X[\phi]$ is understood in the sense that if

$$\phi_n \to \phi \in \mathcal{D},$$

then

$$\mathrm{Var}\,(X[\phi_n] - X[\phi]) = \langle (X[\phi_n] - X[\phi])^2 \rangle \to 0, \tag{4}$$

as $n \to \infty$.

The probabilistic properties of the generalized random function are completely determined by the family of one-dimensional probability distribution functions,

$$F_{X[\phi]}(x) = P(X[\phi] \le x) = \langle 1 - \chi(X[\phi] - x) \rangle, \quad \phi \in \mathcal{D}. \tag{5}$$

or the densities,

$$f_{X[\phi]}(x) = \frac{\partial}{\partial x} P(X[\phi] \le x) = \langle \delta(X[\phi] - x) \rangle, \quad \phi \in \mathcal{D}. \tag{6}$$

In particular, the *covariance bilinear functional*

$$\mathrm{Cov}_X[\phi, \psi] = \langle X[\phi] X[\psi] \rangle,$$

and other higher moments and cross-moments of the random distribution X (if they exist) are directly computable given the information in (5), and (6). Indeed, for example,

$$\text{Cov}_X[\phi, \psi] = \langle X[\phi]X[\psi]\rangle = \left\langle X^2\left[\frac{\phi+\psi}{4}\right]\right\rangle - \left\langle X^2\left[\frac{\phi-\psi}{4}\right]\right\rangle$$

$$= \int x^2 f_{X[\psi/2+\psi/2]}(x)\, dx - \int x^2 f_{X[\psi/2-\psi/2]}(x)\, dx.$$

In a similar fashion,

$$\text{Var}_X[\phi] = \langle X^2[\phi]\rangle = \int x^2 f_{X[\psi]}(x)\, dx.$$

Also, all the finite-dimensional probability distributions of random vectors

$$(X[\phi_1], \ldots, X[\phi_n]), \quad \phi_1, \ldots, \phi_n \in \mathcal{D}, \tag{7}$$

are, in view of the linearity of the functional $X[\phi]$, automatically determined by the family of one-dimensional distributions (5–6) as is clearly seen by considering the *characteristic functional*

$$\Psi_X[\phi] = \langle \exp(iX[\phi])\rangle = \int e^{ix} f_{X[\phi]}(x)\, dx, \quad \phi \in \mathcal{D}.$$

Indeed, the finite-dimensional characteristic function of the random vector (7) is expressible in terms of a one-dimensional characteristic functional of the random distribution X via the formula

$$\left\langle \exp\Big(i(\omega_1 X[\phi_1] + \cdots + \omega_n X[\phi_n])\Big)\right\rangle = \Phi_X[\omega_1\phi_1 + \cdots + \omega_n\phi_n].$$

Remark 1. Classical random functions (stochastic processes). As was the case for deterministic distributions, some of random distributions are generated by regular random functions. Thus, if $\xi(t), t \in \mathbf{R}$, is a random function (i.e., a function with values that are random quantities) then the formula,

$$X_\xi[\phi] = \int \xi(t)\phi(t)\, dt, \tag{8}$$

defines a random distribution whenever the integral in (8) is well defined in the sense of the integral being the limit in the variance (in L^2) of the appropriate discrete approximations. Of course, computation of probabilistic characteristics of discrete approximants of the integral (8) requires, in general, knowledge of all finite-dimensional probability distributions of the random function $\xi(t)$, that is the probability distributions of all real-valued random quantities $\xi(t), t \in \mathbf{R}$, of all 2-D random vectors $(\xi(t_1), \xi(t_2)), t_1, t_2 \in \mathbf{R}$, 3-D random vectors $(\xi(t_1), \xi(t_2), \xi(t_3)), t_1, t_2, t_3 \in \mathbf{R}$, and so on. As we have seen in Chapter 16, these probability distributions can be determined in one of several ways, either via *n-point (n-dimensional) cumulative probability distribution functions*,

$$F_{t_1,\ldots,t_n}(x_1,\ldots,x_n) = P(\xi(t_1) \le x_1,\ldots,\xi(t_n) \le x_n),$$

for all $n = 1,2,\ldots$, and all $t_1,\ldots,t_n,x_1,\ldots,x_n \in \mathbf{R}$, or probability densities

$$f_{t_1,\ldots,t_n}(x_1,\ldots,x_n) = \frac{\partial F_{t_1,\ldots,t_n}(x_1,\ldots,x_n)}{\partial x_1 \ldots \partial x_n},$$

meant in the distributional sense if necessary. Alternative characterizations can be provided by the expectations

$$\Big\langle \phi_1(\xi(t_1)) \cdot \ldots \cdot \phi_n(\xi(t_n)) \Big\rangle, n = 1,2,\ldots,$$

where $\phi(x),\ldots,\phi_n(x)$ range over all bounded continuous test functions, or just over all complex exponential functions $\exp(i\omega x)$, in which case we obtain all the n-dimensional characteristic functions

$$\Phi_{t_1,\ldots,t_n}(\omega_1,\ldots,\omega_n) = \Big\langle \exp i\Big(\omega_1\xi(t_1) + \cdots + \omega_n\xi(t_n)\Big)\Big\rangle.$$

On the other hand, by *Kolmogorov's Existence Theorem*, see, e.g., Billingsley [3], any consistent family

$$F_{t_1,\ldots,t_n}(x_1,\ldots,x_n), n = 1,2,\ldots; t_1,\ldots,t_n,x_1,\ldots,x_n \in \mathbf{R},$$

of finite-dimensional cumulative probability distribution functions determines a random function. The proof of this important result is not trivial (see, e.g., [3]) and beyond the scope of this book. The consistency of the family means that the following two conditions are satisfied:

$$F_{t_1,\ldots,t_n}(x_1,\ldots,x_n) = F_{t_{k_1}\ldots t_{k_n}}(x_{k_1},\ldots,x_{k_n}),$$

for any permutation $k_1,\ldots,k_n$ of indices $1,\ldots,n$, and

$$F_{t_1,\ldots,t_k,t_{k+1},\ldots,t_n}(x_1,\ldots,x_k,\infty,\ldots,\infty) = F_{t_1,\ldots,t_k}(x_1,\ldots,x_k),$$

for any $k < n$, and any $t_1,\ldots,t_n \in \mathbf{R}$.

17.2   Brownian motion, and the white noise

17.2.1.   Basic definitions.   In this section we will consider examples of Gaussian random functions and distributions which, in view of the shape of the finite-dimensional probability densities of Gaussian random vectors (see, Section 16.7), are completely determined by their mean function $\langle \xi(t) \rangle$, and their covariance function,

$$B_\xi(t, s) = \text{Cov}(\xi(t), \xi(s)).$$

If the mean function of ξ is constant (independent of t), and the covariance function depends only on the time lag, $t - s$, that is $B_\xi(t, s) = B_\xi(t - s)$, then we say that the process is *weakly stationary*. For a Gaussian process, weak stationarity is equivalent to strong stationarity, which means the invariance of all finite-dimensional distributions under time translations, and in this case we'll just use the term *stationary* for such processes.

A Gaussian process $b(t), t \geq 0$, with $\langle b(t) \rangle = 0$, and covariance

$$\text{Cov}_b(t, s) := \langle b(t)b(s) \rangle = \min(t, s), \tag{1}$$

is called the *Brownian motion* random function, or the *Wiener stochastic process*. It reflects, in crude approximation, the motion of a Brownian particle. Clearly, its variance (structure) function

$$\text{Var}\, b(t) = \langle b^2(t) \rangle = t,$$

and its increments $b(t_4) - b(t_3)$, and $b(t_2) - b(t_1)$, over disjoint time intervals $t_1 < t_2 < t_3 < t_4$, are uncorrelated (and, in view of their Gaussianness, also independent), since, in view of (1),

$$\left\langle \Big(b(t_4) - b(t_3)\Big)\Big(b(t_2) - b(t_1)\Big) \right\rangle = t_2 - t_2 - t_1 + t_1 = 0.$$

With this information, one can easily compute the finite-dimensional probability distributions of the Brownian motion random function. Indeed, if $0 \leq t_1 < \cdots < t_n$, and $y_1, \ldots, y_n \in \mathbf{R}$, then

$$P\Big(b(t_1) \leq y_1, \ldots, b(t_n) \leq y_n\Big) = \frac{1}{(2\pi)^{n/2}\sqrt{t_1(t_2 - t_1) \cdot \ldots \cdot (t_n - t_{n-1})}}$$

$$\times \int_{-\infty}^{y_n} \cdots \int_{-\infty}^{y_1} \exp\left[-\frac{1}{2}\left(\frac{x_1^2}{t_1^2} + \frac{(x_2 - x_1)^2}{(t_2 - t_1)^2} + \cdots + \frac{(x_n - x_{n-1})^2}{(t_n - t_{n-1})^2}\right)\right] dx_1 \ldots dx_n.$$

In distribution-theoretic terms, the Gaussian random distribution

$$X_b[\phi] = \int b(t)\phi(t)\, dt$$

corresponding to the Brownian motion random function is characterized by zero mean, $\langle X_b[\phi] \rangle = 0$, and the bilinear covariance functional,

$$\text{Cov}_{X_b}[\phi, \psi] = \langle X_b[\phi]X_b[\psi] \rangle = \left\langle \int b(t)\phi(t)\, dt \int b(s)\psi(s)\, ds \right\rangle$$

$$= \int_0^\infty \int_0^\infty \langle b(t)b(s) \rangle \phi(t)\psi(s)\,dt\,ds = \int_0^\infty \int_0^\infty \min(t,s)\phi(t)\psi(s)\,dt\,ds$$

$$= \int_0^\infty \phi(t) \int_0^t s\psi(s)\,ds\,dt + \int_0^\infty \psi(s) \int_0^s t\phi(s)\,dt\,ds.$$

The above integrals can be directly evaluated by integration by parts twice to obtain the final formula,

$$\mathrm{Cov}_{X_b}[\phi, \psi] = \int_0^\infty \left(\int_t^\infty \phi(s)\,ds \int_t^\infty \psi(s)\,ds \right) dt.$$

17.2.2.   Continuity of Brownian motion.   There exists a version of the Brownian motion process $b(t)$ with continuous sample paths. This continuity statement has been proved first by Norbert Wiener in 1923, and we provide here a modern version of his argument based on the ideas of fractional integration of Section 6.8 of Volume 1 (also, see Part VI of the present volume). The Brownian motion process $b(t)$ will be constructed as a sum of a uniformly convergent (almost surely, i.e., with probability 1) random series of continuous functions on $[0,1]$. An extension to the whole real line can be easily provided. More precisely, we will prove the following statement:

If $\xi_1, \xi_2, \ldots$ are independent, identically distributed standard Gaussian random quantities, and $g_1, g_2, \ldots$ are an orthonormal complete sequence in $L^2[0,1]$ (such as the trigonometric system, or the Haar wavelet system), then the series

$$b(t) = \sum_{n=1}^\infty G_n(t)\xi_n,$$

where $G_n(t) = \int_0^t g_n(s)\,ds$, $n = 1, 2, \ldots$, converges a.s. uniformly on $[0,1]$ (i.e., in the norm of $C[0,1]$), and its sum $b(t)$ is a Brownian motion process.

Clearly, $G_1\xi_1, G_2\xi_2, \ldots$ is a sequence of independent random quantities with values in $C[0;1]$ (i.e., with values being continuous functions). Denote by

$$(I^\alpha f)(t) = \frac{1}{\Gamma(\alpha)} \int_0^t (t-s)^{\alpha-1} f(s)\,ds, \qquad 0 < \alpha < 1,$$

the fractional integration operators of order α discussed in detail in Section 6.8 of Volume 1.

Now, let $2 < p < \infty$, and let $\alpha > 1/2$, $\beta > 1/p$, and $\alpha + \beta = 1$. Noting that $G_i = I^1 g_i$, and that $I^\alpha I^\beta = I^1$, inequalities (6.8.7) and (6.8.8), Volume 1, implies that to prove that $\sum_{i=1}^\infty (I^1 g_i)\xi_i$ converges a.s. in $C[0,1]$ it suffices to demonstrate that $\sum_{i=1}^\infty (I^\alpha g_i)\xi_i$ converges with probability 1 in $L^p[0,1]$, or even, by the standard probabilistic argument, that the above series converges in $L^p[0,1]$ in the pth mean. The latter assertion is proved as follows

$$\left\langle \| \sum_{i=1}^\infty (I^\alpha g_i)\xi_i \|_p^p \right\rangle = \int_0^1 \left\langle \left| \sum_{i=1}^\infty (I^\alpha g_i)(t)\xi_i \right|^p \right\rangle dt$$

$$= c(p) \int_0^1 \left(\sum_{i=1}^{\infty} (I^\alpha g_i)^2(t) \right)^{p/2} dt, \qquad (2)$$

where the last equality follows from the fact that for each sequence of scalars $\alpha_1, \alpha_2, \ldots$, the random quantities $\sum_{i=1}^{\infty} \alpha_i \xi_i$ and $(\sum_{i=1}^{\infty} \alpha_i^2)^{1/2} \xi_1$ are equidistributed. So, the constant $c(p) = E|\xi_1|^p$.

By the Bessel inequality, for each t,

$$\left(\sum_{i=1}^{\infty} (I^\alpha g_i)^2(t) \right)^{1/2} = \left(\sum_{i=1}^{\infty} \left(\int_0^1 k_\alpha(t-s) I_{[0,t)}(s) g_i(s) \, ds \right)^2 \right)^{1/2}$$

$$\leq \|k_\alpha(t-.)I_{[0,t)}\|_2 = \frac{1}{\Gamma(\alpha)} \left(\int_0^t s^{(\alpha-1)2} \, ds \right)^{1/2} \leq \frac{1}{\Gamma(\alpha)} (2\alpha-1)^{-1/2},$$

where $k_\alpha(t)$ denotes the kernel $t^{\alpha-1}\chi(t)/\Gamma(\alpha)$, and I_A denotes the *indicator function* of set A (equal to 1 on the set A, and 0 off the set). Hence

$$\int_0^1 \left(\sum_{i=1}^{\infty} (I^\alpha g_i)^2(t) \right)^{p/2} dt < \infty$$

and, by the Lebesgue Theorem, and (2),

$$\lim_{k,m \to \infty} \left\langle \| \sum_{i=k}^{m} (I^\alpha g_i) \xi_i \|_p^p \right\rangle = 0,$$

which gives the desired convergence in the pth mean.

So, the sum $b(t) = \sum_{i=1}^{\infty} G_i(t) \xi_i$ defines, for $t \in [0,1]$, a stochastic process with continuous sample paths, which is a zero-mean Gaussian process as a limit of zero-mean Gaussian processes. So, to prove that $b(t)$ is a Brownian motion, it remains to check that $\langle b(t)b(s) \rangle = \min(t,s)$. Indeed, by the Lebesgue Theorem, and the Parseval Identity,

$$\langle b(t)b(s) \rangle = \sum_{i,j=1}^{\infty} G_i(s) G_j(t) \langle \xi_i \xi_j \rangle = \sum_{i=1}^{\infty} G_i(s) G_i(t)$$

$$= \sum_{i=1}^{\infty} \left(\int_0^1 g_i(u) I_{[0,s]}(u) \, du \right) \cdot \left(\int_0^1 g_i(u) I_{[0,t]}(u) \, du \right)$$

$$= \int_0^1 I_{[0,s]}(u) I_{[0,t]}(u) \, du = \min(t,s).$$

This completes the construction of a version of the Brownian motion process with continuous sample paths a.s.

17.2.3.   Nondifferentiability of Brownian motion. Observe that, in
the mean square sense, the derivative $b'(t)$, of the Brownian motion process,
$b(t)$, does not exist for any $t > 0$, since

$$\lim_{\Delta t \to 0} \text{Var} \frac{b(t + \Delta t) - b(t)}{\Delta t} = \lim_{\Delta t \to 0} \frac{\Delta t}{(\Delta t)^2} = \infty.$$

However, a much stronger property can be demonstrated: *With probability
1, the sample paths of Brownian motion are nowhere differentiable.* The
provided below proof of this results, due to Dvoretzky, Erdös, and Kakutani,
1961,[2] is subtle. More precisely, we will prove that

$$\mathbf{P}(b'(t) \text{ exists for some } t) = 0. \tag{3}$$

To verify (3) note that the differentiability of $b(t)$ at a point $s \in [0,1]$
implies the existence of an integer $N \geq 1$ such that, $|b(t) - b(s)| < N$, as
$t \searrow s$. More precisely, for sufficiently large n, and

$$j = \lfloor ns \rfloor + 2, \lfloor ns \rfloor + 3, \lfloor ns \rfloor + 4,$$

we have

$$\left| b\left(\frac{j}{n}\right) - b\left(\frac{j-1}{n}\right) \right| < \frac{7N}{n},$$

because

$$\left| b\left(\frac{j}{n}\right) - b\left(\frac{j-1}{n}\right) \right| \leq \left| b\left(\frac{j}{n}\right) - b(s) \right| + \left| b\left(\frac{j-1}{n}\right) - b(s) \right|.$$

In other words,

$$\mathbf{P}(b'(t) \text{ exists for some } t \in (0,1))$$

$$\leq \mathbf{P}\left(\exists N \geq 1 \, \exists s \, \exists m \geq 1 \, \forall n \geq m \, \forall j = \lfloor ns \rfloor + 2, \lfloor ns \rfloor + 3, \lfloor ns \rfloor + 4, \right.$$

$$\left. \left| b\left(\frac{j}{n}\right) - b\left(\frac{j-1}{n}\right) \right| < \frac{7N}{n} \right)$$

$$\leq \mathbf{P}\left(\exists N \geq 1 \, \exists m \geq 1 \, \forall n \geq m \, \exists i = 2, \ldots, n+1 \, \forall j = i+1, i+2, i+3, \right.$$

$$\left. \left| b\left(\frac{j}{n}\right) - b\left(\frac{j-1}{n}\right) \right| < \frac{7N}{n} \right)$$

$$\leq \mathbf{P}\left(\bigcup_{N \geq 1} \bigcup_{m \geq 1} \bigcap_{n \geq m} \bigcup_{1 < i \leq n+1} \bigcap_{i < j \leq i+3} \left(\left| b\left(\frac{j}{n}\right) - b\left(\frac{j-1}{n}\right) \right| < \frac{7N}{n} \right) \right)$$

---

[2]The original result was discovered in the 1930s by Paley, Wiener, and Zygmund.

$$\leq \sum_{N=1}^{\infty} \mathbf{P}\left(\liminf_{n\to\infty} \bigcup_{1<i\leq n+1} \bigcap_{i<j\leq i+3} \left(\left|b\left(\frac{j}{n}\right) - b\left(\frac{j-1}{n}\right)\right| < \frac{7N}{n}\right)\right)$$

and, by the classical Fatou Lemma,

$$\leq \sum_{N=1}^{\infty} \liminf_{n\to\infty} n \cdot \left(\mathbf{P}\left(\left|b\left(\frac{1}{n}\right)\right| < \frac{7N}{n}\right)\right)^3$$

$$= \sum_{N=1}^{\infty} \liminf_{n\to\infty} n \cdot \left(\mathbf{P}\left(|b(1)| < \frac{7N}{n} \cdot n^{1/2}\right)\right)^3 \leq \sum_{N=1}^{\infty} \liminf_{n\to\infty} n \cdot \left(\frac{7N}{\sqrt{2\pi}n^{1/2}}\right)^3 = 0.$$

17.2.4. White noise: the distributional derivative of the Brownian motion. Although $W(t) = b'(t)$ does not exist in the classical sense, *the white noise*, the distributional derivative of the Brownian motion is well defined by the usual formula

$$(X_b)'[\phi] = -X_b[\phi'] = -\int b(t)\phi'(t)\, dt, \quad \phi \in \mathcal{D}.$$

and denoted

$$(X_b)' = W.$$

The reasons for the name, "white noise," will become apparent in the next section.

Obviously, W is a Gaussian random distribution in the sense that, for any $\phi \in \mathcal{D}$, the random quantity

$$W[\phi] = -\int b(t)\phi'(t)\, dt, \quad \phi \in \mathcal{D}.$$

is Gaussian, with zero mean. The covariance bilinear functional of the white noise is easily computable from the covariance functional of the Brownian motion:

$$\mathrm{Cov}_W[\phi, \psi] = \langle W[\phi]W[\psi]\rangle$$

$$= \left\langle \int b(t)\phi'(t)\, dt \int b(s)\psi'(s)\, ds \right\rangle = \int \phi(t)\psi(t)\, dt.$$

The latter expression can be rewritten in terms of the Dirac delta of two variables (t, s) so that

$$\mathrm{Cov}_W[\phi, \psi] = \int_0^\infty \delta(t - s)[\phi(t)]\psi(s)\, ds = \delta(t - s)[\phi(t)\psi(s)].$$

For that reason the white noise is often referred to as *delta-correlated (generalized) random function*. Its characteristic functional is now easily calculated:

$$\langle e^{iW[\phi]}\rangle = \exp\left(-\frac{1}{2}\int \phi^2(t)\, dt\right), \quad \phi \in \mathcal{D}.$$

Remark 1. Covariances and the kernel theorem. It is a consequence of a deep result of the theory of distributions (called the *kernel theorem*)[3] that any random distribution has a covariance that can be identified as a distribution in two variables. More precisely, every bilinear continuous functional $B[\phi, \psi], \phi, \psi \in \mathcal{D}$ is of the form

$$B[\phi, \psi] = T_{(t,s)}[\phi(t)\psi(s)],$$

where $T_{(t,s)}$ is a distribution in $\mathcal{D}'(\mathbf{R}^2)$.

17.3   Spectral analysis of stationary generalized processes

A random distribution $X[\phi], \phi \in \mathcal{D}$, is said to be *strictly stationary* if its one-dimensional (and thus any finite-dimensional, see Section 17.1) probability distribution functions are invariant under translations. More precisely, X is strictly stationary if, for any $\tau \in \mathbf{R}$, and any $\phi \in \mathcal{D}$,

$$F_{X[\phi]} = F_{X[\phi_\tau]}, \tag{1}$$

where $\phi_\tau(t) := \phi(t + \tau)$. If the mean value of X is constant (say zero), and the covariance bilinear functional is invariant under translations, that is,

$$\operatorname{Cov}_X[\phi, \psi] = \operatorname{Cov}_X[\phi_\tau, \psi_\tau], \tag{2}$$

then the random distribution X is called *weakly stationary* (or, simply, *stationary*). Obviously, strict stationarity implies the weak stationarity (if second moments exist). For zero-mean Gaussian random distribution, it suffices to verify the second-order condition (2) (i.e., the weak stationarity) to demonstrate their strict stationarity.

Example 1. Stationarity of the white noise. The covariance bilinear functional of the white noise W is of the form,

$$\int \phi(t)\psi(t)\, dt = (\phi * \psi^-)(0) = \delta(t)[\phi * \psi^-], \tag{3}$$

with $\psi^-(t) := \psi(-t)$, so it is obviously invariant under translations. Since the mean $\langle W[\phi] \rangle = 0$, the white noise is an example of a stationary (and strictly stationary) random distribution. Moreover, because the Dirac delta $\delta(t)$ is the Fourier transform of a function identically equal to 1, we arrive at the so-called spectral representation of the covariance bilinear functional of the white noise:

$$\operatorname{Cov}_W[\phi, \psi] = \int \tilde{\phi}(\lambda)(\tilde{\psi}(\lambda))^* \, d\lambda,$$

---

[3]See L. Schwartz [5].

where $\tilde{\phi}$ and $\tilde{\psi}$ are Fourier transforms of test functions ϕ and ψ.

In more generality one can prove (see Remark 1, Subsection 17.2.4) that the covariance bilinear functional of any stationary distribution X is of the form,

$$\mathrm{Cov}_X[\phi, \psi] = T[\phi * \psi^-],$$

where T is a certain semipositive-definite distribution (i.e., $T[\phi * \phi^-] \geq 0$), or equivalently, that

$$\mathrm{Cov}_X[\phi, \psi] = \int \tilde{\phi}(\lambda)(\tilde{\psi}(\lambda))^* \sigma(d\lambda),$$

where σ is the *spectral measure* of the stationary random distribution X. In the special case, when the spectral measure σ is generated by a regular locally integrable nonnegative function φ, then is, when

$$\mathrm{Cov}_X[\phi, \psi] = \int \tilde{\phi}(\lambda)(\tilde{\psi}(\lambda))^* \varphi(\lambda)\, d\lambda,$$

function $\varphi = \varphi(\lambda)$ is called the *spectral density* of the stationary random distribution X. In this terminology, for the white noise the spectral density

$$\varphi_W(\lambda) \equiv 1.$$

The above spectral representation of the covariance bilinear functional of a stationary generalized process can be "lifted up" to the representation of the random distribution itself by introduction of the concept of *random measure*.

Definition 1. An (orthogonally scattered) random measure M is a function mapping Borel sets A on the real line into zero mean, complex-valued random variables with finite variances,

$$A \mapsto M(A) \in L^2(\Omega, \mathcal{F}, \mathcal{P}),$$

such that:
(a) If $A_1, A_2, \ldots$ is a sequence of nonintersecting sets, $A_n \cap A_m = \emptyset$, for $n \neq m$, then

$$M\left(\bigcup_{n=1}^{\infty} A_n\right) = \sum_{n=1}^{\infty} M(A_n),$$

where the convergence of the series on the right-hand side is meant in the mean square (i.e., in $L^2(\Omega, \mathcal{F}, \mathcal{P})$);
(b) If A and B are disjoint Borel subsets of $\mathbf{R}$, then $M(A)$ and $M(B)$ are uncorrelated (orthogonal), that is

$$A \cap B = \emptyset \implies \langle M(A)M^*(B)\rangle = 0.$$

Notice that the above definition implies the existence of a countably additive measure σ on $(\mathbf{R}, \mathcal{B})$ defined by the formula,

$$\sigma(A) = \langle |M(A)|^2 \rangle. \tag{4}$$

Indeed, if $A \cap B = \emptyset$ then, in view of property (b),

$$\sigma(A \cup B) = \langle |M(A \cup B)|^2 \rangle = \langle |M(A) + M(B)|^2 \rangle$$

$$= \langle |M(A)|^2 + 2M(A)M^*(B) + |M(B)|^2 \rangle = \sigma(A) + \sigma(B).$$

The measure σ introduced by (4) permits introduction of the linear isometric mapping

$$L^2(\mathbf{R}, \mathcal{B}, \sigma) \mapsto L^2(\Omega, \mathcal{F}, \mathcal{P})$$

defined as the integral of function f with respect to random measure M,

$$f \mapsto \int f(\lambda) M(d\lambda).$$

Indeed, for a simple function (a linear combination of indicator functions of disjoin Borel sets) $f = \sum a_k I_{A_k}$, the obvious definition is

$$\int f(\lambda) M(d\lambda) = \int \sum_k a_k I_{A_k}(\lambda) M(d\lambda) := \sum_k a_k M(A_k),$$

which assures the linearity of the operation, and the isometry follows from (4) since

$$\left\langle \left| \int f(\lambda) M(d\lambda) \right|^2 \right\rangle = \left\langle \left| \sum_k a_k M(A_k) \right|^2 \right\rangle = \sum_k |a_k|^2 \langle |M(A_k)|^2 \rangle = \int |f(\lambda)|^2 \sigma(d\lambda). \tag{5}$$

The above property extends to the whole space $L^2(\mathbf{R}, \mathcal{B}, \sigma)$ by continuity.

Also, observe that, the above isometry property implies the equality of the corresponding scalar products in spaces $L^2(\mathbf{R}, \mathcal{B}, \sigma)$, and $L^2(\Omega, \mathcal{F}, \mathcal{P})$, that is taking $f_1(\lambda) = I_A(\lambda)$, and $f_2(\lambda) = I_B(\lambda)$, for arbitrary Borel sets A, and B, we have

$$\langle M(A) \cdot M^*(B) \rangle = \left\langle \int f_1(\lambda) M(d\lambda) \cdot \left(\int f_2(\lambda) M(d\lambda) \right)^* \right\rangle$$

$$= \int f_1(\lambda) f_2(\lambda) \sigma(d\lambda) = \sigma(A \cap B).$$

The structure of stationary random distributions (generalized stationary stochastic processes) is described in the following *spectral representation* result:

THEOREM: *The random distribution, $X[\phi], \phi \in \mathcal{D}$, is stationary if, and only if, there exists a random measure M, such that X is the generalized Fourier transform of M, that is*

$$X[\phi] = \int \int \phi(t)e^{i\lambda t} M(d\lambda)dt = \int \tilde{\phi}(\lambda)M(d\lambda), \qquad \phi \in \mathcal{D}, \qquad (6)$$

where, for any Borel sets A, and B, in $\mathbf{R}$

$$\langle M(A) \cdot M^*(B) \rangle = \sigma(A \cap B).$$

Proof: First, we will demonstrate that any random distribution that is a Fourier transform of a random measure, and thus represented by (6), is stationary. Obviously, in view of the assumption, the mean value of M is zero, the mean value of $\int f(\lambda)M(d\lambda)$ is also zero, so that (6) also has zero mean. Next, the covariance bilinear functional,

$$\text{Cov}_X[\phi, \psi] = \langle X[\phi]X^*[\psi] \rangle = \left\langle \int \tilde{\phi}(\lambda)M(d\lambda) \left(\int \tilde{\psi}(\nu)M(d\nu) \right)^* \right\rangle$$

$$= \int \int \tilde{\phi}(\lambda)(\tilde{\psi}(\nu))^* \langle M(d\lambda)(M(d\nu))^* \rangle = \int \tilde{\phi}(\lambda)(\tilde{\psi}(\lambda))^* \sigma(d\lambda),$$

Since

$$\tilde{\phi}_\tau(\lambda) = e^{-i\lambda\tau}\tilde{\phi}(\lambda), \quad \text{and} \quad \tilde{\psi}_\tau(\nu) = e^{-i\nu\tau}\tilde{\psi}(\nu),$$

we get that

$$\text{Cov}_X[\phi_\tau, \psi_\tau] = \int e^{-i\lambda\tau}\tilde{\phi}(\lambda)(e^{-i\lambda\tau}\tilde{\psi}(\lambda))^*\sigma(d\lambda) = \text{Cov}_X[\phi, \psi],$$

Thus the stationarity of (6) has been verified.

The proof of the fact that any stationary random distribution can be represented as the Fourier transform of a random measure is more delicate. We begin by defining two Hilbert spaces, $\mathcal{H}$ and $\mathcal{X}$, by completing, in the first case, the space $\mathcal{D}$ equipped with the scalar product

$$(\phi, \psi)_{\mathcal{H}} = \text{Cov}(X[\phi], X[\psi]),$$

and in the second case, by completing the linear space of random variables $(X[\phi], \phi \in \mathcal{D})$, with the scalar product

$$(X[\phi], X[\psi])_{\mathcal{X}} = \text{Cov}(X[\phi], X[\psi]).$$

It is obvious from the above definitions that the mapping,

$$\mathcal{H} \ni \phi \longmapsto X[\phi] \in \mathcal{X},$$

is a linear isometry preserving the norms generated by the respective scalar products in $\mathcal{H}$, and $\mathcal{X}$.[4]

The stationarity assumption about $X[\phi]$ gives that

$$(\phi, \psi)_{\mathcal{H}} = \mathrm{Cov}_X[\phi, \psi] = \int \tilde{\phi}(\lambda)(\tilde{\psi}(\lambda))^* \sigma(d\lambda),$$

so that the Fourier mapping can be extended to an isometry,

$$\mathcal{I} : \mathcal{H} \ni \phi \longmapsto \tilde{\phi} \in L^2(\mathbf{R}, \mathcal{B}, \sigma),$$

with the well-defined inverse mapping

$$\mathcal{I}^{-1} : L^2(\mathbf{R}, \mathcal{B}, \sigma) \ni \tilde{\phi} \longmapsto \phi \in \mathcal{H},$$

which is also a linear isometry. The complete Hilbert space $L^2(\mathbf{R}, \mathcal{B}, \sigma)$ contains the indicator functions $I_A(t)$ of all the Borel sets $A \subset \mathbf{R}$, and the well-defined mapping

$$\mathcal{B} \ni A \longmapsto M(A) := X(\mathcal{I}^{-1}(I_A)) \in \mathcal{X},$$

is a random measure. Indeed, the countable additivity follows from the linearity and mean square continuity of the mapping, as

$$M\left(\bigcup_n A_n\right) = X\left(\mathcal{I}^{-1}\left(\sum_n I_{A_n}\right)\right) = \sum_n X(\mathcal{I}^{-1}(I_{A_n})) = \sum_n M(A_n),$$

if $A_n \cap A_m = \emptyset$, for $n \neq m$, and from the fact that

$$\langle M(A), M(B)\rangle = \left(X(\mathcal{I}^{-1}(I_A)), X(\mathcal{I}^{-1}(I_B))\right)_{\mathcal{X}} = \int I_A(\lambda)I_B(\lambda)\sigma(d\lambda) = \sigma(A \cap B).$$

So, now, it remains to prove that X is indeed the Fourier transform of the random measure M, that is that

$$X(\phi) = \int\int \phi(t)e^{-i\lambda t}M(d\lambda)dt = \int \tilde{\phi}(\lambda)M(d\lambda), \qquad \phi \in \mathcal{D}.$$

Actually, the equality is true for any $\phi \in \mathcal{H} \supset \mathcal{D}$, because $\mathcal{I}(\phi)$ can be approximated by linear combinations of the disjoint indicator functions $\sum_k a_k I_{A_k}(\lambda)$, so that

$$\int\int e^{-i\lambda t}\phi(t)dt M(d\lambda) = \int \mathcal{I}(\phi)(\lambda)M(d\lambda) \approx \int \left(\sum_k a_k I_{A_k}(\lambda)\right) M(d\lambda)$$

---

[4]We are implicitly assuming here that the covariance function is nondegenerate; otherwise the arguments will have to be carried out on the relevant factor spaces.

$$= \sum_k a_k M(A_k) = \sum_k a_k X(\mathcal{I}^{-1}(I_{A_k})) = X(\mathcal{I}^{-1}(\sum_k a_k I_{A_k})) \approx X(\phi).$$

Thus the proof of the spectral representation is complete.

Example 2. The case of classical stochastic processes. If a weakly stationary random distribution represents a random function (classical stochastic process) $\xi(t)$, then the latter is stationary in the sense that it's mean value is constant (say zero), and its covariance function,

$$\gamma_\xi(t,s) := \mathrm{Cov}(\xi(t), \xi(s)) = \mathrm{Cov}(\xi(t-s), \xi(0)),$$

so that, essentially, it is a function of one variable $\gamma_\xi(t,s) = \gamma_\xi(t-s, 0) = \gamma_\xi(t-s)$, and the above spectral representation for the covariance bilinear functional can be translated into the spectral representation,

$$\gamma_\xi(t) = \int e^{i\lambda t} \, dF(\lambda),$$

where $F(\lambda)$ is a nondecreasing bounded function called the cumulative power spectrum for the process $\xi(t)$.

If $dF(\lambda) = f(\lambda) \, d\lambda$, where $f(\lambda)$ is a nonnegative integrable function, then the covariance function of process $\xi(t)$ has the spectral representation,

$$\gamma_\xi(t) = \int e^{i\lambda t} f(\lambda) \, d\lambda,$$

and f is called the spectral density of the stationary process $\xi(t)$.

If, for example, the weakly stationary stochastic processes (stationary time series) is parametrized by discrete time, $n = \ldots, -2, -1, 0, 1, 2, \ldots$, then it has the spectral representation,

$$\xi(n) = \int_0^1 e^{-i2\pi n\lambda} d\mathcal{W}(\lambda),$$

where $\mathcal{W}(\lambda), 0 \le \lambda \le 1$, is a zero-mean stochastic process with $\mathcal{W}(0) = 0$, and uncorrelated increments such that $\mathcal{C}(\lambda) := \langle \mathcal{W}^2(\lambda) \rangle$ is the cumulative power spectrum of $X(n)$, that is

$$\gamma_\xi(n) = \int_0^1 e^{i2\pi n\lambda} d\mathcal{C}(\lambda).$$

where $\gamma_\xi(n)$ is, as before, the autocovariance sequence of the time series $\xi(n)$.[5]

And, to repeat our previous observation for general random distributions, for general processes the covariance function and the mean function do not completely determine the finite-dimensional distributions of the process, but for the Gaussian processes they do.

---

[5]For more on the spectral theory of classical stationary processes, see W.A. Woyczyński [4].

17.4   Representing generalized processes by sequences of ordinary stochastic processes

In the next three sections we are going to present an alternative approach to random distributions (and distributions themselves) that is less dependent on the functional theoretic background and also may be more appealing and intuitive for the reader working in the physical and engineering sciences. Essentially, the theory relies on the idea that every distribution is a derivative of a certain order of a continuous function and can be also viewed as a "limit" of a sequence of continuous functions. We shall start with the alternative definition of the distributions themselves. To clearly distinguish our alternative theory framework, the notation for distributions introduced in this chapter will also be different. This approach to the ordinary deterministic distributions has been developed by J. Mikusiński and R. Sikorski [7] in 1955–1957 and expanded to the stochastic context by K. Urbanik in a series of papers [8–10] published in 1958.

17.4.1.   Sequential theory of distributions: Representing distributions by sequences of ordinary functions.   We'll start with the basic concepts as the construction depends on the mathematical *abstraction principle* which permits consideration of a class of objects equivalent to each other in a certain sense, as a new mathematical object. To assure that the classes are disjoint (thus defining the new object uniquely) the equivalence relation, say, denoted by $\sim$, must satisfy the following, intuitively obvious, conditions:

(i) *Reflexivity:* $x \sim x$,
(ii) *Symmetry:* If $x \sim y$, then $y \sim x$
(iii) *Transitivity:* If $x \sim y$, and $y \sim z$, then $x \sim z$.

Example 1. Real numbers as equivalence classes of sequences of rational numbers. As an elementary illustration considers the equivalence principle that is also the starting point of the formal introduction of the concept of a real number given the set of rational numbers: A sequence $\{a_n\}$ of rational numbers is called *fundamental* if it satisfies the Cauchy condition, that is if, for every rational number $\epsilon > 0$, there exists an index n_0 such that, for every $m, n > n_0$, we have $|a_m - a_n| < \epsilon$. Then, two fundamental sequences, $\{a_n\}$, and $\{b_n\}$, are called equivalent if the sequence $\{a_n - b_n\}$ converges to zero. The relationship obviously satisfies the conditions (i–iii), and thus each class of equivalent sequences defines a distinct object which we call the real number.

The above abstraction principle permits introduction of the following *sequential* definition of distributions:

Definition 1. A sequence $\{f_n(t)\}$ of continuous functions defined on the (finite or infinite) open interval (A, B) is said to be *fundamental* if there exists a sequence $\{F_n(t)\}$ of functions such that, for a certain integer $k \geq 0$, the kth derivatives,

$$F_n^{(k)}(t) = f_n(t), \quad n = 1, 2, \ldots, \tag{1}$$

and the sequence $\{F_n(t)\}$ converges *almost uniformly* (a.u.); that is, it converges uniformly in each closed and bounded interval contained in (A, B).

We will use the symbol $\Rightarrow$ to denote almost uniform convergence and make a few elementary observations:

(a) Each a.u. convergent sequence of continuous functions is fundamental (take $k = 0$).

(b) If $\{f_n(t)\}$ is fundamental, and the mth derivatives $f_n^{(m)}(t)$ are continuous, then the sequence $\{f_n^{(m)}(t)\}$ is also fundamental.

(c) A sequence $\{p_n(t)\}$ of polynomials of degree $\leq m$,

$$p_n(t) = a_{n0} + a_{n1}t + \cdots + a_{nm}t^m, \quad n = 1, 2, \ldots,$$

is fundamental if, and only if, it is a.u. convergent.

(d) If $\{F_n(t)\}$ is fundamental then, for each integer m, the sequence

$$\bar{F}_n(t) = \int_{t_0}^t dt_1 \int_{t_0}^{t_1} dt_2 \ldots \int_{t_0}^{t_{m-1}} F_n(t_m) dt_m, \quad n = 1, 2, \ldots, \tag{2}$$

is also fundamental and, moreover, if $F_n(t) \Rightarrow F(t)$, then

$$\bar{F}(t) \Rightarrow \int_{t_0}^t dt_1 \int_{t_0}^{t_1} dt_2 \ldots \int_{t_0}^{t_{m-1}} F(t_m) dt_m. \tag{3}$$

Example 2. Fundamental sequences. (a) The following two sequences of functions:

$$f_n(t) = \frac{t}{n} \Rightarrow 0,$$

and

$$f_n(t) = \left(1 + \frac{t}{n}\right)^n \Rightarrow e^t,$$

are examples of fundamental sequences of functions on $(-\infty, \infty)$. Note that both converge to a continuous function.

(b) On the other hand, the sequence

$$f_n(t) = \frac{1}{1 + e^{-nt}}, \quad n = 1, 2, \ldots,$$

is fundamental because $f_n(t) \Rightarrow 0$ in the interval $(-\infty, 0)$, and $f_n(t) \Rightarrow 1$ in the interval $(0, \infty)$, but the limit is not a continuous function.

(c) Finally, the sequences of functions

$$f_n(t) = \sqrt{\frac{n}{2\pi}} e^{-nt^2/2}, \quad n = 1, 2, \ldots,$$

and,

$$f_n(t) = \begin{cases} n(1 - n|t|), & \text{for } -1/n < t < 1/n, \\ 0, & \text{otherwise;} \end{cases} \quad n = 1, 2, \ldots,$$

are also fundamental although they do not converge to any particular classical function. Note that in both cases, for each $n=1, 2, \ldots$, we have $\int_{-\infty}^{\infty} f_n(t)\, dt = 1$.

Definition 2. Two fundamental sequences, $\{f_n(t)\}$, and $\{g_n(t)\}$, are said to be *equivalent* (in brief, $\{f_n(t)\} \equiv \{g_n(t)\}$), if there exist sequences of functions, $\{F_n(t)\}$, and $\{G_n(t)\}$, such that, for a certain integer $k \geq 0$,

$$F_n^{(k)}(t) = f_n(t), \quad and \quad G_n^{(k)}(t) = g_n(t), \tag{4}$$

and $\{F_n(t)\}$, and $\{G_n(t)\}$, converge a.u. to the same function. The latter behavior will be denoted

$$\{F_n(t)\} \Rightarrow\Leftarrow \{G_n(t)\}, \tag{5}$$

Obviously, the integer k appearing in the above definition can be replaced by any greater integer so that functions $F_n(t)$ appearing in (4) can always be replaced by the functions $\bar{F}_n(t)$ defined via indefinite integrals in (2–3).

The above equivalence relation satisfies the reflexivity, symmetry, and transitivity conditions listed at the beginning of this section thus splitting the set of all fundamental sequences into nonintersecting abstraction classes. Two fundamental sequences are in the same class if, and only if, they are equivalent.

Definition 3. The abstraction class of a fundamental sequence $\{f_n(t)\}$ of functions, corresponding to the equivalence relationship $\equiv$ on the interval (A, B), will be called a *distribution* and denoted $f(t) = [f_n(t)]$ in this chapter.[6] The fundamental sequence $\{f_n(t)\}$ will be called a *proxy* for the distribution $[f_n(t)]$. In other words, the notion of distribution is obtained by identification of equivalent fundamental sequences.

Example 3. Equivalence class for Dirac delta. The fundamental sequences defined in Example 1 (c) are in the same equivalence class, and thus determine the same distribution which will be called the *Dirac delta*, and denoted, $\delta(t)$. Indeed, if $\{f_n(t)\}$ denotes either of the these two sequences, then the sequence

$$\int_{-\infty}^{t} dt_2 \int_{-\infty}^{t_2} f_n(t_1)\, dt_1, \quad n = 1, 2, \ldots,$$

---

[6]We deliberately use here a different notation for distributions introduced through the sequential procedure, to distinguish this approach from the functional definition used throughout the rest of the book. This notation is traditional in this part of the distribution theory.

converges a.u. to the same continuous function

$$F(x) = \begin{cases} 0, & \text{for } x < 0; \\ x, & \text{for } x \geq 0. \end{cases}$$

Each classical continuous function determines a distribution via the fundamental sequence $\{f_n(t)\} = \{f(t)\}$, and it is clear that different functions determine different distributions. But the above example of the Dirac delta clearly shows that not all distributions can be identified with continuous functions.

17.4.2.   Algebraic operations and differentiation.   The basic algebraic operations of addition of distributions, and multiplication of a distribution by a number, can now be defined by the corresponding operations on the proxy sequences. Thus,

$$[f_n(t)] + [g_n(t)] = \{f_n(t) + g_n(t)\}, \quad \text{and} \quad \lambda[f_n(t)] = \{\lambda f_n(t)\}, \qquad (6)$$

and these two operations enjoy the usual properties of commutativity and associativity.

Since for every continuous function $F(t)$ there is a sequence of polynomials $\{P_n(t)\}$ such that $P_n(t) \Rightarrow F(t)$, each distribution $f(t)$ has a representation $[p_n(t)]$, for some sequence of polynomials $\{p_n(t)\}$. This fact permits us to introduce the definition of the derivative of an arbitrary order of any distribution.

Definition 4. The *derivative* of order $m \geq 0$ of the distribution $f(t)$ is defined by the formula

$$f^{(m)}(t) = [p_n^{(m)}(t)].$$

Definition 5. A sequence of distributions $\{f_n(t)\}$ is said to be *convergent* to distribution $f(t)$, denoted $f_n(t) \to f(t)$, if there exists an integer $k \geq 0$, and continuous functions, $F(t), F_1(t), F_2(t), \ldots$ such that the derivatives

$$F_n^{(k)}(t) = f_n(t), \quad F^{(k)}(t) = f(t), \quad \text{and} \quad F_n(t) \Rightarrow F(t).$$

Example 4. More equivalence classes for Dirac delta. In addition to fundamental sequences in the Dirac delta equivalence classes displayed in Example 3, we would like to quote three, perhaps less obvious, examples of fundamental sequences of functions distributionally convergent to $\delta(t)$:

$$\frac{\sin nt}{\pi t}, \quad \frac{1}{2} n e^{-n|t|}, \quad \text{and} \quad \frac{n}{\pi(e^{nt} + e^{-nt})}.$$

The proof is left to the reader (see the Exercises section at the end of this chapter).

The above two definitions guarantee that we can interchange the order of the operations of passing to the limit and taking the derivative of arbitrary order, thus giving us the following two useful observations:

For every sequence of distributions $\{f_n\}$, and every $m \geq 0$

$$f_n(t) \to f(t) \qquad implies \qquad f_n^{(m)}(t) \to f^{(m)}(t),$$

and, for every convergent series of distributions $f_n(t)$

$$\left(\sum_{n=1}^{\infty} f_n(t)\right)^{(m)} = \sum_{n=1}^{\infty} f_n^{(m)}(t).$$

The operation of multiplication of two distributions is not well defined, but one can easily define the product of a distribution $f(t) = [f_n(t)]$ with an infinitely differentiable function $w(t)$ by the equation,

$$w(t)f(t) = [w(t)f_n(t)].$$

which leads to the extension of the familiar distributional equality for the derivative of the product,

$$\left(w(t)f(t)\right)' = w'(t)f(t) + w(t)f'(t),$$

and, by induction, to the following identity,

$$w(t)\delta^{(k)}(t) = \sum_{j=0}^{k}(-1)^j \binom{k}{j} w^{(j)}(0)\delta^{(k-j)}(t),$$

which, in particular, yields the following two simple formulas:

$$t\delta(t) = 0, \qquad and \qquad t\delta'(t) = -\delta(t).$$

17.4.3.   Superposition of a distribution with a function.   If $f(t) = [f_n(t)]$ is a distribution in (A, B), and $\phi(t)$ is an infinitely differentiable function on (A_0, B_0) taking values in the interval (A, B), with the derivative $\phi'(t) \neq 0$, for all $t \in (A_0, B_0)$, then, by definition, the distribution

$$f(\phi(t)) := [f_n(\phi(t))], \qquad t \in (A_0, B_0). \tag{7}$$

It follows directly from the definition and the above formula for the derivative of the product of a function and a distribution that

$$(f(\phi(t)))' = f'(\phi(t))\phi'(t). \tag{8}$$

Example 5. Differentiating the Heaviside function. It follows from (8) that

$$\delta(\phi(t)) = \frac{1}{\phi'(t)}\Big(H(\phi(t))\Big)',$$

where $H(t)$ is the usual Heaviside function. Indeed, if $\phi(t) \neq 0$ everywhere (thus, being infinitely differentiable, it is either positive everywhere, or negative everywhere) then $H(\phi(t))$ is either zero everywhere, or 1, everywhere, so that in this case

$$\delta(\phi(t)) = 0.$$

If $\phi(t_0) = 0$ at some point t_0 (which is then unique because the assumption on $\phi(t)$ was that its derivative does not vanish), then

$$H(\phi(t)) = \begin{cases} H(t - t_0), & \text{if } \phi(t) \text{ is increasing;} \\ 1 - H(t - t_0), & \text{if } \phi(t) \text{ is decreasing,} \end{cases}$$

so that, in this case,

$$\delta(\phi(t)) = \frac{1}{|\phi'(t_0)|}\delta(t - t_0).$$

Example 6. Linear substitutions. First, note that for every distribution, and every integer $k \geq 0$, the kth derivative of a linear substitution,

$$\big(f(\alpha x + \beta)\big)^{(k)} = \alpha^k f^{(k)}(\alpha x + \beta),$$

as long as $\alpha \neq 0$. Indeed, if $p_n(x)$ is a polynomial fundamental sequence for the distribution $f(x)$, then

$$\big(f(\alpha x + \beta)\big)^{(k)} = \big[p_n(\alpha x + \beta)\big]^{(k)} = \big[\alpha^k p_n^{(k)}(\alpha x + \beta)\big] = \alpha^k f^{(k)}(\alpha x + \beta).$$

In particular, since for $\alpha \neq 0$ we have the obvious identity,

$$\frac{1}{\alpha}\left(H(\alpha x + \beta) - \frac{1}{2}\right) = \frac{1}{|\alpha|}\left(H(x + \beta/\alpha) - \frac{1}{2}\right),$$

we get the important formulas for linear substitutions for the Dirac delta and its derivatives:

$$\delta^{(k)}(\alpha x + \beta) = \frac{1}{|\alpha|\alpha^k}\delta^{(k)}(x + \beta/\alpha), \qquad k = 0, 1, 2, \ldots . \qquad (9)$$

17.4.4.   The value of a distribution at a point. So far in this section our "function-like" notation for distributions, $f(t), g(t)$, etc., was purely symbolic and we never tried to determine the value of them at a given point. Here, we will break this barrier and introduce the concept of the value of a distribution at a point (a big no-no in the functional framework) which is fairly easy to justify in the sequential framework of the distribution theory. It will play an important role later on in this chapter in the development of random distributions (generalized stochastic processes). We shall start with a seemingly innocuous observation[7]:

If a distribution $f(t)$ on $(-\infty, \infty)$ satisfies the equality,

$$f(\lambda t) = f(t), \tag{10}$$

for each $\lambda \neq 0$, then it is a constant function.

Indeed, since for every distribution,

$$t f'(t) = \lim_{a \to 0} \frac{f(t + at) - f(t)}{a},$$

condition (10) implies that $f'(t) = 0$, for $t \neq 0$, and thus must be of the form (see Exercises)

$$f'(t) = a_0 \delta(t) + a_1 \delta'(t) + \cdots + a_k \delta^{(k)}(t),$$

for a certain integer k, so that

$$f(t) = c + a_0 H(t) + a_1 \delta(t) + \cdots + a_k \delta^{(k-1)}(t),$$

where c is a constant. Since, in view of Example 6,

$$\delta(\alpha t + \beta) = \frac{1}{|\alpha|} \delta\left(t + \frac{\beta}{\alpha}\right)$$

condition (10) implies that

$$a_0(H(\lambda t) - H(t)) + a_1\left(\frac{1}{|\lambda|} - 1\right)\delta(t) + \cdots + a_k\left(\frac{1}{|\lambda|\lambda^{(k-1)}} - 1\right)\delta^{(k-1)}(t) = 0,$$

so that $a_0 = a_1 = \cdots = a_k = 0$ (proof by induction in k), and consequently $g(t) = c$.

The above statement immediately implies that if the distributional limit $f(t) = \lim_{a \to 0} f(at + t_0)$ exists, it must be a constant function. This fact leads us to the following definition of the value of a distribution at a point:

Definition 6. The distributional limit

$$\lim_{a \to 0} f(at + t_0) \tag{11}$$

of a distribution $f(t)$ (if it exists !!!) will be called the *value of the distribution at the point t_0*, and t_0 itself will be called the *regular point* of $f(t)$. Points that are not regular will be called *singular*.

---

[7]This Lemma is due to Z. Zieleźny [11].

Example 6. Value at a point for the Heaviside function, the Dirac delta, and sin$(1/x)$.

(a) For the Heaviside function $H(t)$, all points $t_0 \neq 0$ are regular in the distributional sense, with the obvious values at them corresponding to the classical values. The point $t_0 = 0$ is singular in the distributional sense since the limit

$$H(at) = \frac{a}{|a|}\left(H(t) - \frac{1}{2}\right) + \frac{1}{2},$$

does not exists as $a \to 0$. The same is true for the Dirac delta $\delta(t)$, because $|a|\delta(at) = \delta(t)$.

(b) In the distributional sense the value of the function sin$(1/x)$ at 0 is equal to 0. Indeed,

$$\left(\int_0^t \sin\frac{1}{\tau}\,d\tau\right)'\bigg|_{t=0} = \lim_{t \to 0}\frac{1}{t}\int_0^t \sin\frac{1}{\tau}\,d\tau = 0$$

because, in view of the mean-value theorem, for $\tau \in (1/t, \infty)$,

$$\frac{1}{t}\int_0^t \sin\frac{1}{\tau}\,d\tau = \frac{1}{t}\int_{1/t}^\infty \frac{\sin\tau}{\tau}\cdot\frac{d\tau}{\tau} = \int_\nu^\infty \frac{\sin\tau}{\tau}\,d\tau.$$

We conclude this subsection by quoting a more general characterization of regular points of a distributions which is provided by the following theorem[8]:

THEOREM: *A distribution $f(t)$ has value c at a point t_0 if, and only if, there exists an integer $k \geq 0$, and a continuous function $F(t)$, such that $F^{(k)}(t) = f(t)$, and*

$$\lim_{t \to t_0}\frac{F(t)}{(t - t_0)^k} = \frac{c}{k!}.$$

17.4.5. Representing generalized stochastic processes by sequences of ordinary processes: basic concepts.

From the physical and computational perspective it may be more comfortable to think about random distributions as limits (of classes) of ordinary classical random functions, that is stochastic processes. Such a theory was formally developed by K. Urbanik, see [8–10]. In this subsection we will present an outline of his approach.

A sequence of random functions $\{\xi_n(t)\}$ is said to be convergent to the random function $\xi(t)$ (in short, $\xi_n(t) \Rightarrow \xi(t)$) if, for almost every sample path (realization), the sequence $\{\xi_n(t)\}$ converges almost uniformly to $\xi(t)$, which means that, for all ϵ, there exists a set A of t's of Lebesgue measure less than ϵ such that the sequence $\{\xi_n(t)\}$ converges uniformly on the complement of A. One can prove that the uniform convergence on the set of t's of full Lebesgue measure implies the almost uniform convergence.

---

[8]Its proof, due to S. Łojasiewicz, can be found in [7].

A sequence $\{\xi_n(t)\}$ of stochastic processes with almost surely continuous realizations will be called *fundamental* if there exists a convergent sequence of processes $\{X_n(t)\}$, and a nonnegative integer k, such that the kth derivatives

$$X_n^{(k)}(t) = \xi_n(t),$$

for all $n = 1, 2, \ldots$. Obviously, every convergent sequence of processes is fundamental.

Two sequences of continuous processes, $\{\xi_n(t)\}$, and $\{\theta_n(t)\}$, are called *equivalent*, if there exists sequences of processes $\{X_n(t)\}$, and $\{Y_n(t)\}$, convergent to the same limit, and an integer k such that

$$X_n^{(k)}(t) = \xi_n(t), \qquad and \qquad Y_n^{(k)}(t) = \theta_n(t),$$

for all $n = 1, 2, \ldots$. The above relationship of equivalence is reflexive, symmetric, and transitive, so it splits all the fundamental sequences into disjoint classes which permits the following definition:

Definition 7. Each of the above-defined classes of equivalence will be called a *generalized stochastic process* and will be denoted by capital Greek letters $\Xi(t), \Theta(t)$, etc. If the sequence $\{\xi_n(t)\} \in \Xi(t)$, we shall say that the sequence of continuous processes $\{\xi_n(t)\}$ represents the generalized stochastic process $\Xi(t)$, and write

$$\Xi(t) = [\xi_n(t)].$$

The sequence $\{\xi_n(t)\}$ will be called a *proxy* (or *representative*) of the generalized process $\Xi(t)$.

Operations of addition, and multiplication by a constant, are obviously well defined by analogous operations on the corresponding proxy processes, but the operation of multiplication of two generalized processes is not well defined, although the operation of multiplying a generalized process $\Xi(t)$ by a continuous process $\eta(t)$ is, because for any fundamental sequence $\{\xi_n(t)\}$, the sequence $\{\eta(t)\xi_n(t)\}$ is also fundamental. Note that the above statement does not depend on the choice of the proxy processes.

The derivative of a generalized process is always well defined and differentiation of any order is always possible. Let us start with the following observation.

LEMMA. *Every generalized stochastic process has a proxy composed of processes which are polynomials in variable t, that is, are of the form*

$$\alpha_0 + \alpha_1 t + \cdots + \alpha_k t^k,$$

where $\alpha_0, \alpha_1, \ldots, \alpha_k$ *are random coefficients.*

Proof: Let

$$\Xi(t) = [\xi_n(t)]. \tag{12}$$

Since $\{\xi_n(t)\}$ is a fundamental sequence, there exists a sequence of continuous processes $\{X_n(t)\}$ such that

$$X_n^{(k)}(t) = \xi_n(t), \qquad n = 1, 2, \ldots, \tag{13}$$

with

$$X(t) = \lim_{n \to \infty} X_n(t).$$

Now, for a fixed realization, define the nth Bernstein polynomial of $X(t)$ in a fixed interval, $-T \le t \le T$, as follows:

$$B_{n,T}(t) = \frac{1}{(2T)^n} \sum_{k=0}^{n} \binom{n}{k} X\left(T\left(\frac{2k}{n} - 1\right)\right)(T+t)^k (T-t)^{n-k},$$

whose coefficient is obviously random and dependent on the realization of $X(t)$. Choosing k_n so that

$$B_{k_n,n}(t) \Rightarrow \Xi(t), \tag{14}$$

we see that each of the processes,

$$\eta_n(t) = B_{k_n,n}^{(k)}(t), \qquad n = 1, 2, \ldots \tag{15}$$

is a polynomial in t, of degree not exceeding $n - k$. In view of (14–15) the sequence $\{\eta_n(t)\}$ is fundamental and, by (2) and the definition of $X(t)$, the sequences $\{\eta_n(t)\}$, and $\{\xi_n(t)\}$, are equivalent. Thus we have the equality, $\Xi(t) = [\eta_n(t)]$, which proves the Lemma. ∎

If $\{\xi_n(r)\}$ and $\{\eta_n(t)\}$ are infinitely differentiable proxies of the generalized process $\Xi(t)$, then $\{\xi'_n(t)\}$, and $\{\eta'_n(t)\}$, are also fundamental and equivalent. Hence, the *generalized derivative of the generalized process*

$$\frac{d\Xi(t)}{dt} = [\xi'_n(r)] = [\eta'_n(t)],$$

is always well defined, and the usual rules of differentiation can be proven to apply. Finally, an important observation following from the above Lemma is the following

COROLLARY. *Every generalized stochastic process is a generalized derivative of finite order of a continuous process.*

17.4.6. Convergence of generalized stochastic processes. The sequence $\{\Xi_n(t)\}$ of generalized stochastic processes is said to be *convergent* to a generalized process $\Xi(t)$ (in brief, $\Xi_n(t) \to \Xi(t)$) as $n \to \infty$, if there exist continuous processes $\xi(t), \xi_1(t), \xi_2(t), \ldots$, and a nonnegative integer k, such that

$$\xi_n(t) \Rightarrow \xi(t), \qquad \xi^{(k)}(t) = \Xi(t), \qquad and \qquad \xi_n^{(k)}(t) = \Xi_n(t),$$

for $n = 1, 2, \ldots$

The following properties follow directly from the above definition:

(a) If $\Xi_n(t) \to \Xi(t)$, then $\Xi'_n(t) \to \Xi'(t)$, and $\xi(t)\Xi_n(t) \to \xi(t)\Xi(t)$, for the proper multiplier $\xi(t)$;

(b) If $\Xi_n(t) \to \Xi(t)$, and $\Theta_n(t) \to \Theta(t)$, then $\Xi_n(t) + \Theta_n(t) \to \Xi(t) + \Theta(t)$;

(c) If $\Xi(t) = [\xi_n(t)]$, then $\xi_n(t) \to \Xi(t)$.

Example 7. Derivative of the Poisson process. Consider the standard Poisson process,

$$\Pi(t) = \lim_{n \to \infty} \sum_{k=0}^{n} H\Big(t - (\eta_1 + \cdots + \eta_k)\Big),$$

where $H(t)$ is the Heaviside function equal to 1, for $t \geq 0$, and 0, for $t < 0$, and $\{\eta_i\}$ is a sequence on independent, identically distributed random variables with an exponential probability distribution. Then,

$$\Pi'(t) = \sum_{k=0}^{\infty} \delta\Big(t - (\eta_1 \cdots + \eta_k)\Big).$$

Remark 1. Remembering that any generalized stochastic process $\Xi(t)$ is the distributional derivative of a certain order of a continuous process one can prove that

$$\Xi(t) = \lim_{n \to \infty} \sum_{k=1}^{k_n} \lambda_{k,n} \delta\Big(t - \nu_{k,n}\Big),$$

for a suitably chosen sequences of real numbers $\{\nu_{k,n}\}$, and of random variables $\{\lambda_{k,n}\}$. For a complete proof of this statement, see Urbanik [8], p. 277.

17.4.7.   Derivatives and difference quotients of order k. As observed before, for every generalized stochastic process $\Xi(t)$, there exists a continuous process $\xi(t)$, and a nonnegative integer k, such that $\Xi(t) = \xi^{(k)}(t)$, so that $\theta(t) = \int_0^t \xi(s)\, ds$, has almost all realizations with continuous derivatives, $\Xi(t) = \theta^{(k+1)}(t)$, and

$$\frac{1}{h}\Big(\theta(t+h) - \theta(t)\Big) \Rightarrow \theta'(t),$$

as $h \to 0$. Thus, differentiating k times we get

$$\frac{1}{h}\Big(\Xi(t+h) - \Xi(t)\Big) \to \Xi'(t).$$

Also, if the generalized process $\Xi(t)$ has the representation $[\xi_n(t)]$, then there exists a sequence $h_n \to 0$, such that $[h_n^{-1}(\xi_n(t+h_n) - \xi_n(t))]$ is a proxy for $\Xi'(t)$.

Similarly one can verify (see Exercises) that

$$\frac{1}{h^k}\Delta_h^{(k)}\Xi(t) \to \frac{d^k}{dt^k}\Xi^{(k)}(t),$$

where

$$\Delta_h^{(k)}\Xi(t) := \sum_{j=0}^{k}(-1)^j\binom{k}{j}\Xi(t+(k-j)h)$$

is the usual kth-order difference.

17.4.8.   Expected values of generalized stochastic processes.   Let, as before, $\Xi(t) = \xi^{(k)}(t)$, for a certain nonnegative integer k, and a continuous process $\xi(t)$. We will assume that the expectation $\mathbf{E}\xi(t)$ is well defined and is a locally integrable function. Then the expectation of the generalized process $\Xi(t)$ is a distribution defined by the formula

$$\mathbf{E}\Xi(t) = \big(\mathbf{E}\xi(t)\big)^{(k)}.$$

One can demonstrate that the definition is independent of the choice of k, and the continuous process $\xi(t)$.

Let us observe that, if the expectations of generalized processes $\Xi(t)$ and $\Theta(t)$ exist, then

$$\mathbf{E}\Big(\Xi(t) + \Theta(t)\Big) = \mathbf{E}\Xi(t) + \mathbf{E}\Theta(t),$$

and, if $\mathbf{E}\Xi(t)$ exists, then

$$\mathbf{E}\Xi'(t) = \Big(\mathbf{E}\Xi(t)\Big)'.$$

Example 8. Expected value of a random point process. Consider again the generalized stochastic process

$$\Xi(t) = \sum_{j=1}^{n}\delta(t-\nu_j),$$

where ν_j are random variables with probability densities $g_j(x), j = 1, 2, \ldots n$. It is easy to verify that

$$\Xi(t) = \xi''(t),$$

where

$$\xi(t) = \sum_{j=1}^{n}\Big(\max(0, t-\nu_j) + \min(0, \nu_j)\Big),$$

and that $\xi(t)$ is locally integrable; indeed, $|\xi(t)| \le n|t|$. Thus the process $\Xi(t)$ has a well-defined expected value

$$\mathbf{E}\Xi(t) = \Big(\mathbf{E}\xi(t)\Big)'' = \frac{d^2}{dt^2}\sum_{j=1}^{n}\int_0^t\int_{-\infty}^{y}g_j(x)\,dx\,dy = \sum_{j=1}^{n}g_j(t).$$

Example 9. Expected value of the derivative of Brownian motion with drift. Let $\xi(t)$ be the standard Brownian motion with linear drift, with

$$P(\xi(t) < x) = \frac{1}{\sqrt{2\pi|t|}} \int_{-\infty}^{x} e^{(y-\lambda t)^2/|t|}\, dy.$$

Consider its derivative $\Xi(t) = \xi'(t)$. Noting that $E|\xi(t)|$ is locally integrable (it is of the order $|t| + \sqrt{|t|}$) we obtain that

$$\mathbf{E}\Xi(t) = \Big(\mathbf{E}\xi(t)\Big)' = (\lambda t)' = \lambda.$$

Example 10. Generalized process with Dirac delta expectation. Consider a nonnegative, continuous process

$$\xi(t) = \frac{1 - \cos(\nu t)}{2\pi\nu^2 g(\nu)},$$

where ν is a random variable with positive and continuous density $g(x)$. Then

$$\mathbf{E}\xi(t) = \frac{1}{2\pi} \int_{-\infty}^{\infty} \frac{1 - \cos(tx)}{x^2}\, dx = \frac{|t|}{2},$$

and the generalized process,

$$\Xi(t) = \Big(\xi(t)\Big)'',$$

has the expectation

$$\mathbf{E}\Xi(t) = \Big(E\xi(t)\Big)'' = \delta(t).$$

17.5   Statistical independence concepts in the context of generalized processes

As a natural extension of the concept of classical statistical independence we will call generalized processes, $\Xi_1(t), \ldots, \Xi_r(t)$, *(statistically) independent* if there exist representations $\{\xi_{s,n}(t)\} \in \Xi_s(t), s = 1, 2, \ldots, r$, such that, for arbitrary positive integers, $n_1, \ldots, n_r$, the processes, $\xi_{s,n_1}(t), \ldots, \xi_{s,n_r}(t)$, are independent.

It is easy to see that the independence of generalized processes is preserved under the limit operations and, in particular, under differentiation. Indeed, if $\Xi_1(t), \ldots, \Xi_r(t)$ are independent then, for any combination of derivatives' orders, $m_1, \ldots, m_r$, the generalized processes,

$$\Xi_1^{(m_1)}(t), \ldots, \Xi_r^{(m_r)}(t),$$

are also independent.

17.5.1.   Processes with independent increments. The extension of the concept of independence of increments to a generalized process $\Xi(t)$ is more delicate. The first instinct is to define it by demanding that the proxy continuous processes $[\xi_n(t)]$ have independent increments. But this requirement restricts the theory to processes with Gaussian distributions. Indeed, it is a classical result of the theory of stochastic processes (see Doob [12], p. 420) that a stochastic process with independent increments and continuous trajectories necessarily has Gaussian distributions and, in particular, if the increments have distributions invariant under translations, then the process is the classical Brownian motion (Wiener process).

Thus we have to proceed in a more circumspect fashion and start with introducing a more relaxed concept of ϵ-independence which requires the independence of increments but only if they are separated by a positive time lag $\epsilon > 0$. More precisely we have the following definition for cad-lag (or, regular) processes, that is processes with realizations that are continuous on the right and have left limits:

Definition 1. A stochastic process $\xi(t)$ is said to have *ϵ-independent increments* if for every system $(u_{j,s}, t_{j,s}), j = 1, \ldots, j_s; s = 1, 2, \ldots, r$, of intervals separated from each other by more then ϵ, the random vectors,

$$\Big(\xi(t_{1,1}) - \xi(u_{1,1}), \xi(t_{2,1}) - \xi(u_{2,1}), \ldots, \xi(t_{j_1,1}) - \xi(u_{j_1,1})\Big),$$

$$\cdots$$

$$\Big(\xi(t_{1,r}) - \xi(u_{1,r}), \xi(t_{2,r}) - \xi(u_{2,r}), \ldots, \xi(t_{j_r,r}) - \xi(u_{j_r,r})\Big),$$

are independent.

Since a cad-lag process has independent increments if, and only if, it has ϵ-independent increments for every $\epsilon > 0$, the following definition is justified.

Definition 2. A generalized stochastic process $\Xi(t)$ is said to have *independent increments* if there exists a proxy representation $\{\xi_n(t)\} \in \Xi(t)$ such that for every $\epsilon > 0$ there exists an n_0 such that, for $n > n_0$, the processes $\xi_n(t)$ have ϵ-independent increments.

To better illuminate the relationship between the concepts of independence of increments for generalized processes, and that of ϵ-independence for the classical processes, we present the following

THEOREM. *A generalized stochastic process $\Xi(t)$ has independent increments if, and only if, for a certain k,*

$$\Xi(t) = \xi^{(k)}(t),$$

where $\xi(t)$ is a continuous process such that, for every $k > 0$, the kth difference process $\Delta_h^{(k)} \xi(t)$ has (kh)-independent increments for every $h > 0$.

Sketch of the proof. If $\Xi(t)$ has independent increments then, by definition, for some proxy sequence $\{\xi_n(t)\} \in \Xi(t)$, and for every $\epsilon > 0$, the processes $\xi_n(t)$ have ϵ-independent processes for large enough n's. So, by definition of the fundamental sequence, there exist continuous processes $\eta(t)$, $\eta_1(t), \eta_2(t), \ldots$, and a nonnegative integer k, such that

$$\eta^{(k)}(t) = \Xi(t), \qquad \eta_n^{(k)}(t) = \xi_n(t), \qquad \text{and} \qquad \eta_n(t) \Rightarrow \eta(t). \qquad (1)$$

Now, it is sufficient to demonstrate that, for every $h > 0$, the processes $\Delta_h^{(k)} \eta_n(t)$ (for definition, see Subsection 17.4.7) have (kh)-independent increments for sufficiently larger n's. So, let us take a system $(u_{j,s}, t_{j,s}), j = 1, \ldots, j_s; s = 1, 2, \ldots, r$, of intervals separated from each other by more then kh, and let us denote the minimum of these distances by $kh + \epsilon$. But, in view of (1), we have

$$\Delta_h^{(k)} \eta_n(t_{j,s}) - \Delta_h^{(k)} \eta_n(u_{j,s})$$

$$= \int_{u_{j,s}}^{u_{j,s}+h} \int_{v_k}^{v_k+h} \cdots \int_{v_2}^{v_2+h} \left(\xi_n(v_1) - \xi_n(v_1 + u_{j,s} - t_{j,s}) \right) dv_1 dv_2 \ldots dv_k,$$

so that the above increments are, for different s, integrals of increments of the processes $\xi_n(t)$ over the intervals that are separated by distances greater than, or equal to ϵ, which in turn implies that for sufficiently large n the increments of the process $\Delta_h^{(k)} \xi(t)$ are (kh)-independent.

On the other hand if, for a certain k, $\Xi(t) = \xi^{(k)}(t)$, where $\xi(t)$ is a continuous (or just cad-lag) process such that, for every $k > 0$, the kth difference process $\Delta_h^{(k)} \xi(t)$ has (kh)-independent increments then, denoting $\theta(t) = \int_0^t \xi(u) du$, we see that the process $\theta(t)$ is continuous, and

$$\xi^{(k)}(t) = [n^{k+1} \Delta_{1/n}^{(k+1)} \theta(t)],$$

where the latter fundamental sequence can be proven to have $(k+1)/n$-independent increments. ∎

Remark 1. One can also easily verify that if $\Xi(t)$ has independent increments then so does its generalized derivative $\Xi'(t)$.

17.5.2. Generalized processes with independent values at each point. The concept of a classical cad-lag (or continuous) stochastic process with independent values at different points, that is such that, if $t \neq s$, then $\xi(t)$ and $\xi(s)$ are statistically independent random variables, is vacuous. Indeed, only in the extreme case of completely deterministic (non-random) processes is the condition satisfied. However, the concept makes perfect sense in the case of the distributional theory of generalized stochastic processes. To begin the exposition of this concept we shall adapt the definition of the process with ϵ-independent increments as follows:

Definition 3. A continuous (or cad-lag) stochastic process $\xi(t)$ is said to have *ε-independent values* if, for every system $t_{j,s}, j = 1, \ldots, j_s; s = 1, 2, \ldots, r$, of points separated from each other by more then ϵ, the random vectors

$$\Big(\xi(t_{1,1}), \ldots, \xi(t_{j_1,1})\Big), \ldots, \Big(\xi(t_{1,r}), \ldots, \xi(t_{j_r,r})\Big)$$

are independent.

Observe that if $\xi(t)$ is a cad-lag process with independent increments, then the process

$$\eta(t) = \xi(t + \epsilon) - \xi(t)$$

has ϵ-independent values.

Definition 4. A generalized stochastic process $\Xi(t)$ is said to have *independent values* if there exists a proxy representation $\{\xi_n(t)\} \in \Xi(t)$ such that, for every $\epsilon > 0$, the processes $\xi_n(t)$ have ϵ-independent values for large enough n's.

Example 1. White noise. The white noise $\Xi(t)$, the distributional derivative of the Brownian motion process $\xi(t)$, has independent values. Indeed it is sufficient to choose as the needed proxy representation

$$\Big\{ n\Big(\xi(t + 1/n) - \xi(t)\Big) \Big\}.$$

In general, it is easy to see that if $\Xi(t)$ has independent values, and $\{\xi_n(t)\}$ is its proxy representation such that, for every $\epsilon > 0$, the processes $\xi_n(t)$ have ϵ-independent values for large enough n's, then for every multiplier process $\eta(t)$, the generalized process $\eta(t)\Xi(t)$ has independent values, and $\{\eta(t)\xi_n(t)\}$ is its proxy representation composed of processes with ϵ-independent values for sufficiently large n's.

The following properties of generalized processes with independent values are direct consequences of Definition 4:

(i) If $\Xi(t)$ has independent values and $\xi(t)$ is a continuous process such that $\xi^{(k)}(t) = \Xi(t)$ for a certain integer k then, for every $h > 0$, the process $\Delta_h^{(k)} \xi(t)$ has (kh)-independent values.

(ii) If the cad-lag process $\xi(t)$ is such that, for every $h > 0$, the process $\Delta_h^{(k)} \xi(t)$ has (kh)-independent values, then the generalized kth-order derivative $\xi^{(k)}(t)$ has independent values.

More comprehensively, we have the following structural result:

THEOREM. *A generalized process $\Xi(t)$ has independent values if, and only if, $\Xi(t) = \Theta'(t)$, where the generalized process $\Theta(t)$ has independent increments.*

Sketch of the proof. Choose the proxy representation $\{\xi_n(t)\}$ of $\Xi(t)$ required in the definition of the process with independent values. Since it is a fundamental sequence, there exist continuous processes $\eta(t), \eta_1(t), \eta_2(t), \ldots$; $\eta_n(t) \Rightarrow \eta(t)$, and a positive integer k, such that

$$\eta^{(k)}(t) = \Xi(t), \qquad \eta_n^{(k)} = \xi_n(t).$$

If we define

$$\Theta(t) = [\eta_n^{(k-1)}], \tag{2}$$

then, clearly, $\Theta'(t) = \Xi(t)$, and $\Theta(t)$ has independent increments. Indeed, let $(u_{j,s}, t_{j,s}), j = 1, \ldots, j_s; s = 1, 2, \ldots, r$, be a system of intervals separated from each other by more then ϵ. Then

$$\eta_n^{(k-1)}(t_{j,s}) - \eta_n^{(k-1)}(y_{j,s}) = \int_{u_{j,s}}^{t_{j,s}} \xi_n(u) du,$$

so that the increments are integrals of the values of the process at points separated from each other by more than ϵ, which implies the independence of the increments of the process $\eta_n^{(k-1)}$ for sufficiently large n, and thus, in view of (2), the independence of increments of $\Theta(t)$.

The reverse implication follows from the observation that if $\Theta(t)$ has independent increments with proxy representation $\xi_n(t)$ with ϵ-independent increment, as required by the definition, then if $h_n \to 0$, the processes $(\xi_n(t + h_n) - \xi_n(t))/h_n$ have, for every $\epsilon > 0$, ϵ-independent values for large enough n's, and we can choose the sequence $\{h_n\}$ so that

$$\Theta'(t) = \left[(\xi_n(t + h_n) - \xi_n(t))/h_n \right],$$

which proves the independence of values of the process $\Theta'(t)$. ∎

17.6   Local characteristics of generalized processes; generalized "densities"

17.6.1.   Basic concepts. For a deterministic distribution the definition of the value at a fixed point is tricky. We have discussed this issue briefly in the first section of this chapter and discovered that distributions such as $\delta(t)$ and $\delta'(t)$ do not have well-defined values at $t = 0$. So, obviously, the idea of the probability distribution of values of a random distribution is challenging. Our discussion will be restricted here to the class $\mathcal{M}$ of measurable stochastic processes $\xi(t)$ for which all the moments are finite, that is $\mathbf{E}|\xi(t)|^n < \infty, n = 1, 2, \ldots$. We will also assume that the moment functions $\mathbf{E}|\xi(t)|^n$ are locally integrable and satisfy the condition $\lim_{h \to 0} \mathbf{E}|\Delta_h \xi(t)|^n = 0$. The theory has been developed by K. Urbanik in [9].

The intuition behind what follows is based on the fact that generalized stochastic processes have realizations that are distributions which at each point take either numerical values, or "infinite values." But those infinite values can be of different orders. For example the order of "infinity" at $t = 0$ of $\delta'(t)$ is stronger than the of $\delta(t)$, because $t\delta(t) = 0$, but $t\delta'(t) = -\delta(t)$. Moreover, in addition to simple infinities $+\infty$ and $-\infty$, one has to take into account the "dipole infinity" $\pm\infty$ of $\delta'(t)$, at $t = 0$. So our local characteristics (generalized probability distribution functions) of a generalized process have to take into account the distribution of the "values" of the process over those different infinities.

Definition 1. The sequence of pairs $\{a_n, L_n\}, n = 1, 2, \ldots$, where $a_n = a_n(\lambda)$ are positive functions defined for $\lambda > 0$, and $L_n = L_n(\varphi)$ are positive operators (i.e., $\varphi \geq 0$ implies $L_n(\varphi) \geq 0$) on the following Banach space of functions,

$$\mathcal{X}_n := \left\{ \varphi : G_n^+(\varphi) := \lim_{x \to \infty} \frac{\varphi(x)}{x^n}, \quad and \quad G_n^-(\varphi) := \lim_{x \to -\infty} \frac{\varphi(x)}{|x|^n} \quad exist \right\},$$

equipped with the norm,

$$\|\varphi\| := \sup_{-\infty < x < \infty} \frac{|\varphi(x)|}{1 + |x|^n},$$

is said to be the *local characteristic* of the generalized stochastic process $\Xi(t)$ at time $t = t_0$, if there exist an integer k, a system of positive functions $A_{1n}(\lambda), \ldots, A_{kn}(\lambda)$, and a continuous stochastic process $\xi(t) \in \mathcal{M}$, such that

$$\xi^{(k)}(t) = \Xi(t), \quad a_n(\lambda) = \prod_{j=1}^{k} A_{jn}(\lambda), \quad n = 1, 2, \ldots,$$

and such that, for each $\varphi \in \mathcal{X}_n$,

$$L_n(\varphi) = \lim_{h_1, \ldots, h_k \downarrow 0} \prod_{j=1}^{k} A_{jn}(h_j) \int_{-\infty}^{\infty} \varphi(x) dP(D_{h_1, \ldots, h_k} \xi(t_0) < x).$$

where

$$D_{h_1, \ldots, h_k} \xi(t) := \frac{\Delta_{\lambda_1} \ldots \Delta_{\lambda_k} \xi(t)}{\lambda_1 \cdots \lambda_k}$$

is the usual kth order difference ratio with $\Delta_h \xi(t) := \xi(t + h) - \xi(t)$,

$$\lambda_k = h_k, \quad and \quad \lambda_j = h_j + \sum_{s=1}^{k-j} 2^{s-1} h_{j+s}.$$

We will normalize L_n so that $L_n(1 - |x|^n) = 1$.

The definition may seem opaque at the first sight but we will see its usefulness in the next couple of pages and in Part VI, where we'll discuss the theory of Lévy processes and anomalous diffusions from the perspective of generalized stochastic processes. One can prove that the local characteristic of a generalized stochastic process at fixed time is uniquely determined (see Urbanik [9]).

Example 1. Local characteristics of the Brownian motion and its derivatives. Let $\xi(t)$ be the standard Brownian motion process with variance $|t|$, and let $\Xi_k(t) = \xi^{(k)}(t)$, $k = 1, 2, \ldots$, be generalized derivatives of $\xi(t)$ of order k. Each of these stationary generalized processes has independent values at each time point, and its local characteristics at time $t = t_0$ are given by the following formulas:

$$a_0(\lambda) = \frac{1}{2}, \quad a_n(\lambda) = \sqrt{\pi}\lambda^{n/2+k-1}2^{-k^n}/\Gamma((n+1)/2), \quad n \geq 1,$$

and

$$L_0(\varphi) = \frac{1}{4}(G_0^+(\varphi) + G_0^-(\varphi)), \quad and \quad L_n(\varphi) = \frac{1}{2}(G_n^+(\varphi) + G_n^-(\varphi)), \quad n \geq 1.$$

Note that the characteristic operators $L_n(\phi)$ are independent of the order k of the derivative $\Xi_k(t)$ of the Brownian motion $\xi(t)$, and both local characteristics are independent of the time t_0. The derivatives are generalized stationary processes.

Example 2. Local characteristics of the derivative of the Poisson process. If $\xi(t) = \int_0^t \eta(s)ds$ is the integral of the standard Poisson process $\eta(t)$ with mean $|t|$, then $\xi(t)$ is continuous and its second distributional derivative,

$$\Xi(t) = \xi''(t) = \eta'(t),$$

has the following local characteristics at an arbitrary time $t = t_0$:

$$a_0(\lambda) = a_1(\lambda) = \frac{1}{2}, \quad a_n(\lambda) = \lambda^{n-1}, \quad n \geq 2,$$

and

$$L_0(\varphi) = \frac{1}{2}\varphi(0), \quad L_1(\varphi) = \frac{1}{2}(\varphi(0) + G_1^+(\varphi)), \quad and \quad L_n(\varphi) = G_0^+(\varphi), \quad n \geq 2.$$

Again, the local characteristics are independent of the time t_0.

Example 3. Local characteristics of continuous processes. For any continuous process $\xi(t)$ in $\mathcal{M}$ (which, in particular, means that all its moments are finite) with the cumulative probability distribution function (CDF) $p_{t_0}(x) = P(\xi(t_0) < x)$, the local characteristics are directly controlled by the the CDF. Indeed, for $n = 1, 2, \ldots$,

$$a_n(\lambda) = \frac{1}{1 + \int_{-\infty}^{\infty} |x|^n dp_{t_0}(x)}$$

and

$$L_n(\varphi) = \frac{\int_{-\infty}^{\infty} \varphi(x)dp_{t_0}(x)}{1 + \int_{-\infty}^{\infty} |x|^n dp_{t_0}(x)}$$

17.6.2. Local characteristics of transformed processes. The behavior of local characteristics under the usual operations on generalized stochastic processes is easy to obtain. Thus

(i) *Multiplication by a constant.* If $\Xi_c(t) = c \cdot \Xi(t)$, with the constant $c \neq 0$, then its local characteristics $(a_{n,c}(\lambda), L_{n,c}(\varphi))$ are expressed in terms of the local characteristics $(a_n(\lambda), L_n(\varphi))$ of $\Xi(t)$ as follows:

$$(a_{n,c}(\lambda), L_{n,c}(\varphi)) = \frac{(a_n(\lambda), L_n(\varphi))}{|c|^n + (1 - |c|^n)L_n(1)}, \quad n = 0, 1, 2, \ldots .$$

(ii) *Differentiation.* If $\xi(t)$ is a continuous process in $\mathcal{M}$ and $\xi_k(t) = \eta^{(k)}(t)$ then, for each $\varphi \in \mathcal{X}_n, n = 0, 1, 2, \ldots,$

$$\lim_{h_1,\ldots,h_k \downarrow 0} \sup_{-\infty < y < \infty} \frac{1}{1 + |y|^n} \int_{-\infty}^{\infty} \varphi(x+y) d\Big(P(D_{h_1,\ldots,h_k}\eta(t) < x) - P(\xi(t) < x)\Big) = 0.$$

(iii) *Addition of independent processes.* Let $\Xi(t)$ be a generalized process with local characteristics $(a_n(\lambda), L_n(\varphi))$, and $\xi(t)$ be an independent continuous process in $\mathcal{M}$. Then the local characteristics $(a_{n,+}(\lambda), L_{n,+}(\varphi))$ of the sum $\xi(t) + \Xi(t)$ are as follows:

$$(a_{n,+}(\lambda), L_{n,+}(\varphi)) = (a_n(\lambda), L_n(\varphi * P)), \quad n = 0, 1, 2, \ldots,$$

where

$$(\varphi * P)(x) = \int_{-\infty}^{\infty} \varphi(x+y) dP(\xi(t) < y).$$

17.7 Exercises

1. Prove that the three function sequences

$$\frac{\sin nt}{\pi t}, \qquad \frac{1}{2}ne^{-n|t|}, \qquad \text{and} \qquad \frac{n}{\pi(e^{nt} + e^{-nt})}.$$

listed in Example 4, in Subsection 17.4.2, are in the equivalence class defining the Dirac delta.

2. Verify the formula (8) in Subsection 17.4.3 involving differentiation of the superposition of a distribution and an infinitely differentiable function.

3. Demonstrate that if a distribution $f(t)$ is equal to 0 for $t \neq 0$ then it must be of the form

$$a_0\delta(t) + a_1\delta'(t) + \cdots + a_k\delta^{(k)}(t)$$

See, Subsection 17.4.4.

4. Prove that

$$\frac{1}{h^k}\Delta_h^{(k)}\Xi(t) \;\to\; \frac{d^k}{dt^k}\Xi(k)(t),$$

where

$$\Delta_h^{(k)}\Xi(t) := \sum_{j=0}^{k}(-1)^j \binom{k}{j}\Xi(t+(k-j)h),$$

is the usual kth-order difference.

5. Prove that if $\Xi(t)$ has independent increments then so does its generalized derivative $\Xi'(t)$.

Chapter 18

Dynamical and Statistical Characteristics of Random Fields and Waves

The time dependence of statistical characteristics of fields and waves in random media is often significantly different from the behavior of their realizations. In this chapter[1] we discuss this phenomenon employing the tool of lognormal distributions. A preliminary analysis of parametric stochastic resonance, dynamical and statistical energy localization for wavefields in randomly layered media, wave-beam propagation in random parabolic waveguides, and, finally, diffusing passive and active tracers in random velocity fields is included, but some of these topics will be explored in more detail in the following chapters. We will also take a look at the appearance of certain singularities in the dynamics of individual realizations, accompanied by their absence in the statistical description; the phase fluctuations of plane waves in a randomly layered medium provide an example of such behavior.

18.1 Lognormal processes

The time-dependent, 1-D probability distributions of a random process $Y(t)$ are determined by their *cumulative distribution functions (CDF)*

$$F_Y(y; t) = \mathbf{P}\Big(Y(t) \leq y\Big) = \int_{-\infty}^{y} f_Y(y'; t) dy' = \Big\langle \chi(y - Y(t)) \Big\rangle,$$

where $\chi(z)$ is the usual Heaviside unit jump function equal to 0, for $z < 0$, and to 1, for $z \geq 0$, function $f_Y(y; t) = \langle \delta(Y(t) - y) \rangle$ is the probability density

---

[1]The material discussed in this chapter is based on the papers by V.I. Klyatskin and W.A. Woyczyński [1], and V.I. Klyatskin and A. Saichev [2].

© Springer International Publishing AG, part of Springer Nature 2018
A. I. Saichev and W. A. Woyczynski, *Distributions in the Physical and Engineering Sciences, Volume 3*, Applied and Numerical Harmonic Analysis, https://doi.org/10.1007/978-3-319-92586-8_18

of the process $Y(t)$ at time t, and the brackets $\langle \, . \, \rangle$ stand for averaging over the ensemble of realizations of the stochastic process $Y(t)$.

For a given t, the inverse function $F_Y^{-1}(p;t)$ to the (nondecreasing !!) CDF, $F_Y(y)$, defined for p in the interval $(0,1)$ by the equation,

$$F_Y(F_Y^{-1}(p;t));t) = p, \tag{1}$$

is immediately recognizable as a concrete representation of the random quantity $Y(t)$ defined on the sample space $[0,1]$ equipped with the uniform Lebesgue probability measure.

For a fixed $0 < p < 1$, the deterministic function,

$$Z_Y(t;p) := F_Y^{-1}(p;t),$$

of the variable t will define the so-called *p-isoprobable curve* of the process $Y(t)$. Integrating the equality (1) over an arbitrary time interval (t_1, t_2), we obtain that

$$\left\langle \int_{t_1}^{t_2} \chi\Big(Z_Y(t;p) - Y(t) \Big) \right\rangle dt = \langle T_Y(t_1,t_2) \rangle = p(t_2 - t_1),$$

where $T_Y(t_1, t_2)$ is the total time, inside the interval (t_1, t_2), spent by process $Y(t)$ underneath the isoprobable curve $Z_Y(t;p)$.

Intuitively, if $p = 1/2$, the realization of the process $Y(t)$ will weave around the isoprobable curve, being on the average half of the time above, and a half of the time below that curve. For these reasons, it is natural to think of the isoprobable curve $Z_Y(t;1/2)$ as a sort of substitute of a *"typical" realization* of the process $Y(t)$, although the plot of $Z_Y(t;1/2)$ can differ significantly from the plot of any realization of the process $Y(t)$.

If p is selected close to 1, the plots of the realization of the process $Y(t)$ inside an any interval (t_1, t_2) are, with high probability, below the isoprobable curve. More precisely, for any small $\epsilon > 0$,

$$p = \frac{\langle T_Y(t_1,t_2) \rangle}{t_2 - t_1}$$

$$\leq (p - \epsilon) \cdot \mathbf{P}\left(\frac{T_Y(t_1,t_2)}{t_2 - t_1} \leq p - \epsilon \right) + 1 \cdot \mathbf{P}\left(\frac{T_Y(t_1,t_2)}{t_2 - t_1} > p - \epsilon \right)$$

$$\leq 1 - (1 + \epsilon - p) \cdot \mathbf{P}\left(\frac{T_Y(t_1,t_2)}{t_2 - t_1} \leq p - \epsilon \right)$$

and, as a consequence, necessarily,

$$\mathbf{P}\left(\frac{T_Y(t_1,t_2)}{t_2 - t_1} > p - \epsilon \right) > \frac{\epsilon}{1 + \epsilon - p}.$$

In particular, if we select $p = 0.999$, and $\epsilon = 0.1$, then the above estimate assures that, with probability better than 0.99, the realizations of $Y(t)$ will stay below the isoprobable curve more than 89.9 percent of time.

We will illustrate the above phenomena on the special example of the *lognormal random process* (also called the *geometric Brownian motion*, especially in the context of mathematical finance), which arises in many fields of physics and econometrics. Let us begin by an observation that the *Brownian motion (Wiener) random process* $b(t)$ is a solution of the stochastic equation

$$\frac{d}{dt}b(t) = W(t), \qquad W(0) = 0, \tag{2}$$

where $W(t)$ is a Gaussian white noise generalized process, delta-correlated in time with parameters $\langle W(t)\rangle = 0$, $\langle W(t)W(t')\rangle = 2\delta(t - t')$.[2] The Brownian motion process, $b(t)$, introduced in Chapter 16, is a continuous *nonstationary* (but with stationary increments) Gaussian random process (with almost surely continuous realizations, see Chapter 16) determined by parameters

$$\langle b(t)\rangle = 0, \qquad \langle b(t)b(t')\rangle = 2\min(t, t'). \tag{3}$$

In addition to the usual Brownian motion process, we will consider a Brownian motion process $b(t; \alpha)$ with drift $\alpha > 0$, defined by the formula,

$$b(t; \alpha) = -\alpha t + b(t).$$

The process $b(t; \alpha)$ is also a Markov process, and it has Gaussian 1-D probability densities,

$$f_b(x; t, \alpha) = \langle \delta(b(t; \alpha) - x)\rangle = \frac{1}{2(\pi t)^{1/2}} \exp\left[-\frac{(x + \alpha t)^2}{4t}\right]. \tag{4}$$

By the usual reflection principle we can find (see, also, Subsection 18.3.1) that the cumulative probability distribution of the absolute maximum

$$b_{max}(\alpha) = \max_{t \in (0, \infty)} b(t; \alpha)$$

of the process $b(t; \alpha)$ is of the form

$$F(h; \alpha) = P(b_{max}(\alpha) \le h) = 1 - e^{-\alpha h}. \tag{5}$$

In this context, the *lognormal process*,

$$Y(t; \alpha) := e^{-b(t; \alpha)}, \tag{6}$$

---

[2] We assume that the time t is a dimensionless quantity. Also, note that in the above definition of the Brownian motion we have introduced the multiplier 2 in the covariance function, a common choice in the physical sciences.

constructed with the help of the Brownian motion process, has the 1-D densities,

$$f_Y(y; t, \alpha) = \frac{1}{2(\pi t)^{1/2} y} \exp\left[-\frac{1}{4t} \ln^2(ye^{\alpha t})\right]. \tag{7}$$

Its moments are easily found to be

$$\langle Y^n(t; \alpha) \rangle = e^{n(n-\alpha)t}, \quad n = 1, 2, \dots,$$

and, in particular, for the process

$$Y(t) = Y(t; 1) = e^{b(t)-t}$$

they are given by the formula

$$\langle Y^n(t) \rangle = e^{n(n-1)t}, \quad n = 1, 2, \dots.$$

Thus, the mean value of process $Y(t)$ is constant and equal to 1, while all the higher moments of $Y(t)$ grow exponentially with t.

The exponential increase of higher moments of the lognormal process $Y(t)$ is caused by a slow decrease of *tails* of the probability density (7) for $y \gg 1$. As far as the realizations of the process $Y(t)$ are concerned, this means that rare but high peaks will appear while t increases. Thus, most of the time, realizations of the process $Y(t)$ will remain below the level of its mean value $\langle Y(t) \rangle = 1$, although its statistical moments will be mainly determined by its large jumps. That apparent contradiction between the behavior of statistical moments of the process $Y(t)$ and of its realizations provides a motivation for a more detailed study of the dynamics of realizations of process $Y(t)$. And, in the interpretation described above, the isoprobable curve $Z(t; 1/2) = e^{-t}$ can be seen as a "typical" realization of process $Y(t)$ introduced in (6).

A *p-majorant curve* $M_Y(t; p)$ for process $Y(t)$ is defined by the condition,

$$P\left(Y(t) \leq M_Y(t; p), \text{ for all } t\right) = p.$$

The shape of the distribution (5) of the absolute maximum of process $b(t; \alpha)$ suggests a large class of majorant curves (see, e.g., [2])

$$M_Y(t; , p, \beta) = (1 - p)^{-1/\beta} e^{-(1-\beta)t} \tag{8}$$

parameterized by parameter $0 < \beta < 1$. Note that, despite the facts that the statistical mean of $\langle Y(t) \rangle = 1$, and that the higher moments of process $Y(t)$ grow exponentially, it is always possible to find an exponentially decreasing ($\beta < 1$) majorant curve (8) such that the realization of process $Y(t)$ will stay below it at all times, with any previously set probability $p < 1$. In particular, one-half of realizations of $Y(t)$ will lie below the exponentially decreasing majorant curve

$$M_Y(t; 1/2, 1/2) = 4e^{-t/2}. \tag{8'}$$

Existence of exponentially decreasing majorant curves implies the conclusions regarding the statistical and the dynamical behavior of realizations of the process $Y(t)$:

• The exponential growth of higher moments of $y(t)$ is a purely statistical effect, caused by averaging over the whole ensemble of realizations.

• The area under an exponentially decreasing majorant curve is finite. Hence, very large jumps of process $Y(t)$ do not influence the areas under its realizations, which, in practice, are also finite.

The random area,

$$S = \int_0^\infty Y(t)\,dt,$$

under realizations of the process $Y(t)$ has the probability density

$$f_S(s) = s^{-2}e^{-1/s}, \tag{9}$$

and the cumulative distribution function,

$$F_S(s) = P(S \le s) = e^{-1/s}.$$

Observe that all the moments of the random variable S are infinite.

18.2   Lognormal law in physical phenomena

In this section we will discuss several examples of physical phenomena where the lognormal processes play an important role. They include the stochastic parametric resonance, localization of plane waves in layered random media, wave-beam propagation in random parabolic waveguides, and, finally, statistical description of diffusing tracers in random velocity fields. The last topic will be studied in greater depth in the following two chapters.

18.2.1.   Stochastic parametric resonance.   The dynamical system under consideration describes the stochastic parametric excitation in an oscillatory system with small linear friction due to parameter fluctuations and is represented by the following system of equations:

$$\frac{d}{dt}X(t) = Y(t), \qquad \frac{d}{dt}Y(t) = -2\gamma Y(t) - \omega_0^2[1 + Z(t)]X(t), \tag{1}$$

with initial values $X(0) = x_0$ and $Y(0) = y_0$. Process $Z(t)$ is assumed to be stationary, Gaussian, with zero mean, variance $\sigma_z^2 = \langle Z^2(t)\rangle$, and correlation time τ_0. In the delta-correlated approximation $Z(t)$ would be just the white noise generalized process and assumed to have parameters

$$\langle Z(t)\rangle = 0, \qquad \langle Z(t)Z(t')\rangle = 2\sigma_z^2\tau_0\delta(t - t').$$

System (1) appears in many areas of physics, and it permits a parametric excitation because $Z(t)$ contains harmonic components of all frequencies, including values $2\omega_0/n$, $n = 1, 2, 3, \ldots$, which exactly correspond to *parametric resonance* in a system where $Z(t)$ is just a periodic function, in which case it corresponds to the classical Mathieu equation (assuming $\gamma = 0$).

Let us rewrite the solutions of system (1) in the form,

$$X(t) = A(t)\sin(\omega_0 t + \phi(t)), \quad Y(t) = \omega_0 A(t)\cos(\omega_0 t + \phi(t)), \qquad (2)$$

where $A(t)$ is the oscillation amplitude function and $\phi(t)$, the phase function.

If parameter ω_0/D, where $D = \sigma_z^2 \omega_0^2 \tau_0$ is the diffusion coefficient in the corresponding Fokker–Planck equation, is large ($\omega_0/D \gg 1$), the relatively slow change (in t) of statistical characteristics of the solution system (2) is accompanied by ordinary oscillations with frequency ω_0. If the latter are eliminated by averaging the appropriate statistical characteristics over the time period $T = 2\pi/\omega_0$, then the probability density of the oscillation amplitude $A(t)$ is lognormal (see, e.g., [3]). Its moment functions satisfy the equality

$$\langle A^n(t)\rangle = A_0^n \exp\left[-n\gamma t + \frac{1}{8}n(2+n)Dt\right].$$

If, for a certain n, the condition $8\gamma < (2+n)D$ is satisfied, then the stochastic system (1) is subject to stochastic parametric excitation. Indeed, note that the "typical" realization function of the random process $A(t)$ is of the form,

$$A_0 e^{-(\gamma - D/4)t}.$$

Thus, if the friction in the dynamical system (1) is small, but $D < 4\gamma < (1 + n/2)D$, then the "typical" realization function decays exponentially. However, at the same time, the moment functions of the order $\geq n$ have an exponential growth.

18.2.2. Localization of plane waves in layered random media.

In this model, a layered, randomly inhomogeneous medium occupies a layered slab $L_0 < x < L$ in the space, and an inclined plane wave

$$U(x, \boldsymbol{\rho}) = \exp[ip(L - x) + i\mathbf{q}\boldsymbol{\rho}],$$

where $\boldsymbol{\rho} = (y, z)$, and $p = (k^2 - q^2)^{1/2} = k\cos\theta$, is incident on it from the region $x > L$. Within the layer, the wavefield is of the form,

$$U(x, \boldsymbol{\rho}) = u(x)\exp[i\mathbf{q}\boldsymbol{\rho}],$$

where function $u(x)$ is a solution of the boundary-value problem,

$$\frac{d^2}{dx^2}u(x) + \left[p^2 + k^2\varepsilon(x)\right]u(x) = 0, \qquad (3)$$

$$\frac{i}{p}\frac{d}{dx}u(x) + u(x)\Big|_{x=L} = 2, \quad \frac{i}{p}\frac{d}{dx}u(x) - u(x)\Big|_{x=L_0} = 0,$$

where $\varepsilon(x)$ is a random process describing random layering of the medium (e.g., fluctuations of the refraction coefficient for acoustic waves, or of the dielectric permittivity for the electromagnetic waves). In the case of wave's normal incidence to the boundary $x = L$ the problem simplifies because then $\theta{=}0$ and $p{=}k$.

The wavefield $u(x)$ also depends on parameter L, and in the framework of the so-called *imbedding method*, the boundary problem (3) can be reformulated as the following initial value problem for $u(x) = u(x; L)$ with respect to parameter L (see, [4]):

$$\frac{\partial}{\partial L}u(x; L) = ipu(x; L) + \frac{ik^2}{2p}\varepsilon(L)(1 + R_L))u(x; L),$$

$$\frac{d}{dL}R_L = 2ipR_L + \frac{ik^2}{2p}\varepsilon(L)(1 + R_L)^2, \tag{3'}$$

where R_L is the reflection coefficient.

In the half-space limit $(L_0 \rightarrow -\infty)$, the reflection coefficient $|R_L| = 1$ with probability 1, and it has the structure $R_L = \exp(i\phi_L)$ ([4]). Now, the phase ϕ_L and the wavefield intensity

$$I(x; L) = |u(x; L)|^2 = 2Y(x; L)(1 + \cos\phi_L)$$

satisfy the imbedding equations,

$$\frac{d}{dL}\phi_L = 2p + \frac{k^2}{p}\varepsilon(L)(1 + \cos\phi_L), \tag{4}$$

$$\frac{\partial}{\partial L}I(x; L) = -\frac{k^2}{p}\varepsilon(L)I(x; L)\sin\phi_L.$$

In the case of wave's normal incidence function, $Y(x; L)$ has the structure ([4]),

$$Y(x; L) = e^{-[q(L)-q(x)]}, \tag{5}$$

where the function, $q(L)$, is described by the stochastic equation,

$$\frac{d}{dL}q(L) = k\varepsilon(L)\sin\phi_L.$$

If the random process, $\varepsilon(L)$, is a Gaussian process with variance σ_ε^2, and correlation radius l_0, then, in the delta-correlated process approximation, random functions $q(L)$, and ϕ_L, are statistically independent, and the random quantity,

$$\eta(x; L) = \ln Y(x; L)$$

has a Gaussian distribution with parameters

$$\langle\eta(x; L)\rangle = -D(L - x), \quad \sigma_\eta^2(x; L) = D(L - x),$$

where $D = k^2 \sigma_\varepsilon^2 l_0/2$ is the diffusion coefficient in the corresponding Fokker–Planck equation. Consequently, the quantity $Y(x; L)$ has a lognormal distribution with mean $\langle Y(x; L) \rangle = 1$, and higher order moments

$$\langle Y^n(x; L)) \rangle = e^{n(n-1)D(L-x)}, \quad n = 2, \ldots,$$

growing exponentially as we move deeper into the medium thus again displaying the *stochastic wave parametric resonance*.

This case was discussed in detail in the previous subsection. The lognormal distribution indicates the existence in each realization of rare, but strong discontinuities of the wavefield intensity. These jumps take place against the background of exponential decay described by the function,

$$Y(x; L)) = \exp[-(L-x)l_{loc}^{-1}], \tag{6}$$

where

$$l_{loc}^{-1} = -\frac{d}{dL}\langle \eta(x; L) \rangle.$$

Usually, such a phenomenon is associated with the so-called *dynamic localization* property of physical disordered systems, and the quantity l_{loc} is then called the *localization length*. In the case considered above $l_{loc} = D^{-1}$. However, all statistical moments of quantities $Y(x; L)$ increase, so the *statistical energetic localization* does not occur in this case.

Each realization of random function $Y(x; L)$ represents an exponentially decreasing function with possibly large peaks and, in the physics of disordered systems, function $\exp(-\xi)$, $\xi = D(L - x)$, is called a "typical" realization of the random function $Y(x; L)$. This term is justified because it is the *p*-isoprobable curve for random function $Y(x; L)$, corresponding to $p = 1/2$. In other words, for any finite interval on the axis $\xi = D(L - x)$, on the average, over half of the interval (measure-wise) function $Y(x; L)$ majorizes the typical realization, and over the other half it is dominated by it.

It is also possible to find a global majorant of the wavefield intensity $Y(x; L)$. In particular, one can show that, with probability $p = 1/2$,

$$Y(x; L) < 4e^{-\xi/2}$$

for all values of ξ (see, [2]).

Next, consider the situation where the source of plane waves is located inside the medium layer at a point x_0, $L_0 < x_0 < L$. Then, within the layer, the wavefield is described by the boundary-value wave problem (see [4]),

$$\frac{d^2}{dx^2}G(x; x_0) + k^2\left[1 + \varepsilon(x)\right]G(x; x_0) = 2ik\,\delta(x - x_0), \tag{7}$$

$$\frac{i}{k}\frac{d}{dx}G(x; x_0) + G(x; x_0)\Big|_{x=L} = 0, \quad \frac{i}{k}\frac{d}{dx}G(x; x_0) - G(x; x_0)\Big|_{x=L_0} = 0,$$

where the random process, $\varepsilon(x)$, is, in general, complex-valued. Outside the layer $\varepsilon(x) = 0$, and inside the medium slab $\varepsilon(x) = \varepsilon_1(x) + i\gamma$, where $\varepsilon_1(x) = \varepsilon_1^*(x)$ is real-valued, and the attenuation parameter $\gamma \ll 1$ describes wave absorption. Introduction of a small attenuation parameter is essential here since, as we shall see later, the solution of this statistical problem is singular with respect to parameter γ.

Note that the problem (3) of wave incidence on a layered medium corresponds, in problem (7), to the location of the source on the boundary $x_0 = L$. In other words, $u(x) = G(x; L)$.

The solution of the boundary-value problem (7) is of the form,

$$G(x; x_0) = \frac{1 + R_2(x_0)}{1 - R_1(x_0)R_2(x_0)} u(x; x_0), \qquad x \le x_0, \tag{8}$$

where the quantity, $R_1(L) = R_L$, is the reflection factor of the plane wave incident from region $x > L$ on the medium layer, and $u(x; L)$ is a solution of the problem (3'). Quantity $R_2(x_0)$ has a similar meaning. Moreover, in the region $x < x_0$, the field of a point source is proportional to the field of the plane wave incident from the homogeneous space $x > x_0$ on the medium layer (L_0, x_0). The influence of the sublayer (x_0, L) is reflected only by the quantity $R_2(x_0)$.

18.2.3.   Wave-beam propagation in random parabolic waveguides.

In the quasioptics approximation, wave-beam propagation in the direction of the x-axis is described by a parabolic equation (see, e.g., [3]),

$$2ik\frac{\partial}{\partial x}u(x, \boldsymbol{\rho}) + \Delta_\perp u(x, \boldsymbol{\rho}) + k^2\varepsilon(x, \boldsymbol{\rho})u(x, \boldsymbol{\rho}) = 0, \tag{9}$$

where $\boldsymbol{\rho} = (y, z)$, and

$$\Delta_\perp = \frac{\partial^2}{\partial y^2} + \frac{\partial^2}{\partial z^2}.$$

Random field $\varepsilon(x, \boldsymbol{\rho})$ represents the fluctuating part of the dielectric permittivity, or of the refractive index, and $\boldsymbol{\rho}$ designates coordinates in the plane perpendicular to the x-axis.

In several important cases describing propagation of waves in natural waveguides (such as acoustic waves in the ocean, or radio waves in the atmosphere) one can assume that there are no random fluctuations and $\varepsilon(x, \boldsymbol{\rho}) = -\alpha^2\boldsymbol{\rho}^2$. Note that, in the absence of random inhomogeneities, wavefield $u(x, \boldsymbol{\rho})$ assumes the form,

$$u_0(x, \boldsymbol{\rho}) = u_0 \exp\left\{-\frac{\boldsymbol{\rho}^2}{2a^2} - i\alpha x\right\},$$

where parameter a satisfies the condition, $k\alpha a^2 = 1$. This wavefield corresponds to an eigenmode of the wave equation, and its amplitude does not vary during the wave propagation.

In the case when the fluctuation field is random, and given by the formula

$$\varepsilon(x, \boldsymbol{\rho}) = -\alpha^2 \boldsymbol{\rho}^2 + Z(x)\boldsymbol{\rho}^2,$$

where $Z(x)$ is a Gaussian, delta-correlated white noise generalized random process with parameters

$$\langle Z(x) \rangle = 0, \qquad \langle Z(x)Z(x') \rangle = 2\sigma^2 l \delta(x - x'),$$

the solution of the wave-beam problem (9) can be written in the form,

$$u(x, \boldsymbol{\rho}) = u_0 \exp\left\{ -\frac{\boldsymbol{\rho}^2}{2a^2} A(x) + B(x) \right\},$$

so that we obtain the following expression for the wavefield intensity,

$$I(x, \boldsymbol{\rho}) = I(x, 0) \exp\left\{ -\frac{\boldsymbol{\rho}^2}{a^2} I(x, 0) \right\},$$

where

$$I(x, 0) = \frac{1}{2}[A(x) + A^*(x)]$$

is the variation along the x-axis of wave intensity of the unperturbed waveguide.

Thus the statistical characteristics of wave intensity are determined by the statistical characteristics of quantity A which satisfies the equation,

$$\frac{d}{dx}A(x) = -\frac{i}{\kappa a^2}[A^2(x) - \alpha^2 \kappa^2 a^4] - i\kappa^2 Z(x), \quad A(0) = 1,$$

similar to equation for the wave reflection coefficient in a one-dimensional problem [5], and if parameters of the beam are matched to parameters of the waveguide, that is, if $\alpha k a^2 = 1$. As a result we get that

$$\langle I^n(x, 0) \rangle = e^{Dn(n-1)x}, \tag{10}$$

where $D = \sigma^2 l / 2\alpha^2 \kappa^2$ is the diffusion coefficient in the corresponding Fokker–Planck equation. This indicates that quantity $I(x, 0)$ has a lognormal distribution. Moments of the wave intensity field grow exponentially along the waveguide axis. However, as we have seen earlier, the "typical" realization of wavefield intensity is of the form,

$$I(x, 0) \sim e^{-Dx}, \quad \text{for } Dx \gg 1,$$

so that radiation escapes from the waveguide axis for each concrete realization; this indicates existence of the dynamical localization in the x-direction.

18.2.4. Statistical description of diffusing tracers in random velocity fields.

In this subsection we begin a preliminary study of the phenomenon of transport of passive tracers in randomly moving media. The topic will be investigated in much greater detail in the next two chapters. Governing equations for the problem of diffusing tracers in a random velocity field,

$$U(r, t) = (u_1(r, t), u_2(r, t), u_3(r, t)),$$

are of the following two types:

$$\left(\frac{\partial}{\partial t} + U(r, t) \frac{\partial}{\partial r} \right) q(r, t) = \kappa \frac{\partial^2}{\partial r^2} q(r, t), \quad q(r, 0) = q_0(r), \tag{11}$$

$$\left(\frac{\partial}{\partial t} + U(r, t) \frac{\partial}{\partial r} \right) p_i(r, t) = -\frac{\partial u_k(r, t)}{\partial r_i} p_k(r, t) + \kappa \frac{\partial^2}{\partial r^2} p_i(r, t),$$

$$p(r, 0) = p_0(r). \tag{12}$$

Equation (11) describes a scalar field $q(r, t)$, and equation (12) describes its spatial gradient $p(r, t) = (p_1(r, t), p_2(r, t), p_3(r, t)) = \partial q(r, t)/\partial r$. Let us also note the equation

$$\left(\frac{\partial}{\partial t} + \frac{\partial}{\partial r} U(r, t) \right) \rho(r, t) = \kappa \frac{\partial^2}{\partial r^2} \rho(r, t), \quad \rho(r, 0) = \rho_0(r), \tag{13}$$

for the density $\rho(r, t)$ of the passive "matter." In the above equations, κ denotes the *molecular diffusion coefficient*.

The fluid flow can be either *compressible* or *incompressible* ($\nabla U(r, t) = 0$). In the latter case equations (11) and (13) are identical. In the one-dimensional case, equations (12) and (13) are also identical, and the fluid flow is always compressible.

Equations (11)–(13) give the *Eulerian description* of the system. The probability distribution of $q(r, t)$ cannot be solved explicitly as (11) contains the second-order (diffusion) term depending on r. To remedy this situation we will introduce an auxiliary field $\tilde{q}(r, t)$ described by a stochastic equation

$$\left(\frac{\partial}{\partial t} + U(r, t) \frac{\partial}{\partial r} \right) \tilde{q}(r, t) = -\alpha(t) \frac{\partial}{\partial r} \tilde{q}(r, t), \quad \tilde{q}(r, 0) = q_0(r), \tag{14}$$

where $\alpha(t)$ is a delta-correlated Gaussian random vector generalized process (independent of U) with parameters

$$\langle \alpha(t) \rangle = 0, \quad \langle \alpha_i(t) \alpha_j(t') \rangle = 2\kappa \delta_{ij} \delta(t - t'); \quad i, j = 1, 2, 3.$$

Then, the solution of (11) corresponds to ensemble averaging of (14) relative to the process $\alpha(t)$, that is (see, [5])

$$q(r, t) = \langle \tilde{q}(r, t) \rangle_\alpha. \tag{15}$$

Formula (15) gives what is often called the *path integral representation* of the solution of (11).

The first-order stochastic partial differential equation (14) can be solved by the *method of characteristics* which replaces (14) by a system of ordinary differential equations

$$\frac{d}{dt}\boldsymbol{r}(t) = \boldsymbol{U}(\boldsymbol{r}(t), t) + \boldsymbol{\alpha}(t), \qquad \boldsymbol{r}(0) = \boldsymbol{\xi};$$

$$\frac{d}{dt}\tilde{q}(t) = 0, \qquad \tilde{q}(0) = q_0(\boldsymbol{\xi}). \tag{16}$$

Solution of the equations (16) depends on the initial parameter $\boldsymbol{\xi}$, that is,

$$\boldsymbol{r}(t) = \boldsymbol{r}(t|\boldsymbol{\xi}); \qquad \tilde{q}(t) = \tilde{q}(t|\boldsymbol{\xi}),$$

which corresponds to the so-called *Lagrangian description*. Eliminating parameter $\boldsymbol{\xi}$ in system (16) we obtain the *Eulerian description* of the concentration field:

$$\boldsymbol{\xi} = \boldsymbol{\xi}(t, \boldsymbol{r}); \qquad \tilde{q}(\boldsymbol{r}, t) = \tilde{q}\Big(t|\boldsymbol{\xi}(t, \boldsymbol{r})\Big).$$

To begin our analysis of statistical solutions of the diffusing tracer problem we will restrict ourselves initially to the simplest, *one-dimensional problem* with zero-mean velocity, and without molecular diffusion. In this case our equations take the form,

$$\left(\frac{\partial}{\partial t} + u(x, t)\frac{\partial}{\partial x}\right) q(x, t) = 0, \qquad q(x, 0) = q_0(x), \tag{17}$$

$$\left(\frac{\partial}{\partial t} + \frac{\partial}{\partial x}u(x, t)\right) \rho(x, t) = 0, \qquad \rho(X, 0) = \rho_0(x), \tag{18}$$

replacing equations (11) and (13). The above one-dimensional problem always describes a compressible fluid flow, and spatial concentration gradient $p(x, t) = \partial q(x, t)/\partial x$ is also described by equation (18).

The method of characteristics applied to equation (17) gives the corresponding Lagrange description

$$\frac{d}{dt}x(t|\xi) = u(x(t|\xi), t), \qquad x(0|\xi) = \xi;$$

$$\frac{d}{dt}q(t|\xi) = 0, \qquad q(0|\xi) = q_0(\xi). \tag{19}$$

Hence, $q(t|\xi) = q_0(\xi)$. The divergence, $j(t|\xi) = |\partial u(t|\xi)/\partial \xi|$, plays an important role if transferred to the Eulerian description, and, in view of (19), it satisfies the equation,

$$\frac{d}{dt}j(t|\xi) = \frac{\partial u(x, t)}{\partial x}j(t|\xi), \quad j(0|\xi) = 1. \tag{20}$$

The statistical properties of $x(t|\xi)$, and $j(t|\xi)$, will be described in terms of the function,

$$\Phi_t(x, j|\xi) = \delta\Big(x(t|\xi) - x\Big)\delta\Big(j(t|\xi) - j\Big),$$

satisfying Liouville's equation,

$$\frac{\partial}{\partial t}\Phi_t(x, j|\xi) = -\left[\frac{\partial}{\partial x}u(x, t) + \frac{\partial}{\partial j}j\frac{\partial u(x, t)}{\partial x}\right]\Phi_t(x, j|\xi),$$

$$\Phi_0(x, j|\xi) = \delta(x - \xi)\delta(j - 1). \tag{21}$$

For the sake of simplicity we shall assume that random velocity field $u(x, t)$ is Gaussian, homogeneous, and isotropic in space, and stationary in time, with parameters

$$\langle u(x, t)\rangle = 0, \qquad \langle u(x, t)u(x', t')\rangle = B(x - x', t - t').$$

We shall also keep the assumption that the field is delta-correlated in time, in which case the correlation function

$$B(x, t) = 2B^{eff}(x)\delta(t), \quad \text{where} \quad 2B^{eff}(x) := \int_{-\infty}^{\infty} d\tau\, B(x, \tau). \tag{22}$$

Averaging equation (22) over the realization ensemble of the random field $u(x, t)$ we obtain the Fokker–Planck equation for the joint probability density, $f_t(x, j|\xi) = \langle \Phi_t(x, j|\xi)\rangle$, of the "particle" position, and of its divergence (see, also, [6])

$$\frac{\partial}{\partial t}f_t(x, j|\xi) = D_1\frac{\partial^2}{\partial x^2}f_t(x, j|\xi) + D_2\frac{\partial^2}{\partial j^2}j^2 f_t(x, j|\xi),$$

$$f_0(x, j|\xi) = \delta(x - \xi)\delta(j - 1), \tag{23}$$

where diffusion coefficients D_i are determined by the equalities,

$$D_1 = B^{eff}(0), \qquad D_2 = -\frac{\partial^2}{\partial x^2}B^{eff}(x)\Big|_{x=0}.$$

It is clear from equation (23) that the diffusion of a "particle" does not depend upon divergence statistics, and that it is described by a Gaussian probability distribution with parameters

$$\langle x(t|\xi)\rangle = \xi, \qquad \sigma_x^2(t) = \left\langle [x(t|\xi) - \langle x(t|\xi)\rangle]^2\right\rangle = 2D_1 t,$$

i.e., it corresponds to the usual Brownian motion. Probability distribution of the divergence j is lognormal, and

$$\langle j(t)\rangle = 1, \qquad \langle j^n(t)\rangle = e^{D_2 n(n-1)t}, \tag{24}$$

i.e., the mean value of divergence is constant and moments of order ≥ 2 grow exponentially in time.

Note that the quantity, $\tilde{\rho}(t) = 1/j(t)$, which represents the particle density, satisfies in the Lagrangian description the equation,

$$\frac{d}{dt}\tilde{\rho}(t) = -\frac{\partial u(x,t)}{\partial x}\tilde{\rho}(t), \quad \tilde{\rho}(0) = 1.$$

One can prove that it has a lognormal probability distribution with the moment functions

$$\langle \tilde{\rho}^n(t) \rangle = e^{D_2 n(n+1)t}.$$

Thus, the mean density of passive tracer grows exponentially in time together with its higher moments.

This "paradoxical" behavior of statistical characteristics of divergence and of particle density, which both have moment functions simultaneously growing in time, is due to properties of the lognormal probability distribution. Thus a "typical" realization of the random divergence field may be represented by an exponentially decaying curve

$$j(t) = e^{-D_2 t},$$

and there exist exponential majorants for realizations of random process $j(t)$. In particular, with probability $p = 1/2$,

$$j(t) < 4e^{-D_2 t/2}$$

for any time t. Similarly, a "typical" realization of the density is of the form,

$$\tilde{\rho}(t) = e^{D_2 t},$$

and there exist exponential minorants so that, with probability $p = 1/2$,

$$\tilde{\rho}(t) > \frac{1}{4}e^{D_2 t/2}.$$

The above estimates for the statistics of $j(t)$, and $\tilde{\rho}(t)$, indicate the presence in their realizations of jumps over the "typical" realization as the particles are being compressed and form clusters located in mostly low-density zones.

Next, let us consider the Eulerian description of our problem and introduce the probability density functions,

$$f_{t,x}(q) = \langle \delta(q(x,t) - q) \rangle, \qquad f_{t,x}(\rho) = \langle \delta(\rho(x,t) - \rho) \rangle,$$

which are described by the Fokker–Planck equations,

$$\frac{\partial}{\partial t}f_{t,x}(q) = D_1 \frac{\partial^2}{\partial x^2}f_{t,x}(q), \qquad f_{0,x}(q) = \delta(q_0(x) - q), \qquad (25)$$

$$\frac{\partial}{\partial t}f_{t,x}(\rho) = D_1 \frac{\partial^2}{\partial x^2}f_{t,x}(\rho) + D_2 \frac{\partial^2}{\partial \rho^2}\rho^2 f_{t,x}(\rho),$$

$$f_{0,x}(\rho) = \delta(\rho_0(x) - \rho). \tag{26}$$

The solution of (25) corresponds to the spatial diffusion of the initial distribution. In the simplest case of homogeneous initial condition, the probability distribution does not depend on x, and $f_t(q) = \delta(q - q_0)$.

If the initial density $\rho_0(x) = \rho_0$ is constant, then the probability distribution does not depend on x either, and equation (26) takes a simplified form,

$$\frac{\partial}{\partial t} P_t(\rho) = D_2 \frac{\partial^2}{\partial \rho^2} \rho^2 P_t(\rho), \qquad P_0(\rho) = \delta(\rho_0 - \rho). \tag{27}$$

The solution of (27) corresponds to the lognormal distribution and

$$\langle \rho(x,t) \rangle = \rho_0, \qquad \langle \rho^n(x,t) \rangle = \rho_0^n e^{D_2 n(n-1)t}. \tag{28}$$

From (27), and (28), one can obtain a "typical" realization,

$$\rho(x,t) = \rho_0 e^{-D_2 t},$$

of the random field at any fixed point in space. The Eulerian statistics are related to density fluctuations relative to this curve, which confirms their cluster character.

The spatial gradient of concentration $p(x,t) = \partial q(x,t)/\partial x$ is described by an equation which coincides with the equation for media density. In this case, the joint probability density $f_{t,x}(q,p) = \langle \delta(q(x,t) - q)\delta(p(x,t) - p) \rangle$, for values of $q(x,t)$, and $p(x,t)$, is also described by the equation,

$$\frac{\partial}{\partial t} f_{t,x}(q,p) = D_1 \frac{\partial^2}{\partial x^2} f_{t,x}(q,p) + D_2 \frac{\partial^2}{\partial p^2} p^2 f_{t,x}(q,p),$$

$$f_{0,x}(q,p) = \delta\left(q_0(x) - q\right)\delta\left(\frac{\partial}{\partial x} q_0(x) - p\right),$$

from which one obtains the joint moment functions

$$\left\langle q^n(x,t)p^m(x,t) \right\rangle \sim e^{D_2 m(m-1)t}.$$

Hence, at a fixed point in space, the statistics of concentration gradient are formed by jumps with respect to a "typical," exponentially decaying realization.

The peculiarities of the statistical solutions described above are entirely caused by the compressibility of the one-dimensional flow. The situation also changes completely in the case of a *multidimensional incompressible fluid flow*. The above discussion indicates that the knowledge of the behavior of moment functions for concentration and its gradient, or for the spatial density, is insufficient for a detailed description of the passive tracer diffusion. What is needed here is a more complete study of the full probability distributions. In the general, three-dimensional case, the molecular diffusion term is the cause of principal difficulties and one has to resort to approximate methods. These issues will be addressed in later chapters.

18.3 Boundary-value problems for Fokker–Planck equation

18.3.1. Probability distribution of the maximum of the Brownian motion process. As a simple but instructive example of the boundary-value problem for the Fokker–Planck equation we will consider derivation of the probability distribution (see, Section 18.1) of the maximum of the Brownian motion process.

For the purposes of this subsection, the Brownian motion process $b(t; \alpha)$ is described by a stochastic differential equation

$$\frac{db(t; \alpha)}{dt} = -\alpha + W(t), \quad b(0, \alpha) = 0, \tag{1}$$

where $W(t)$ is a Gaussian generalized white noise process, delta-correlated in dimensionless time, with parameters

$$\langle W(t) \rangle = 0, \quad \langle W(t) W(t') \rangle = 2\delta(t - t').$$

Process $b(t, \alpha)$ is a Markov process, and its probability density $f_b(x; t) = \langle \delta(b(t, \alpha) - x \rangle$ satisfies the Fokker–Planck equation

$$\frac{\partial}{\partial t} f_b(x; t) = \alpha \frac{\partial}{\partial x} f_b(x; t) + \frac{\partial^2}{\partial x^2} f_b(x; t), \quad f_b(x; 0) = \delta(x). \tag{2}$$

Its solution is a Gaussian probability density function (18.1.4).

Now, in addition to the initial condition, let us complement equation (2) with a boundary condition

$$f_b(x; t) \big|_{x=h} = 0, \quad t > 0, \tag{3}$$

"killing" the realizations of the process $b(t; \alpha)$ at the moment when they reach the boundary h. The solution of the boundary problem (2–3) will be denoted by $f_b(x; t, h)$. For $x < h$, it describes the probability distribution of values of those realizations of the process $b(t; \alpha)$ that have survived up to time t; that is, during the entire time they have never reached the boundary h. Consequently, the probability density $f_b(x; t, h)$ integrates not to 1 but to the probability of $t > t^*$, where t^* is the time when process $b(t; \alpha)$ reaches the boundary h for the first time:

$$\int_{-\infty}^{h} dx \, f_b(x; t, h) = \mathbf{P}(t < t^*). \tag{4}$$

Let us introduce the cumulative probability distribution function and the probability density of the random time moment when the boundary is attained for the first time:

$$F(t; \alpha, h) = 1 - P(t < t^*) = 1 - \int_{-\infty}^{h} dx \, f_b(x; t, h), \tag{5}$$

$$f(t; \alpha, h) = \frac{\partial}{\partial t} F = -\frac{\partial}{\partial x} f_b(x; t, h) \Big|_{x=h}.$$ (5')

If $\alpha > 0$, then, on the average, process $b(t; \alpha)$ moves away from the boundary h as t increases. When $t \to \infty$, the probability $\mathbf{P}(t < t^*)$ in (4) converges to the probability of an event that process $b(t; \alpha)$ never reaches the boundary h. In other words,

$$\lim_{t \to \infty} \int_{-\infty}^{h} dx\, f_b(x; t, h) = \mathbf{P}(b_m(\alpha) < h),$$ (6)

where

$$b_m(\alpha) = \max_{t \in (0, \infty)} b(t; \alpha)$$ (7)

is the global maximum of the process $b(t; \alpha)$. Thus, it follows from (6), and (4), that the cumulative distribution function of the global maximum $b_m(\alpha)$ has the form,

$$F(h; \alpha) = P(b_m(\alpha) \le h) = \lim_{t \to \infty} \int_{-\infty}^{h} dx\, f_b(x; t, h).$$ (8)

For example, having solved the boundary problem (2–3) by the reflection method, we obtain that

$$f_b(x; t, h) = \frac{1}{2\sqrt{\pi t}} \left(\exp\left[-\frac{(x + \alpha t)^2}{4t} \right] - \exp\left[-h\alpha - \frac{(x - 2h + \alpha t)^2}{4t} \right] \right).$$ (9)

Substituting this expression into (5), we find that the probability density function of the first hitting time t^* of the boundary h by process $b(t; \alpha)$ is given by the formula,

$$f_b(t; \alpha, h) = \frac{1}{2t\sqrt{\pi t}} \exp\left[-\frac{(h + \alpha t)^2}{4t} \right]$$ (10)

Integrating (9) with respect to x and allowing $t \to \infty$ we obtain the cumulative distribution function

$$F(h; \alpha) = 1 - e^{-\alpha h}$$

of the global maximum b_m, as claimed in (4), and (18.1.5).

18.3.2. Phase fluctuations of plane wave in layered random medium. Localization of plane waves with normal incidence on the layer boundary was already discussed in Subsection 18.2.2, and we use the same notation in the present subsection. The situation is different in the case of inclined incidence. To study this case, instead of phase angle ϕ_L, we shall employ the random function $Z_L = \tan \phi_L/2$ which has singular points. The dynamical equation,

$$\frac{d}{dL} Z_L = p(1 + Z_L^2) + \frac{k^2}{p} \varepsilon(L),$$ (11)

for Z_L follows from (18.2.4). As before, function $\varepsilon(L)$ is assumed to be a Gaussian, delta-correlated generalized white noise random process. Then, in the half-space case, the steady-state probability density

$$f_Z(z) = \lim_{L_0 \to -\infty} f_L(z),$$

is described by the equation,

$$-\kappa \frac{d}{dz}(1 + z^2) f_Z(z) + \frac{d^2}{dz^2} f_Z(z) = 0,$$

where parameter

$$\kappa = p^3/2k^2 D = \frac{\alpha}{2} \cos^3 \theta, \quad \alpha = k/D.$$

Further analysis of this equation largely depends on the specific boundary conditions with respect to z, which determine the type of problems to be investigated. In the case of discontinuous function Z_L, defined for all values of L, the divergence of its values to $-\infty$ on one side of some points is accompanied by their divergence to $+\infty$ on the other side of the same points. In this situation the boundary condition satisfies the condition,

$$J(z)|_{z=-\infty} = J(z)|_{z=+\infty},$$

where

$$J(z) := \lim_{L_0 \to -\infty} J_L(z) = -\kappa(1 + z^2) f_Z(z) + \frac{d}{dz} f_Z(z)$$

is the steady state of the flux of probability density, with $f_L(z)$, and $J_L(z)$, satisfying the Fokker–Planck equation,

$$\frac{\partial}{\partial z} f_Z(z) = \frac{\partial}{\partial z} J_L(z).$$

Hence (see, e.g., [7]),

$$f_Z(z) = J(\kappa) \int_z^\infty d\xi \, \exp\left\{-\kappa\xi[1 + \xi^2/3 + z(z + \xi)]\right\}, \quad (12)$$

$$J^{-1}(\kappa) = \left(\frac{\pi}{k}\right)^{1/2} \int_0^\infty d\xi \, \xi^{-1/2} e^{-\kappa(\xi + \xi^3/12)}.$$

Asymptotically, for $\kappa \gg 1$, the sought probability density of Z is the Cauchy density, $f_Z(z) = 1/\pi(1 + z^2)$, $-\infty < z < \infty$, and it corresponds to the uniform probability distribution of ϕ : $f_\Phi(\phi) = 1/2\pi$, $0 \le \phi \le 2\pi$.

Probability distribution (12) allows us to calculate various quantities connected with fluctuations of the phase of the reflection coefficient. For example, on the boundary $x = L$ the mean value of the wavefield intensity

$$\langle I(L; L) \rangle = 2\langle 1 + \cos\phi_L \rangle = 2(3)^{1/6}\Gamma(2/3)\kappa^{1/3}, \quad \kappa \ll 1.$$

For the sliding incidence, when $\theta \to \pi/2$, the reflection coefficient $R_L \to -1$, and on the boundary $x = L$ wavefield $u(L) = 1 + R_L$ tends to zero. This result indicates that the medium behaves in a mirror-like fashion—an effect that is linked to a discontinuity of function $\varepsilon(x)$ at the boundary $x = L$. If the jump is small then it contributes little to the statistics of the phase in incidence angles θ ($\kappa \gg 1$). On the other hand, for the sliding incidence, this jump appears to be like an infinite barrier and contributes significantly to the statistics. Hence, the probability distribution of the reflection coefficient's phase contains information about how the wave scatters on a jump of $\varepsilon(L)$ at the boundary, and on random inhomogeneities inside the medium, without separating these effects.

Now, consider a medium with an adjusted right-hand boundary so that $\varepsilon(x) = \varepsilon(L)$ for $x > L$. In this case $k_L = p$, and the effect of discontinuity at $x = L$ can be excluded. The boundary problem is then reduced to the following initial-value problem with respect to L:

$$\frac{\partial}{\partial L} u(x; L) = ipu(x; L) + \frac{k^2 \xi(L)}{2p^2}(1 - R_L)u(x; L),$$

$$\frac{d}{dL} R_L = 2ip R_L + \frac{k^2 \xi(L)}{2p^2}(1 - R_L^2),$$

where $\xi(L) = \partial\varepsilon(L)/\partial L$ [8]. It is easy to see what is the impact of the nonlinear term in the equation for the reflection coefficient R_L. In the case of a random medium occupying half-space $x < L$, for the sliding incidence, we see that values of the phase of reflection coefficient $R_L = \exp[i\varphi_L] \to \pm 1$. Hence, these two points contribute most to the statistical characteristics linked with the phase.

For the finite slab with $\varepsilon(x)$ with no discontinuity at L, substituting $R_L = \exp[i\varphi_L]$, we obtain equations

$$\frac{d}{dL}\varphi_L = 2p - \frac{k^2 \xi(L)}{2p^2}\sin\varphi_L, \quad \frac{d}{dL}Z_L = p(1 + Z_L^2) - \frac{k^2 \xi(L)}{2p^2}Z_L,$$

for φ_L, and $Z_L = \tan\varphi_L/2$. And in the case of half-space, there is the steady-state probability density, which, for $\kappa \gg 1$, becomes the uniform probability distribution of the phase. In this case, for the mean value of the wave intensity on the boundary $x = L$, we have $\langle I(L; L) \rangle = 2$, as $\kappa \ll 1$, which means that the statistical weights of phase values for which $R_L = \pm 1$ are equal, although the probability distribution of φ differs considerably from the uniform one [8].

Finally, let us consider the statistical moment functions of the wavefield intensity $\langle I^n(x, L) \rangle = \langle |u(x; L)|^{2n} \rangle$. From (18.2.4), we obtain that

$$\langle I^n(\xi) \rangle = 2^n \pi^{1/2} \frac{(2n-3)!!}{(n-1)!}\xi^{-1/2}, \quad \xi \gg 1, \kappa \ll 1, \tag{13}$$

where $\xi = D(L - x)$ [7]. Hence, for the sliding incidence of the plane wave on a layered medium, quantities $\langle I^n(\xi) \rangle$ decrease according to the universal $\xi^{-1/2}$ law as we move deeper inside the medium.

For the adjusted boundary, one obtains the expression,

$$\langle I^n(\xi) \rangle \cong \frac{2^{n+1}(2n-3)!!}{(n-1)!} \frac{e^{-n\xi}}{\pi[1 + 4\pi^{-2}(\ln\kappa + \gamma)^2]\kappa}, \tag{14}$$

where γ is the Euler constant, and which is valid for $\kappa \ll 1$, and $\xi \gg 1$ (see [8]). This primary difference in the behavior of statistical characteristics for continuous and discontinuous $\varepsilon(x)$ is explained by the discontinuity at $x = L$. Thus, the asymptotic behavior of (14) is a "purely statistical" effect. We excluded the influence of boundary $x = L$ linked with a discontinuity of $\varepsilon(x)$ and have taken into account scattering of the wave by random inhomogeneities.

18.3.3.   Appearance of caustics in random media.

In the parabolic equation approximation the wavefield is described by equation (18.2.9) and, for the purpose of this subsection, we will write it in the form,

$$u(x, \rho) = A(x, \rho)e^{iS(x,\rho)}.$$

In the geometric optics approximation, when $k \to \infty$, analysis of the amplitude and phase fluctuations is greatly simplified, and equations for the phase and wave intensity assume the following form:

$$2k\frac{\partial}{\partial x}S(x, \boldsymbol{\rho}) + (\nabla_\perp S)^2 = k^2\varepsilon(x, \boldsymbol{\rho}), \quad k\frac{\partial}{\partial x}I(x, \boldsymbol{\rho}) + \nabla_\perp(I\nabla_\perp S) = 0, \tag{15}$$

where $(\nabla_\perp = \partial/\partial\boldsymbol{\rho})$. Solving these partial differential equations by the method of characteristics (rays), we can introduce a function $\boldsymbol{R}(x)$ such that, substituting $\boldsymbol{p}(x, \boldsymbol{\rho}) = \frac{1}{k}\nabla_\perp S$, we obtain a system,

$$\frac{d}{dx}\boldsymbol{R}(x) = \boldsymbol{p}(X), \quad \frac{d}{dx}\boldsymbol{p}(x) = \frac{1}{2}\frac{\partial}{\partial\boldsymbol{R}}\varepsilon(x, \boldsymbol{R}), \tag{16}$$

of closed equations. It must be solved with the initial condition at $x = 0$, determining parameterization of rays.

The equation for intensity can be rewritten in the form of an equation along the characteristics:

$$\frac{d}{dx}I(x) = -\frac{1}{k}I(x)\Delta_{\boldsymbol{R}}S(x, \boldsymbol{R}). \tag{17}$$

Let us introduce functions

$$u_{ij}(x, \boldsymbol{\rho}) = \frac{1}{k}\frac{\partial^2}{\partial\rho_i\partial\rho_j}S(x, \boldsymbol{\rho}),$$

which describe the *curvature of the phase front* $S(x, \boldsymbol{\rho}) = $ const. These functions satisfy equations

$$\frac{d}{dx}u_{ij}(x) + u_{ik}(x)u_{kj}(x) = \frac{1}{2}\frac{\partial^2}{\partial R_i\partial R_j}\varepsilon(x, \boldsymbol{R}), \tag{18}$$

$$\frac{d}{dx}I(x) = -I(x)u_{ii}(x).$$

In the two-dimensional case $(R = y)$, equations (16)–(18) become much simpler and take the form,

$$\frac{d}{dx}y(x) = p(x), \qquad \frac{d}{dx}p(x) = \frac{1}{2}\frac{\partial}{\partial y}\varepsilon(x, y), \tag{19}$$

$$\frac{d}{dx}I(x) = -I(x)u(x), \qquad \frac{d}{dx}u(x) = -u^2(x) + \frac{1}{2}\frac{\partial^2}{\partial y^2}\varepsilon(x, y).$$

Note that in absence of medium inhomogeneities $(\varepsilon = 0)$, rays appear to be straight lines, and integration of equations (19) gives

$$u(x) = \frac{u_0}{1 + u_0 x}, \qquad I(x) = \frac{I_0}{u_0}u(x). \tag{20}$$

If the initial condition u_0 is < 0, then $u(x_0) = -\infty$, and $I(x_0) = \infty$ at the point $x_0 = -1/u_0$; this means that the solution has an *explosive character*. In presence of inhomogeneities, $\varepsilon(x, y)$, such singular points exist regardless of the sign of u_0, which means that in the statistical problem singular points appear at finite distances. Their appearance is a result of random focusing of the wavefield which is caused by the fact that the phase front curvature and the wavefield intensity become infinite. It is this *random focusing* of the wavefield in the randomly inhomogeneous medium that causes appearance of *caustics*.

In the two-dimensional case, the phase curve curvature is described by equation (3.19). For a Gaussian, homogeneous, isotropic, and delta-correlated fluctuation $\varepsilon(x, y)$ with

$$\langle\varepsilon(x, y)\varepsilon(x', y'))\rangle = \delta(x - x')A(y - y'),$$

its probability density, $f(x, u)$, is described by the Fokker–Planck equation,

$$\frac{\partial}{\partial x}f(x, u) = \frac{\partial}{\partial u}u^2 f(x, u) + \frac{D}{2}\frac{\partial^2}{\partial u^2}f(x, u), \qquad P_0(u) = \delta(u - u_0), \tag{21}$$

where the diffusion coefficient

$$D = \frac{1}{4}\frac{\partial^4}{\partial y^4}A(y)\Big|_{y=0}.$$

Equation (21) has been studied in [9,10], where it was shown that random process $u(x)$ becomes infinite at finite distance $x(u_0)$, which is determined by the initial condition u_0. In this case, the mean value

$$\langle x(u_0)\rangle = \frac{2}{D}\int_{-\infty}^{u_0}d\xi\,e^{2\xi^3/3D}\int_{\xi}^{\infty}d\eta\,e^{-2\eta^3/3D} \tag{22}$$

and, consequently,

$$D^{1/3}\langle x(\infty)\rangle = 6.27, \quad D^{1/3}\langle x(0)\rangle = \frac{2}{3}D^{1/3}\langle x(\infty)\rangle = 4.18.$$

Quantity $\langle x(0)\rangle$ describes the mean distance of the focus appearance for the initial plane wave, and quantity $\langle x(\infty)\rangle$ describes the mean distance between two subsequent foci.

Further analysis of equation (21) depends in an essential way on the specific boundary condition with respect to u. Thus, as in Subsection 18.3.1, if we consider function $u(x)$ to be discontinuous, and defined for all values of x, and if its blowup to $-\infty$, for $x \to x_0 - 0$, accompanied by a blowup to $+\infty$, for $x \to x_0 + 0$, then the boundary condition is

$$J(x,u)\big|_{u\to\infty} = J(x,u)\big|_{u\to-\infty},$$

where

$$J(x,u) = u^2 f(x,u)) + \frac{D}{2}\frac{\partial}{\partial u}f(x,u) \tag{21'}$$

is the flux of probability density.

In this case, there exists a steady-state probability distribution,

$$f(\infty,u) = Je^{-2u^3/3D}\int_0^\infty d\xi\, e^{2u^3/3D},$$

defined by the equation,

$$J = u^2 f(\infty,u) + \frac{D}{2}\frac{d}{du}f(\infty,u),$$

where the steady-state probability flux density is

$$J = 1/\langle x(\infty)\rangle.$$

The asymptotic behavior of $f(\infty,u)$ for large values of u can be shown to be as follows:

$$f(\infty,u) \sim \frac{1}{\langle x(\infty)\rangle}\frac{1}{u^2},$$

and it depends on the jumps of function $u(x)$, and on its behavior near the discontinuity (caustic) points x_k which is of the form (20),

$$u(x) \sim \frac{1}{x - x_k}.$$

Note that according to (20) the wave intensity normalized by the intensity of incident wave I_0 has the following structure in the neighborhood of the discontinuities:

$$I(x) \sim \frac{x_k}{|x - x_k|}.$$

In this case, the asymptotics of probability distribution $z(x) = I^2(x)$, for sufficiently large x and z, is described by the expression,

$$f(x, z) \sim \sum_{k=0}^{\infty} \left\langle \delta \left(\frac{x_k^2}{(x - x_k)^2} - z \right) \right\rangle = \frac{x}{z\sqrt{z}} \sum_{k=0}^{\infty} \langle \delta(x - x_k) \rangle$$

$$= \frac{x}{z\sqrt{z}} \frac{1}{2\pi} \int_{-\infty}^{\infty} d\kappa \, e^{-i\kappa x} \frac{\langle e^{i\kappa x_0} \rangle}{1 - \langle e^{i\kappa x} \rangle},$$

where $\langle e^{i\kappa x_0} \rangle$ is the characteristic function of the distance to the first caustic, and $\langle e^{i\kappa x} \rangle$ is the characteristic function of the distance between two subsequent caustics. Thus, if $x \gg \langle x(\infty) \rangle$, then the probability density of Z has asymptotic form

$$f(x, z) \sim \frac{x}{z\sqrt{z}} \frac{1}{\langle x(\infty) \rangle},$$

and, as a result, probability density of the intensity itself has the asymptotics,

$$f(x, I) = \frac{2x}{I^2} \frac{1}{\langle x(\infty) \rangle},$$

so that, in particular, it has a power decay as I increases.

Another type of boundary conditions arises when the curve $u(x)$ blows up to $-\infty$ as x approaches point x_0. This corresponds to the condition that the probability flux density $J(x, u)$ must converge to zero when $u \to \infty$, which can be rephrased by the following conditions,

$$J(x, u) \to 0, \text{ as } u \to \infty; \qquad f(x, u) \to 0, \text{ as } u \to -\infty.$$

The probability density of the caustic location is then ([9])

$$p(x) = \lim_{u \to -\infty} J(x, u). \tag{23}$$

To derive the asymptotic dependence of probability density $p(x)$ on parameter $D \to 0$, we shall use for equation (21) the standard procedure (see, e.g., [11]) for analysis of parabolic equations with a small parameter in the highest derivative. Let us begin by writing the solution of equation (21) in the form,

$$f(x, z) = C(D) \exp \left\{ -\frac{1}{D} A(x, u) - B(x, u) \right\}. \tag{24}$$

Constant $C(D)$ is determined by the condition that, for $x \to 0$, the probability distribution of the plane incident wave must have the form,

$$f(x, z) \sim \frac{1}{(2\pi Dx)^{1/2}} \exp \left\{ -\frac{u^2}{2Dx} \right\}. \tag{23'}$$

This gives that $C(D) \simeq D^{1/2}$.

Note that expressing $f(x, z)$ in the form (23) immediately permits determination of structural dependence of $p(x)$ on x by dimensional analysis. Indeed, quantities u, D, and $P_x(u)$ have, respectively, dimensions $[u] = x^{-1}$, $[D] = x^{-3}$, and $[P_x(u)] = x$. Consequently, in view of (21'), function $p(x)$ has the structure,

$$p(x) \sim C_1 D^{-1/2} x^{-5/2} e^{-C_2/Dx^3},$$

and the only remaining task is to calculate positive constants C_1 and C_2. This has been done in [9], and the resulting formula is

$$p(x) \sim 3\alpha^2 (2\pi D)^{-1/2} x^{-5/2} \exp\{-\alpha^4/6Dx^3\}, \tag{25}$$

where $\alpha = 1.85$. The condition of applicability of (25) is $Dx^3 \ll 1$. However, as was shown in [9] by numerical modeling, expression (25) also accurately describes the probability distribution of appearance of random foci when $Dx^3 \simeq 1$.

Note that for a three-dimensional problem, dimensional considerations give the following form of the probability density of appearance of caustics: $p(x) = \alpha D^{-1} x^{-4} \exp(-\beta/Dx^3)$, where α and β are numerical constants. Their values, $\alpha = 1.74$ and $\beta = 0.66$, have been computed in [10].

18.4 Fokker–Planck equation for dynamical systems

Assume that a vector-valued function $\boldsymbol{\xi}(t)$ satisfies the dynamical equation,

$$\frac{d}{dt}\boldsymbol{\xi}(t) = \boldsymbol{v}(\boldsymbol{\xi}, t) + \boldsymbol{z}(\boldsymbol{\xi}, t), \qquad \boldsymbol{\xi}(0) = \boldsymbol{\xi}_0, \tag{1}$$

where function $\boldsymbol{v}(\boldsymbol{\xi}, t)$ is a deterministic function, and $\boldsymbol{z}(\boldsymbol{\xi}, t)$ is a random field of $(n + 1)$ variables, which satisfies two conditions:

(a) $\boldsymbol{z}(\boldsymbol{x}, t)$ is a Gaussian random field in the $(n + 1)$-dimensional $(\boldsymbol{x}, t)$-space;

(b) $\langle \boldsymbol{z}(\boldsymbol{x}, t) \rangle = 0$.

Statistical characteristics of the field $\boldsymbol{z}(\boldsymbol{x}, t) = (z_i(\boldsymbol{x}, t))$ are fully determined by the correlation tensor

$$B_{ij}(\boldsymbol{x}, t; \boldsymbol{x}', t') = \langle z_i(\boldsymbol{x}, t) z_j(\boldsymbol{x}', t') \rangle.$$

The quantities $\xi_i(t)$ represent nonanticipating functionals that depend only on the values of $f_j(\boldsymbol{x}, t')$, for $t' \leq t$, that is,

$$\frac{\delta \xi_i(t)}{\delta z_j(\boldsymbol{x}, t')} = 0, \text{ if } t' > t. \tag{2}$$

However, for $t'' > t$, statistical dependence between $\xi_i(t)$ and the values of $z_j(\boldsymbol{x}, t'')$ is possible. Indeed, the latter are correlated with values $f_j(\boldsymbol{x}, t')$ for $t' < t$. Obviously, a correlation between $\xi_i(t)$ and the subsequent values $z_i(\boldsymbol{x}, t'')$ exists only if $t'' - t \sim \tau_0$, where τ_0 denotes the correlation radius of the field $z_i(\boldsymbol{x}, t)$ with respect to variable t.

For a large number of actual physical processes the characteristic correlation radius of $\xi_i(t)$ has order of magnitude T, where T reflects the natural scales of the dynamical problem without the fluctuation term. In such cases, parameter τ_0/T is small and can be used in the construction of an asymptotic solution. In the first-order approximation with respect to the small parameter τ_0/T we then take $\tau_0 \to 0$ and obtain the so-called delta-correlation approximation, where there is neither a functional dependence nor a statistical dependence between the values of $\xi_i(t')$ for $t' < t$, and the values of $z_j(\boldsymbol{x}, t'')$ for $t'' > t$. This first-order approximation leads to introduction of the effective correlation tensor

$$B_{ij}(\boldsymbol{x}, t; \boldsymbol{x}', t') = 2\delta(t - t')\beta_{ij}(\boldsymbol{x}, \boldsymbol{x}', t), \tag{3}$$

where

$$\beta_{ij}(\boldsymbol{x}, \boldsymbol{x}', t) = \frac{1}{2}\int_{-\infty}^{\infty} dt' \, B_{ij}(\boldsymbol{x}, t; \boldsymbol{x};, t').$$

Thus the original field z now becomes a Gaussian random field which is delta-correlated with respect to t.

Consider the probability density

$$f_{\boldsymbol{\xi}}(t, \boldsymbol{x}) = \langle \delta(\boldsymbol{x} - \boldsymbol{\xi}(t)) \rangle \tag{4}$$

of the solution $\boldsymbol{\xi}(t)$ of the system of equations (26). If we differentiate (2) with respect to t, taking into account (1), we get the equation,

$$\frac{\partial}{\partial t} f_{\boldsymbol{\xi}}(t, \boldsymbol{x}) = -\frac{\partial}{\partial x_k}\left(v_k(\boldsymbol{x}, t) f_t(\boldsymbol{x})\right) - \frac{\partial}{\partial x_k}\left\langle z_k(\boldsymbol{x}, t)\delta(\boldsymbol{x} - \boldsymbol{\xi}(t)) \right\rangle, \tag{5}$$

which can be rewritten in the form

$$\frac{\partial}{\partial t} f_{\boldsymbol{\xi}}(t, \boldsymbol{x}) = -\frac{\partial}{\partial x_k}\left(v_k(\boldsymbol{x}, t) f_{\boldsymbol{\xi}}(t, \boldsymbol{x})\right) \tag{6}$$

$$+\frac{\partial}{\partial x_k}\int d\boldsymbol{x}' \int_0^t d\tau \left\langle z_k(\boldsymbol{x}, t) z_1(\boldsymbol{x}', \tau) \right\rangle \frac{\partial}{\partial x_m}\left\langle \delta(\boldsymbol{x} - \boldsymbol{\xi}(t))\frac{\delta \xi_m(t)}{\delta z_1(\boldsymbol{x}, \tau)} \right\rangle.$$

Here, we have used the *Furutsu–Novikov formula* [12, 13],

$$\left\langle z_k(\boldsymbol{x}, t) R[\boldsymbol{z}(\boldsymbol{x}', \tau)] \right\rangle$$

$$= \int d\boldsymbol{x}' \int_0^t d\tau \left\langle z_k(\boldsymbol{x}, t) z_1(\boldsymbol{x}', \tau) \right\rangle \left\langle \frac{\delta}{\delta z_1(\boldsymbol{x}, \tau)} R[\boldsymbol{z}(\boldsymbol{x}', \tau)] \right\rangle,$$

which gives the covariance of a Gaussian random field $z(x, t)$ and its arbitrary functional $R[z]$. Notice that condition (2) permitted us to restrict the integration over τ to the interval $(0, t)$. The above equation implies that the probability density for the solution ξ at instant t is determined by the interdependence between $\xi(t)$ and the field $z(x', \tau)$ for all values of τ in the interval $(0, t)$. In general, $f_\xi(t, x)$ does not satisfy a closed differential equation.

If the correlation function of field $z(x, t)$ is asymptotically approximated by (3), then the integration over τ reduces to the substitution $\tau = t$, and only values $\delta\xi_m(t)/\delta z_1(x', t)$ appear. In this case, these values can be expressed in terms of $\xi(t)$:

$$\frac{\delta\xi_m(t)}{\delta z_l(x', t - 0)} = \delta_{ml}\delta(\xi - x'),$$

and we get the Fokker–Planck equation

$$\frac{\partial}{\partial t} f_\xi(t, x) + \frac{\partial}{\partial x_k}\left\{\left[v_k(x, t) + A_k(x, t)\right] f_\xi(t, x)\right\}$$

$$-\frac{\partial^2}{\partial x_k \partial x_1}\left[\beta_{kl}(x, x; t) f_\xi(t, x)\right] = 0, \tag{7}$$

where

$$A_k(x, t) = \frac{\partial}{\partial x_1'}\beta_{kl}(x, x')\Big|_{x'=x}.$$

Equation (7) has to be solved with the initial condition $f_\xi(0, x) = \delta(x - \xi_0)$, or with a more general initial condition $f_\xi(t, x) = W_0(x)$.

Let us now return to the dynamical system (1), and derive an equation for the joint probability density

$$f_m(x_1, t_1; \ldots; x_m, t_m) = \left\langle\delta(\xi(t_1) - x_0)\cdot\ldots\cdot\delta(\xi(t_m) - x_m)\right\rangle, \tag{8}$$

which depends on m different moments of time $t_1 < t_2 < \ldots < t_m$. After differentiation of (8) with respect to t_m, and subsequent application of the dynamical equation (1), the definition of the β_{ij} yields the following equation:

$$\frac{\partial}{\partial t_m} f_m(x_1, t_1; \ldots; x_m, t_m) + \sum_{i=1}^n \frac{\partial}{\partial x_{mi}}\left\{\left[V_i(x_m, t_m) + A_i(x_m, t_m)\right] f_m\right\}$$

$$= \sum_{i=1}^n \frac{\partial}{\partial x_{mi}} \sum_{j=1}^n \frac{\partial}{\partial x_{mj}}\left[\beta_{ij}(x_m, x_m; t_m) f_m\right], \tag{9}$$

similar to the Fokker–Planck equation (6).

The initial condition for (9) can be derived from (8). If we suppose that $t_m = t_{m-1}$ in (8), then we get that

$$f_m(x_1, t_1; \ldots; x_{m-1}, t_{m-1}; x_m, t_{m-1})$$

$$= \delta(\boldsymbol{x}_m - \boldsymbol{x}_{m-1}) f_{m-1}(\boldsymbol{x}_1, t_1; \ldots; \boldsymbol{x}_{m-1}, t_{m-1}), \tag{10}$$

and we can look for a solution of (9) of the form

$$f_m(\boldsymbol{x}_1, t_1; \ldots; \boldsymbol{x}_m, t_m) = p(\boldsymbol{x}_m, t_m | \boldsymbol{x}_{m-1}, t_{m-1}) f(\boldsymbol{x}_1, t_1; \ldots; \boldsymbol{x}_{m-1}, t_{m-1}). \tag{11}$$

Since all the derivatives in (9) are with respect to $t_m, \boldsymbol{x}_m$, a substitution of expression (11) into (9), and (10), yields the following equation for the transition probability density:

$$\frac{\partial}{\partial t} p(\boldsymbol{x}, t | \boldsymbol{x}_0, t_0) + \frac{\partial}{\partial x_k} \left\{ \left[v_k(\boldsymbol{x}, t) + A_k(\boldsymbol{x}, t) \right] p(\boldsymbol{x}, t | \boldsymbol{x}_0, t_0) \right\}$$

$$= \frac{\partial}{\partial x_k \partial x_1} \left[\beta_{kl}(\boldsymbol{x}, \boldsymbol{x}; t) p(\boldsymbol{x}, t | \boldsymbol{x}_0, t_0) \right], \tag{12}$$

with the initial condition,

$$p(\boldsymbol{x}, t | \boldsymbol{x}_0, t_0) \bigg|_{t \to t_0} = \delta(\boldsymbol{x} - \boldsymbol{x}_0). \tag{13}$$

Here, variables $\boldsymbol{x}_m, t_m$ are replaced by $\boldsymbol{x}$, t, and variables $\boldsymbol{x}_{m-1}$, t_{m-1}, by $\boldsymbol{x}_0, t_0$ (i.e., $p(\boldsymbol{x}, t | \boldsymbol{x}_0, t_0) = \langle \delta(\boldsymbol{x} - \boldsymbol{\xi}(t) | \boldsymbol{\xi}(t_0) = \boldsymbol{x}_0) \rangle)$).

Applying (11) $(m-1)$ times, we arrive at the relation,

$$f_m(\boldsymbol{x}_1, t_1; \ldots; \boldsymbol{x}_m, t_m)$$

$$= p(\boldsymbol{x}_m, t_m | \boldsymbol{x}_{m-1}, t_{m-1}) \ldots p(\boldsymbol{x}_2, t_2 | \boldsymbol{x}_1, t_1) f_{t_1}(\boldsymbol{x}_1), \tag{14}$$

where $f_{t_1}(\boldsymbol{x}_1)$ is the probability density determined by equation (12), and depending on the single time instant t_1. Formula (14) expresses the multi-point probability density as a product of transition probability densities so that $\boldsymbol{\xi}(t)$ is a *Markov process*.

In order to determine conditions of applicability of the Fokker–Planck equation we assume that the time correlation radius τ_0 of the field $f_j(\boldsymbol{x}, t)$ is finite. Then, equation (6) for the probability density can be replaced by the equation,

$$\hat{E} f_{\boldsymbol{\xi}}(t, \boldsymbol{x}) = -\frac{\partial}{\partial x_k} S'_k(\boldsymbol{x}, t), \tag{15}$$

where $\hat{E}$ denotes the operator appearing in the left-hand side of equation (9). Instead of $\beta_{kl}(\boldsymbol{x}, \boldsymbol{x}', t)$ we set

$$\tilde{\beta}_{kl}(\boldsymbol{x}, \boldsymbol{x}', t) = \int_0^t dt' \, B_{kl}(\boldsymbol{x}, t; \boldsymbol{x}', t'). \tag{16}$$

Term $S'_k(\boldsymbol{x}, t)$ contains corrections of the probability density due to the finite positive value of τ_0 and it imposes, in general, limitations on the intensity of fluctuations of field $\boldsymbol{z}$. If $\tau_0 \to 0$, then the right-hand side of (15) tends to zero and we obtain equation (6) again.

So, in order to describe the statistical characteristics of the solution of equation (1) in the delta-correlation random field approximation (Fokker–Planck equation), it is necessary, but in general not sufficient, to assume that parameter τ_0/T is small. However, in each concrete problem, a more detailed study can be carried out to determine the validity of the delta-correlated approximation.

We would like to reemphasize that the delta-correlated generalized random process approximation is by no means a simple replacement of $z_j(\boldsymbol{\xi}, t)$ in (1) by a random field with the correlation function (3). It means a limiting procedure where the radius of correlation τ_0 of field $\boldsymbol{z}(\boldsymbol{\xi}, t)$ tends to zero, and mean quantities like $\langle z_j(\boldsymbol{\xi}, t)\phi(\boldsymbol{\xi})\rangle$ tend asymptotically to the values of the corresponding functionals of the delta-correlated in time field. The demand $t \gg \tau_0$ points to the fact that the delta-correlated process approximation is suitable for a time behavior description of such mean values only on time intervals when system (1) is influenced by a great number of independent random sources. Average description of the system behavior during times at most equal to τ_0 can be considered as satisfactory. At the same time, an analysis of stochastic equations (such as (1)) containing random forces like $\boldsymbol{z}(\boldsymbol{\xi}, t)$ requires their continuity to guarantee smooth time dependence of solution $\boldsymbol{\xi}(t)$.

Example. Following [14], consider the *Langevin equation*

$$\frac{d}{dt}\xi(t) = -\lambda\xi(t) + z(t), \quad \xi(t_0) = 0, \tag{17}$$

where $z(t)$ is assumed to be a zero-mean Gaussian random process with covariance function $\langle z(t)z(t')\rangle = B_z(t-t')$. For a given realization of external forces $z(t)$, the solution of the equation (17) has the form

$$\xi(t) = \int_{t_0}^{t} d\tau \, z(\tau)e^{-\lambda(t-\tau)}.$$

Consequently, $\xi(t)$ is a Gaussian random process. Its one-point probability density $f_\xi(t, x) = \langle \delta(\xi(t) - x)\rangle$ satisfies an exact equation

$$\frac{\partial}{\partial t}f_\xi(t, x) = \lambda\frac{\partial}{\partial x}xf_\xi(t, x) + \int_{t_0}^{t} d\tau \, B_z(t - \tau)e^{-\lambda(t-\tau)}\frac{\partial^2}{\partial x^2}f_\xi(t, x), \tag{18}$$

with the initial condition $f_\xi(0, x) = \delta(x)$. When $t_0 \to -\infty$, then process $\xi(t)$ converges to a Gaussian random function with parameters

$$\langle\xi(t)\rangle = 0, \quad \sigma_\xi^2 = \langle\xi^2(t)\rangle = \frac{1}{\lambda}\int_0^\infty d\tau \, B_z(\tau)e^{-\lambda\tau}. \tag{17'}$$

In particular, for covariance function of the form

$$B_z(\tau) = \sigma_z^2 e^{-|\tau|/\tau_0},$$

we obtain

$$\langle \xi(t) \rangle = 0, \quad \langle \xi^2(t) \rangle = \sigma_z^2 \tau_0 / \lambda (1 + \lambda \tau_0) \tag{19}$$

and, in the limit $\tau_0 \to 0$,

$$\langle \xi^2(t) \rangle = 0. \tag{20}$$

Now, multiply equation (17) by $\xi(t)$. Assuming that $\xi(t)$ is a nice enough function of time, we obtain equality

$$\xi(t) \frac{d}{dt} \xi(t) = \frac{1}{2} \frac{d}{dt} \xi^2(t) = -\lambda \xi^2(t) + z(t) \xi(t).$$

Averaging it over the ensemble of realizations of random functions $z(t)$ we obtain the equation,

$$\frac{1}{2} \frac{d}{dt} \langle \xi^2(t) \rangle = -\lambda \langle \xi^2(t) \rangle + \langle z(t) \xi(t) \rangle. \tag{21}$$

A direct calculation shows that the stationary solution of (21),

$$\langle \xi^2(t) \rangle = \frac{1}{\lambda} \langle z(t) \xi(t) \rangle,$$

corresponding to $t_0 \to -\infty$, coincides with (20). This gives that

$$\langle z(t) \xi(t) \rangle = \frac{\sigma_z^2 \tau_0}{1 + \lambda \tau_0} \lambda.$$

Since $\delta \xi(t) / \delta z(t - 0) = 1$, the same result is obtained for $\langle z(t) \xi(t) \rangle$ if one utilizes the formula

$$\langle z(t) \xi(t) \rangle = \int_{-\infty}^{t} d\tau \, B_z^{eff}(t - \tau) \left\langle \frac{\delta}{\delta z(\tau)} \xi(t) \right\rangle$$

including the "effective" correlation function

$$B_f^{eff}(\tau) = 2 \sigma_z^2 \tau_0 \delta(\tau).$$

As we mentioned before, some of probabilistic properties of the above equations in the delta-correlated process approximation coincide with probabilistic properties of a Markov process. However, it should be understood that this is true only for statistical means and equations for them. In particular, for the Langevin equation (17), realizations of the process $\xi(t)$ and the corresponding Markov process are qualitatively different. The latter satisfies the equation (17) with ideal "white noise" $z(t)$ on the right-hand side which has the correlation function $B_z(\tau) = 2\sigma_z^2 \tau_0 \delta(\tau)$; the equation itself should be understood in the generalized sense because the ideal Markov process is not differentiable in the ordinary sense.

At the same time, process $\xi(t)$, whose statistical properties are sufficiently described by a "white noise" approximation, remains differentiable in the ordinary sense and the ordinary rules of calculus apply. For example,

$$\xi(t)\frac{d}{dt}\xi(t) = \frac{1}{2}\frac{d}{dt}\xi^2(t), \tag{22}$$

and, in particular,

$$\langle\xi(t)\frac{d}{dt}\xi(t)\rangle = 0. \tag{23}$$

On the other hand, if $\xi(t)$ is an ideal Markov process satisfying, in the generalized sense, the Langevin equation (17) with the "white noise" on the right-hand side, equality (22) loses its validity, and the relation

$$\langle\xi(t)\frac{d}{dt}\xi(t)\rangle = -\lambda\langle\xi^2(t)\rangle + \langle z(t)\xi(t)\rangle \tag{24}$$

depends on the definition of the mean values. Indeed, if we treat (24) as the limit of the equality,

$$\langle\xi(t+\eta)\frac{d}{dt}\xi(t)\rangle = -\lambda\langle\xi(t)\xi(t+\eta)\rangle + \langle z(t)\xi(t+\eta)\rangle \tag{25}$$

when $\eta \to 0$, the result will be qualitatively different depending on whether $\eta \to 0+$, or $\eta \to 0-$.

If $\eta \to 0+$, then we have

$$\lim_{\eta\to 0+} \langle z(t)\xi(t+\eta)\rangle = 2\sigma_z^2 \tau_0,$$

and, taking into account (20), equality (25) yields the equation

$$\langle\xi(t+0)\frac{d}{dt}\xi(t)\rangle = \sigma_z^2 \tau_0. \tag{26}$$

If $\eta \to 0-$, then $\langle z(t)\xi(t+0)\rangle = 0$, because of the dynamic causality condition, and equality (25) takes the form

$$\langle\xi(t-0)\frac{d}{dt}\xi(t)\rangle = -\sigma_z^2 \tau_0. \tag{27}$$

Comparing (23) with (26), and (27), we see that for the ideal Markov process described by the Langevin equation with the "white noise" (usually called the *Ornstein–Uhlenbeck process*), we have

$$\langle\xi(t+0)\frac{d}{dt}\xi(t)\rangle \neq \langle\xi(t-0)\frac{d}{dt}\xi(t)\rangle \neq \langle\frac{1}{2}\frac{d}{dt}\xi^2(t)\rangle.$$

This concludes our discussion of the question of the delta-correlated approximation of a random process. Throughout this section the phrase "dynamical system (equation) with generalized delta-correlated random parameter fluctuations" was meant to denote the asymptotic case in which the correlation radius of these parameters is small in comparison with all characteristic scales of the problem.

18.5   Exercises

1. Demonstrate that the amplitudes of oscillation $A(t)$ appearing in solutions (2) of the stochastic parametric resonance equations (1) of Section 18.2 follow the lognormal distribution under the conditions described in that Section.

2. Complete the discussion in Subsection 18.2.2 for the case of reflecting boundaries, where $\partial G / \partial x = 0$. If a source is placed on the reflecting boundary then $R_2(x_0) = 1$ and, consequently,

$$G_{ref}(x; x_0) = \frac{2}{1 - R_1(x_0)} u(x; x_0). \tag{8'}$$

Focus on statistical problems in the case of a source in the infinite space (8), and a source in the infinite half-space (8'), when the attenuation is small ($\gamma \to 0$). Assume that fluctuations $\varepsilon_1(x)$ are delta-correlated, so that quantities $R_1(x_0)$ and $R_2(x_0)$ are statistically independent. Find the mean intensity of a point source (8).

3. Show that equalities (26), and (27), of Section 18.4 can be also obtained from the correlation function

$$\langle \xi(t)\xi(t + \tau) \rangle = \frac{\sigma_z^2 \tau_0}{\lambda} e^{-\lambda|\tau|},$$

of process $\xi(t)$.

Chapter 19

Forced Burgers Turbulence and Passive Tracer Transport in Burgers Flows

In this chapter we will discuss the problem of random flows governed by the Burgers equation for the velocity field $\boldsymbol{v}(\boldsymbol{x}, t)$, $\boldsymbol{x} \in \mathbf{R}^d, d \geq 1$,

$$\frac{\partial \boldsymbol{v}}{\partial t} + (\boldsymbol{v} \cdot \boldsymbol{\nabla})\boldsymbol{v} = \mu \Delta \boldsymbol{v} + \boldsymbol{f}(\boldsymbol{x}, t), \quad \boldsymbol{v}(\boldsymbol{x}, t = 0) = \boldsymbol{v}_0(\boldsymbol{x}),$$

where $\mu > 0$, and the initial velocity $\boldsymbol{v}_0$ along with the force $\boldsymbol{f}$ are known and random. Additionally we will study how statistical properties of passive tracers suspended in Burgers flows evolve in time. The first problem is often referred to as "Burgers turbulence" because Burgers equation, discussed in detail in Chapter 14 of Volume 2 of this monograph series, can be seen as a "toy model" of real turbulence problems described by the Navier–Stokes equation. The material is based on a series of authors' articles [1–5]. Also, the second-named author's book *Burgers-KPZ Turbulence: Göttingen Lectures* addresses the general problems in this area mostly in the classical context.

In the first part of the present chapter we'll concentrate on random forced Burgers flows in the context of distributional tools. Statistical properties of multidimensional Burgers turbulence evolving in presence of a force field with random potential, which is *delta-correlated in time and smooth in space*, are studied in the inviscid limit and at the physical level of rigorousness. The solution algorithm reduces the problem to finding multistream fields describing the motion of an auxiliary gas of interacting particles in a force field. Consequently, the statistical description of forced Burgers turbulence

© Springer International Publishing AG, part of Springer Nature 2018
A. I. Saichev and W. A. Woyczynski, *Distributions in the Physical and Engineering Sciences, Volume 3,* Applied and Numerical Harmonic Analysis, https://doi.org/10.1007/978-3-319-92586-8_19

is obtained by finding the largest possible value of the least action for the auxiliary gas. The exponential growth of the number of streams is found to be a necessary condition for the existence of stationary regimes.

In the second part of this chapter a model analytical description of the density field advected in a velocity field governed by the multidimensional Burgers' equation is investigated. This model field satisfies the mass conservation law and, in the zero viscosity limit, coincides with the generalized solution of the continuity equation. A numerical and analytical study of the evolution of such a model density field is much more convenient than the standard method of modeling of transport of passive tracer particles in the fluid which will be studied in the next chapter.

In the 1-D case, a more general KdV-Burgers equation is suggested as a model which permits an analytical treatment of the density field in a strongly nonlinear model of compressible gas which takes into account dissipative and dispersive effects as well as pressure forces, the former not being accounted for in the standard Burgers' framework.

The dynamical and statistical properties of the density field are studied. In particular, utilizing the above model in the 2-D case, and in the most interesting for us situation of small viscosity, we can follow the creation and evolution of the cellular structures in the density field and the subsequent creation of the "quasi-particles" clusters of matter of enormous density. In addition, it is shown that in the zero viscosity limit the density field spectrum has a power tail $\propto k^{-n}$, with different exponents in different regimes.

19.1   Furutsu–Novikov–Donsker formula and mechanism of energy dissipation

In this introductory section we begin with the relatively simple 1-D forced Burgers model and initially determine the evolution of the correlation between between the random forces and the solution velocity field. The second subsection will explore the mechanism of energy dissipation in the inviscid 1-D Burgers turbulence. We'll start with the detailed discussion of the Furutsu–Novikov formula which already has been briefly exploited in Chapter 18.

19.1.1.   Furutsu–Novikov–Donsker formula in application to the forced Burgers turbulence.   Consider an arbitrary zero-mean Gaussian field $\boldsymbol{f}(\boldsymbol{x}) = (f_i(\boldsymbol{x}))_i$, and an analytic functional R acting on it. The Furutsu–Novikov–Donsker formula, introduced by Furutsu (1963) in the context of the statistical theory of electromagnetic waves in a fluctuating medium, and by Novikov in 1964 in a study of randomly forced turbulence,[1] explicitly

---

[1]See, references [12], and [13], in the bibliography to Chapter 18. Donsker obtained it independently in 1964 while studying mathematical theory of path integrals.

calculates the correlation of $\boldsymbol{f}(\boldsymbol{x}) = (f_i(\boldsymbol{x}))_i$ and $R[\boldsymbol{f}]$:

$$\langle f_i(\boldsymbol{x})R[\boldsymbol{f}]\rangle = \int \langle f_i(\boldsymbol{x})f_k(\boldsymbol{x}')\rangle \left\langle \frac{\delta R[\boldsymbol{f}]}{\delta \boldsymbol{f}(\boldsymbol{x}')} \right\rangle d\boldsymbol{x}', \tag{1}$$

where $\delta R[\boldsymbol{f}]/\delta \boldsymbol{f}$ is the variational derivative of the functional R. It is obtained by a straightforward formal comparison of the functional power series expansions of the left-hand side and the right-hand side.

We will illustrate its usefulness by applying it to evaluate the correlation $\langle f(x,t)v(x,t)\rangle$, where $f(x,t)$ is a Gaussian random field with zero mean and correlation function

$$\langle f(x,t)f(x+z,t+\tau)\rangle = \Gamma_f(z)\delta(\tau), \tag{2}$$

and $v(x,t)$ is the solution of 1-D Burgers' equation,

$$\frac{\partial v}{\partial t} + v\frac{\partial v}{\partial x} = \mu\frac{\partial^2 v}{\partial x^2} + f(x,t), \tag{3}$$

for the velocity field $v(x,t)$, where $v_0(x)$ is a statistically homogeneous stochastic process with zero mean and correlation function

$$\Gamma_0(z) = \langle v_0(x)v_0(x+z)\rangle. \tag{4}$$

Above, and in what follows, the angled brackets denote the statistical averaging over the ensemble of realizations of the force and (if necessary) of the random initial data, which are assumed to be independent of each other.

In this case, the Furutsu–Novikov–Donsker formula yields the following exact equality:

$$\langle f(x,t)v(x+z,t)\rangle = \int dy \int_0^t d\tau \langle f(x,t)f(y,\tau)\rangle \left\langle \frac{\delta v(x+z,t)}{\delta f(y,\tau)} \right\rangle. \tag{5}$$

Applying the variational derivative to the Burgers equation (3) gives

$$\frac{\partial}{\partial t}\left(\frac{\delta v(x,t)}{\delta f(y,\tau)}\right) + \frac{\partial}{\partial x}\left(v(x,t)\frac{\delta v(x,t)}{\delta f(y,\tau)}\right) = \mu\frac{\partial^2}{\partial x^2}\frac{\delta v(x,t)}{\delta f(y,\tau)} + \delta(x-y)\delta(t-\tau).$$

Now, taking into account the causality principle, one can replace the above linear equation for the sought variational derivative by the following Cauchy problem for the homogeneous equation:

$$\frac{\partial}{\partial t}\left(\frac{\delta v(x,t)}{\delta f(y,\tau)}\right) + \frac{\partial}{\partial x}\left(v(x,t)\frac{\delta v(x,t)}{\delta f(y,\tau)}\right) = \mu\frac{\partial^2}{\partial x^2}\frac{\delta v(x,t)}{\delta f(y,\tau)},$$

$$\frac{\delta v(x,t=\tau)}{\delta f(y,\tau)} = \delta(x-y). \tag{6}$$

Substituting into (5) the correlation function (4), we obtain the expression,

$$\langle f(x,t)v(x+z,t)\rangle = \int dy\, \Gamma_f(y-x) \int_0^t \delta(t-\tau) \left\langle \frac{\delta v(x+z,t)}{\delta f(y,\tau)} \right\rangle d\tau,$$

which, in view of the probing property of the Dirac delta, gives

$$\langle f(x,t)v(x+z,t)\rangle = \frac{1}{2} \int dy\, \Gamma_f(y-x) \left\langle \frac{\delta v(x+z,t)}{\delta f(y,t)} \right\rangle.$$

So, finally, in view of equality (6),

$$\langle f(x,t)v(x+z,t)\rangle = \frac{1}{2}\Gamma_f(z).$$

19.1.2. Mechanism of energy dissipation in the inviscid 1-D Burgers turbulence. In this subsection we discuss the mechanism of energy dissipation in the inviscid 1-D Burgers turbulence, and the related problems of steady-state regimes maintained in presence of external forces.

Let us begin by considering the homogeneous Burgers' equation

$$\frac{\partial v}{\partial t} + v\frac{\partial v}{\partial x} = \mu\frac{\partial^2 v}{\partial x^2}, \qquad v(x,t=0) = v_0(x),$$

where $v_0(x)$ is a stationary and homogeneous random field. Then, obviously, the solution $v(x,t)$ of this equation is also a statistically homogeneous function of x. This means that the average energy

$$\langle u(x,t)\rangle = \frac{1}{2}\left\langle v^2(x,t)\right\rangle$$

obeys to the equation

$$\frac{d\langle u\rangle}{dt} = -\bar{\epsilon}, \tag{7}$$

where the energy dissipation rate is defined by the formula

$$\bar{\epsilon} = \mu\left\langle g^2(x,t)\right\rangle, \tag{8}$$

where

$$g(x,t) = \frac{\partial v(x,t)}{\partial x}$$

is the Burgers turbulence velocity gradient. It is clear from (8) that, in the inviscid limit $\mu \to 0+$, dissipation occurs only in the infinitesimal vicinities of Burgers' velocity shock fronts, where the velocity gradient has big jumps of size $\sim 1/\mu$. These large peaks balance the influence of the vanishing coefficient μ at the right-hand side of (8).

To recover the detailed mechanism of energy dissipation in the inviscid limit let us recall[2] the universal shape of Burgers' equation's solution in a small vicinity of the shock front of size a, moving with velocity V, and situated at the point $x^* = x - Vt + C$:

$$v_s(x - x^*, a) = V - \frac{a}{2} \tanh\left(\frac{a(x - x^*)}{4\mu}\right).$$

The corresponding velocity field gradient has, in the vicinity of this shock, the form

$$g_s(x - x^*, a) = -\frac{a^2}{8\mu} \cdot \frac{1}{\cosh^2\left(\frac{a(x-x^*)}{4\mu}\right)}. \tag{9}$$

It is physically natural to assume that, for sufficiently small viscosity μ, the gradient is of the same shape in the case of forced Burgers' velocity field. So, neglecting contribution to the dissipation rate of the gradient field realizations between shocks, we can write these realizations in the form of a series of nonoverlapping peaks:

$$g(x, t) = \sum_k g_s(x - x_k, a_k), \tag{10}$$

where x_k and a_k are, respectively, the coordinates and amplitudes of successive shocks. Substituting (10) into (8), and taking into account (9), we get that

$$\bar{\epsilon} = \left\langle \frac{\vartheta a^4}{64\mu} \int_{-\infty}^{\infty} \frac{dx}{\cosh^4(ax/4\mu)} \right\rangle,$$

where $\vartheta(a, t)$ denotes the average spatial frequency of shocks with amplitude a at the time t, and angle brackets denote the statistical averaging over random shock amplitudes a_k. Evaluating the integral we get

$$\bar{\epsilon} = \frac{\langle \vartheta a^3 \rangle}{12}.$$

In the case of forced 1-D Burgers' equation (3), and delta-correlated Gaussian forces (5), the average energy obeys an equation similar to (2):

$$\frac{d\langle u \rangle}{dt} = -\bar{\epsilon} + \frac{1}{2}\Gamma_f,$$

where $\Gamma_f = \Gamma_f(z = 0)$. At the initial stage, when shocks are virtually absent ($\vartheta \approx 0$), we get

$$\frac{d\langle u \rangle}{dt} = \frac{1}{2}\Gamma_f,$$

---

[2]See, e.g., Gurbatov, Malakhov, Saichev (1993), and Woyczyński (1998).

and the energy of turbulence is increasing linearly:

$$< u > \approx \frac{t}{2}\Gamma_f.$$

afterwards, the growth rate of $\bar{\epsilon}$ is reduced due the appearance of shock fronts in Burgers' velocity field realizations. Eventually, for the steady-state regime of forced Burgers turbulence, the frequency of shocks, their amplitudes, and the statistical properties of external forces are tied by the equality:

$$\langle \vartheta a^3 \rangle = 6\Gamma_f.$$

19.1.3. Existence of stationary regimes in forced 1-D Burgers turbulence. One of the goals of the present chapter is to provide a quantitative study of the statistically stationary regimes in Burgers turbulence. To begin with, let us review some conditions for existence of such equilibria.

Since dissipation leads to a decay of turbulence, to sustain it one needs a supply of energy from outside. In the hydrodynamic turbulence in nature such an "engine" is often the solar energy, which generates large-scale convective eddies. Their nonlinear descending cascade maintains in the dynamic equilibrium even smaller-scale, turbulent rotational motions.

In the general, multidimensional Burgers turbulence

$$\frac{\partial \boldsymbol{v}}{\partial t} + (\boldsymbol{v} \cdot \boldsymbol{\nabla})\boldsymbol{v} = \mu \Delta \boldsymbol{v} + \boldsymbol{f}(\boldsymbol{x}, t), \quad \boldsymbol{v}(\boldsymbol{x}, t = 0) = \boldsymbol{v}_0(\boldsymbol{x}),$$

the necessary input of energy is provided by the external random force field $\boldsymbol{f}(\boldsymbol{x}, t)$. Observe, however, that not all force fields $\boldsymbol{f}(\boldsymbol{x}, t)$, even if they are stationary in time and homogeneous in space, will lead to a stationary regime in Burgers turbulence. For that reason one would like to know conditions on forces $\boldsymbol{f}(\boldsymbol{x}, t)$ which would guarantee the establishment of a stationary regime as $t \to \infty$.

A significant result in this direction has been obtained by Sinai in 1991 (see, [7–8]) who gave a rigorous proof of the fact that (in the 1-D case) there exists a broad class of random force potentials $U(\boldsymbol{x}, t)$, periodic in space and delta-correlated in time, for which the solution $\boldsymbol{v}(\boldsymbol{x}, t)$ of the Burgers' equation converges (as $t \to \infty$) to a solution $\boldsymbol{v}_\infty(\boldsymbol{x}, t)$ which is independent of the initial condition, stationary in time and periodic in space.

Below, we display the case of random forces $\boldsymbol{f}(\boldsymbol{v}, t)$ for which the stationary regime is impossible in principle. We shall restrict ourselves here to the 1-D Burgers' equation (2–4).

The spatial autocorrelation function

$$\Gamma(z; t) = \langle v(x, t)v(x + z, t) \rangle,$$

of the 1-D Burgers turbulence satisfies equation

$$\frac{\partial}{\partial t}\Gamma(z;t) + \frac{1}{2}\frac{\partial}{\partial z}\Big[\Gamma_{12}(z;t) - \Gamma_{12}(-z;t)\Big]$$

$$= 2\mu\frac{\partial^2}{\partial z^2}\Gamma(z;t) + \langle f(x,t)\,v(x+z,t)\rangle + \langle f(x+z,t)\,v(x,t)\rangle, \qquad (11)$$

with the initial condition $\Gamma(z;t=0) = \Gamma_0(z)$, and $\Gamma_{12}(z;t) := \langle v(x,t)\,v^2(x+z,t)\rangle$. Applying the Furutsu–Novikov–Donsker formula to the cross-correlations in (11) we obtain the equation,

$$\langle f(x,t)\,v(x+z,t)\rangle = \langle f(x+z,t)\,v(x,t)\rangle = \frac{1}{2}\Gamma_f(z).$$

As a result, equation (11) assumes the form,

$$\frac{\partial}{\partial t}\Gamma(z;t) + \frac{1}{2}\frac{\partial}{\partial z}\Big[\Gamma_{12}(z;t) - \Gamma_{12}(-z;t)\Big] = 2\mu\frac{\partial^2}{\partial z^2}\Gamma(z;t) + \Gamma_f(z). \qquad (12)$$

Now, let us introduce the spatial spectral density,

$$G(\kappa;t) = \frac{1}{2\pi}\int \Gamma(z;t)e^{i\kappa z}dz,$$

of the Burgers turbulence v, and the spatial spectral density

$$G_f(k) = \frac{1}{2\pi}\int \Gamma_f(z)e^{i\kappa z}dz,$$

of the force field f. It follows from (12), and from a natural from the physical viewpoint assumption,

$$\lim_{|z|\to\infty}\Gamma_{12}(z,t) = 0,$$

that, at $\kappa = 0$, the G satisfies the equation,

$$\frac{d}{dt}G(0;t) = G_f(0), \qquad (13)$$

with the initial condition

$$G(0;t=0) = \frac{1}{2\pi}\int \Gamma_0(z)dz, \qquad (14)$$

which has the obvious solution,

$$G(0;t) = G(0,t=0) + G_f(0)t. \qquad (15)$$

Thus, if $G_f(0) \neq 0$ then the spectral density of the Burgers turbulence grows linearly in time at $\kappa = 0$, which is clearly impossible in a stationary regime. Thus, we have arrived at the following result:

(i) A necessary condition for the existence of a stationary regime in forced Burgers turbulence is:

$$G_f(k=0) = \frac{1}{2\pi}\int \Gamma_f(z)\,dz = 0, \tag{16}$$

i.e., that the spectral density of the external force vanishes for $\kappa = 0$.

This condition and its multidimensional analog are fulfilled, in particular, if the *random force's potential $U(x,t)$ is statistically homogeneous in space,* and we will make this assumption in the remainder of this chapter.

It also follows from (15) that the spectral density of the Burgers turbulence depends on $G(0, t = 0)$. This means that if $G(0, t = 0) \neq 0$ then the Burgers turbulence "always remembers" the initial field. Consequently:

(ii) A necessary condition for the stationary regime in forced Burgers turbulence to be ergodic, by which we mean here that it is independent of the initial field, is:

$$G(0; t=0) = \frac{1}{2\pi}\int \Gamma_0(z)dz = 0. \tag{17}$$

Observe that the necessary conditions (13–14) of the existence of an ergodic stationary regime are clearly satisfied for the class of forces and initial conditions studied by Sinai [7].

Equation (12) also permits us to formulate the following, somewhat less obvious, result about statistical properties of stationary regimes in the forced Burgers turbulence. Its validity follows directly from (15–17).

(iii) Assume that there exists an ergodic stationary regime of Burgers turbulence and that the limits,

$$\Gamma^\infty(z) = \lim_{t\to\infty}\Gamma(z,t), \qquad \Gamma_{12}^\infty(z) = \lim_{t\to\infty}\Gamma_{12}(z,t),$$

exist. Then, its spectral density vanishes at $\kappa = 0$, i.e.,

$$G^\infty(\kappa = 0) = \frac{1}{2\pi}\int \Gamma^\infty(z)\,dz = 0. \tag{18}$$

At this point a natural question is whether a Gaussian stationary regime is feasible. To answer it observe that in the stationary regime, equation (12) takes the form

$$\frac{d}{dz}\Gamma_{12:odd}^\infty(z) = 2\mu\frac{d^2}{dz^2}\Gamma^\infty(z) + \Gamma_f(z), \tag{19}$$

where $\Gamma_{12:odd}^\infty(z)$ is the odd part of function $\Gamma_{12}^\infty(z)$. Multiplying the last equation by z^2, integrating it term-by-term over all z's, and taking into account (18), we get that

$$\int z\Gamma_{12:odd}^\infty(z)\,dz = -2\pi\frac{d^2}{d\kappa^2}G_f(\kappa)\Big|_{\kappa=0}, \tag{20}$$

where $G_f(\kappa)$ is the, defined previously, spatial spectral density of the force. Since, for a Gaussian field, necessarily $\Gamma^\infty_{12:odd}(z) \equiv 0$, formula (20) implies the following proposition.

(iv) For the existence of a Gaussian ergodic stationary regime in the forced Burgers turbulence it is necessary that

$$G_f(\kappa) = o(\kappa^2), \quad (\kappa \to 0).$$

Now, from (19), we obtain another result:

(v) If a stationary regime in forced Burgers turbulence is Gaussian then its spectral density satisfies condition

$$G^\infty(\kappa) = \frac{1}{2\mu} \frac{G_f(\kappa)}{\kappa^2}. \tag{21}$$

Other problems related to the energy dissipation mechanism in the inviscid limit ($\mu \to 0+$) and steady-state Burgers turbulence have been addressed in Subsection 19.1.2. But, (21) also implies the following statement:

(vi) If an ergodic stationary regime exists for the inviscid forced Burgers turbulence then it is non-Gaussian.

In a sense, a solution of the Burgers' equation in the inviscid limit gives a good mathematical illustration of the Cheshire cat's mystery: the viscosity disappears ($\mu = 0$), but the dissipation remains. Perhaps this is one of the reasons the study of the inviscid limit keeps attracting researchers. The subtlety of the situation is that in the inviscid limit the solutions of the Burgers' equation exist only in the generalized sense. Thus, right away one runs into the problem of selection of a suitable class of "physically meaningful" generalized solutions and the question of uniqueness of these solutions in the selected class. Although the Hopf–Cole substitution

$$v(x, t) = -2\mu\nabla \ln \phi(x, t) \tag{22}$$

transforms the *nonlinear* Burgers' equation (3) into a *linear* Schrödinger-type diffusion equation

$$\frac{\partial \phi}{\partial t} = \mu\Delta\phi - \frac{1}{2\mu}U(x, t)\phi, \tag{23}$$

the inverse passage from its solutions back to the solutions of the Burgers' equation in the inviscid limit is a formidable mathematical problem. The difficulty of the rigorous mathematical analysis of the forced Burgers turbulence becomes more acute if the forces are assumed to be delta-correlated in time. In that case, equation (23), even for $\mu > 0$, looses its classical meaning and has to be considered as a stochastic partial differential equation in

a generalized sense, like Ito or Stratonovich ordinary stochastic differential
equations (see, e.g., Da Prato, Zabczyk [9]). Similar difficulties arise with
the definition of nonlinear terms in the original Burgers' equation (3).

19.2 Multidimensional forced Burgers turbulence

19.2.1. General issues. In the remainder of this chapter we provide
an approximate method of analysis of the statistical properties of Burgers
turbulence in the inviscid limit, and for an arbitrary spatial dimension, $d \geq 1$,
for a random force field potential which is *delta-correlated in time and smooth
in space*. A special attention is paid to verifying the feasibility of stationary
regimes.

In Subsection 19.2.2, we provide a detailed exposition of the structure
of solutions of the forced inviscid Burgers turbulence., i.e., the velocity field
$v(x, t)$, $x \in \mathbf{R}^d, d \geq 1$, satisfying the multidimensional Burgers' equation

$$\frac{\partial v}{\partial t} + (v \cdot \nabla)v = \mu \Delta v + f(x, t), \qquad v(x, t = 0) = v_0(x), \tag{1}$$

where $\mu > 0$, and the initial velocity v_0 along with force f are known and
random. In this case the solution velocity field corresponds to the "least-
action stream" velocity among the multivalued velocity fields describing the
multistream motion of noninteracting particles in the force field.

Section 19.3 is devoted to the statistical description of the above-mentioned
multistream particle motion and proposes an approximate method of finding
the desired statistical properties of Burgers turbulence based on searching for
the *largest value of least action*. In Section 19.4, we verify the efficacy of this
method on the relatively simple case of the 2-D homogeneous Burgers' equation with random initial conditions[3].

In Section 19.5, we study statistical properties of Burgers turbulence
and, in particular, its average kinetic energy, under the assumption that
the force field has a random potential which is homogeneous in space and
delta-correlated in time.[4] At the same time, we also explain the important
role that the average stream-number in the associated gas of noninteracting
particles plays in the analysis of Burgers turbulence. It turns out that the
exponential growth in time of the average number of streams is a necessary
condition for the existence of a stationary regime. In Section 16.6, using the
theory of Markov processes, we show that the average number of streams
indeed grows exponentially, at least in the 1-D case.

---

[3]See also Funaki, Surgailis, Woyczyński [10], and Molchanov, Surgailis, Woyczyński
[11–12] for other rigorous results in this direction.

[4]A curious maximum principle for the mean kinetic energy in 1-D Burgers turbulence
was discovered in Hu and Woyczyński [13–14].

Finally, we would like to mention, that in the case of simple degenerate random potentials, a rigorous analysis of the forced Burgers turbulence has been done by Molchanov, Surgailis, Woyczynski [11–12] using the variational method and spectral analysis for the Schrödinger equation with a random potential.

The Burgers turbulence is considered as an adequate model of certain aspects of the hydrodynamic turbulence since, as was observed by Burgers himself (see, e.g., Burgers [15]), it takes into account the competition of two most important mechanism for the real turbulence: inertial nonlinearity and viscous dissipation. There exist, however, some discrepancies between the Burgers' and the hydrodynamic turbulences which are aggravated by the fact that the hydrodynamic turbulence has, primarily, a rotational character, whereas by the Burgers turbulence we usually mean the potential velocity field

$$v(x, t) = \nabla S(x, t), \tag{2}$$

generated by potential S, which then satisfies the Hamilton–Jacobi type equation

$$\frac{\partial S}{\partial t} + \frac{1}{2}(\nabla S)^2 = \mu \Delta S + U(x, t), \tag{3}$$

where U is the potential of external forces, i.e.,

$$f(x, t) = \nabla U(x, t). \tag{4}$$

Nevertheless, fondness of theoreticians for potential solutions of the Burgers' equation can be justified not only by existence of the explicit formulas and thus, a realistic hope of quantitative analysis of a strongly nonlinear phenomenon; the differences between the Burgers' and the hydrodynamic turbulence are as instructive as the similarities. Also, by now, the Burgers equation has become one of the common nonlinear model equations of mathematical physics and over a period of time it was discovered that such, or similar, models describe various physical phenomena displaying shock formation; the astrophysicists studying the large-scale distribution of matter in the Universe have convincingly demonstrated that the Burgers turbulence provides an adequate description of the process of formation of cellular structures (see, e.g., Shandarin, Zeldovich [16] , and Molchanov, Surgailis and Woyczynski [12], where the large-scale structure of the Universe and quasi-Voronoi tessellation of shock fronts in forced inviscid Burgers turbulence in $\mathbf{R}^d$ were examined).

19.2.2.   Least-action principle for forced Burgers turbulence.
We devote the present section to a detailed discussion of solutions of the nonhomogeneous Burgers' equation (1) with the potential force (4) in an arbitrary d-dimensional space ($x \in \mathbf{R}^d$, $d \geq 1$).

For the sake of simplicity we will assume in this subsection that the potential $U(\boldsymbol{x}, t)$ is a sufficiently smooth function in both the space variable $\boldsymbol{x}$ and time variable t. Additionally, we will complement equation (1) by the zero initial condition

$$\boldsymbol{v}(\boldsymbol{x}, t = 0) = 0. \tag{5}$$

The nonzero initial conditions can be taken into account by a special choice of the external force's potential $U(\boldsymbol{x}, t)$.

As we observed in Section 19.1, by the Hopf–Cole transformation (19.1.22), equation (1) with the initial condition (5) is reduced to a linear Schrödinger-type diffusion equation (19.1.23) with initial condition

$$\varphi(\boldsymbol{x}, t = 0) = 1. \tag{6}$$

Its solution can be written out in the form of the well-known Feynman–Kac formula

$$\phi(\boldsymbol{x}, t) = \left\langle \exp\left(-\frac{1}{2\mu} \int_0^t U\left(\boldsymbol{x} - \boldsymbol{w}(t) + \boldsymbol{w}(\tau), \tau\right) d\tau\right)\right\rangle, \tag{7}$$

where the averaging is with respect to the ensemble of realizations of the vector-valued Wiener process $\boldsymbol{w}(t) = (w_l(t))$ whose statistical properties are determined by the conditions,

$$\boldsymbol{w}(0) = 0, \quad \langle w_l(t) w_m(t)\rangle = 2\mu t \delta_{lm}, \quad l, m = 1, 2, \ldots, d.$$

To make the further analysis more transparent, let us write (7) in the form of a paths integral. For this purpose, consider a discretized form

$$U(\boldsymbol{x}, t) = \varepsilon \sum_{p=0}^{\infty} U(\boldsymbol{x}, p\varepsilon)\delta(t - p\varepsilon) \tag{8}$$

of the external force potential (4). Substituting it into (7) and assuming, for simplicity, that the time $t = (q + 1)\varepsilon - 0$, $q = 0, 1, 2, \ldots$, is also discrete, we obtain that

$$\phi(\boldsymbol{x}, t) = \left\langle \exp\left[-\frac{\varepsilon}{2\mu} \sum_{p=0}^{q} U\left(\boldsymbol{x} - \sum_{r=p}^{q} \boldsymbol{\Omega}_r, p\varepsilon\right)\right]\right\rangle, \tag{9}$$

where

$$\boldsymbol{\Omega}_r = \boldsymbol{w}((r + 1)\varepsilon) - \boldsymbol{w}(r\varepsilon), \qquad r = 0, 1, 2, \ldots,$$

are mutually independent Gaussian random vectors with the correlation tensor

$$\langle \Omega_{rl} \Omega_{rm}\rangle = 2\mu\varepsilon\delta_{lm}, \qquad l, m = 1, 2, \ldots, d.$$

Writing explicitly the average in (9) with respect to the Gaussian ensemble $\{\boldsymbol{\Omega}_0, \boldsymbol{\Omega}_2, \ldots, \boldsymbol{\Omega}_q\}$ we get

$$\phi(\boldsymbol{x}, t) =$$

$$\int \cdots \int \exp\left[-\frac{1}{2\mu}\left(\varepsilon\sum_{p=0}^{q}U\left(\boldsymbol{x}-\sum_{r=p}^{q}\boldsymbol{z}_r,p\varepsilon\right)+\sum_{p=0}^{q}\frac{\boldsymbol{z}_p^2}{2\varepsilon}\right)\right]\mathcal{D}_{q+1}(\boldsymbol{z}), \quad (10)$$

where each of the above integrals denotes integration over the d-dimensional space and

$$\mathcal{D}_{q+1}(\boldsymbol{z}) = \left(\frac{1}{4\pi\mu\varepsilon}\right)^{d(q+1)/2} d^d\boldsymbol{z}_0\,d^d\boldsymbol{z}_1\ldots d^d\boldsymbol{z}_q. \qquad (11)$$

Remember that our final goal is to find not the auxiliary field $\phi(\boldsymbol{x},t)$ but the solution $\boldsymbol{v}(\boldsymbol{x},t)$ of the nonhomogeneous Burgers' equation (1), expressed through the former via the Hopf–Cole formula (19.1.22). In that solution, in addition to $\phi(\boldsymbol{x},t)$ itself, there also appears its gradient which we shall find by acting with the operator $\boldsymbol{\nabla}$ on the right-hand side of equality (10). Putting the derivatives under the integral signs, noticing that

$$\frac{\partial}{\partial x_l}\exp\left[-\frac{\varepsilon}{2\mu}\sum_{p=0}^{q}U\left(\boldsymbol{x}-\sum_{r=p}^{q}\boldsymbol{z}_r,p\varepsilon\right)\right] = -\frac{\partial}{\partial z_{ql}}\exp\left[-\frac{\varepsilon}{2\mu}\sum_{p=0}^{q}U\left(\boldsymbol{x}-\sum_{r=p}^{q}\boldsymbol{z}_r,p\varepsilon\right)\right],$$

and integrating by parts the integral with respect to $\boldsymbol{z}_q$, we obtain that

$$-2\mu\boldsymbol{\nabla}\phi(\boldsymbol{x},t) = \qquad (12)$$

$$\int \cdots \int \frac{\boldsymbol{z}_q}{\varepsilon}\exp\left[-\frac{\varepsilon}{2\mu}\left(\sum_{p=0}^{q}U\left(\boldsymbol{x}-\sum_{r=p}^{q}\boldsymbol{z}_r,p\varepsilon\right)+\frac{1}{2}\left(\frac{\boldsymbol{z}_p}{\varepsilon}\right)^2\right)\right]\mathcal{D}_{q+1}(\boldsymbol{z}).$$

Let us change variables in integrals (10–11) from $\{\boldsymbol{z}_p\}$ to

$$\boldsymbol{X}_p = \boldsymbol{x} - \sum_{r=p}^{q}\boldsymbol{z}_r, \qquad p = 0,1,\ldots,q, \qquad \boldsymbol{X}_{q+1} = \boldsymbol{x},$$

so that $\boldsymbol{z}_p = \boldsymbol{X}_{p+1} - \boldsymbol{X}_p$, $p = 0,1,\ldots,q$, and, as a result, equalities (10), and (12) take the form,

$$\phi(\boldsymbol{x},t) = \qquad (13)$$

$$\int \cdots \int \exp\left[-\frac{\varepsilon}{2\mu}\sum_{p=0}^{q}\left(U\left(\boldsymbol{X}_p,p\varepsilon\right)+\frac{1}{2}\left(\frac{\boldsymbol{X}_{p+1}-\boldsymbol{X}_p}{\varepsilon}\right)^2\right)\right]\mathcal{D}_{q+1}(\boldsymbol{X}),$$

$$-2\mu\boldsymbol{\nabla}\phi(\boldsymbol{x},t) = \qquad (14)$$

$$\int \cdots \int \frac{\boldsymbol{x}-\boldsymbol{X}_q}{\varepsilon}\exp\left[-\frac{\varepsilon}{2\mu}\sum_{p=0}^{q}\left(U\left(\boldsymbol{X}_p,p\varepsilon\right)+\frac{1}{2}\left(\frac{\boldsymbol{X}_{p+1}-\boldsymbol{X}_p}{\varepsilon}\right)^2\right)\right]\mathcal{D}_{q+1}(\boldsymbol{X}).$$

Now, let us pass to the limit,

$$\varepsilon \to 0, \qquad q = (t-\varepsilon)/\varepsilon \to \infty,$$

in the formulas (13–14). Remark that $\boldsymbol{X}_p$ can be naturally regarded as values, for $\tau = p\varepsilon$, of a certain vector-valued process $\boldsymbol{X}(\tau)$: $\boldsymbol{X}_p = \boldsymbol{X}(p\varepsilon)$, so that the multiple integrals (13) can be interpreted as discretized functional integrals

$$\phi(\boldsymbol{x},t) = \int \cdots \int \exp\left(-\frac{1}{2\mu}S[\boldsymbol{X}(\tau)]\right) \mathcal{D}[\boldsymbol{X}(\tau)], \tag{15}$$

$$-2\mu\boldsymbol{\nabla}\phi(\boldsymbol{x},t) = \int \cdots \int \frac{d\boldsymbol{X}(\tau)}{d\tau}\bigg|_{\tau=t} \exp\left(-\frac{1}{2\mu}S[\boldsymbol{X}(\tau)]\right) \mathcal{D}[\boldsymbol{X}(\tau)], \tag{16}$$

over all the sample paths $\boldsymbol{X}(\tau), \tau \in [0,t]$, satisfying the obvious condition

$$\boldsymbol{X}(\tau = t) = \boldsymbol{x}. \tag{17}$$

In (15), there appears the *action functional*

$$S[\boldsymbol{X}(\tau)] = \int_0^t \left[U(\boldsymbol{X}(\tau),\tau) + \frac{1}{2}\left(\frac{d\boldsymbol{X}}{d\tau}\right)^2\right] d\tau. \tag{18}$$

Substituting (15) in the Hopf–Cole formula (19.1.22), we obtain a solution of the nonhomogeneous Burgers' equation (1), expressed through the functional integrals

$$\boldsymbol{v}(\boldsymbol{x},t) = \frac{\int \frac{d\boldsymbol{X}(\tau)}{d\tau}\big|_{\tau=t} \exp\left(-\frac{1}{2\mu}S[\boldsymbol{X}(\tau)]\right) \mathcal{D}[\boldsymbol{X}(\tau)]}{\int \cdots \int \exp\left(-\frac{1}{2\mu}S[\boldsymbol{X}(\tau)]\right) \mathcal{D}[\boldsymbol{X}(\tau)]}. \tag{19}$$

For arbitrary $\mu > 0$, the above functional form of the nonhomogeneous Burgers' equation's solution is poorly suited for analytic calculations. Nevertheless, for $\mu \to 0+$, expression (19) supplies the following, geometrically helpful, Lagrangian picture of the corresponding generalized solution which is an analogue of the Feynman's least-action principle in quantum electrodynamics.

Least-Action Principle for Forced Burgers Turbulence. *In the inviscid limit,*

$$\boldsymbol{v}(\boldsymbol{x},t) = \frac{d\boldsymbol{X}(\tau)}{d\tau}\bigg|_{\tau=t}, \tag{20}$$

where $\boldsymbol{X}(\tau)$ is the vector-valued process on which the action functional (18) takes the minimal absolute value.

Note, that analogous constructions of generalized solutions of first-order nonlinear partial differential equations can be found in the mathematical literature (see, e.g. Oleinik [17], in the 1-D case, and Lions [18], in the multidimensional case).

The extremals of functional (18) fulfill the equations,

$$\frac{d\boldsymbol{X}}{d\tau} = \boldsymbol{V}, \qquad \frac{d\boldsymbol{V}}{d\tau} = \boldsymbol{f}(\boldsymbol{X},\tau), \tag{21}$$

together with boundary condition (17) combined with another obvious condition at $\tau = 0$:

$$V(\tau = 0) = 0, \qquad X(\tau = t) = x. \tag{22}$$

Equations (21), along equations

$$\frac{dS}{dt} = U(X, \tau) + \frac{1}{2}V^2, \qquad S(\tau = 0) = 0, \tag{23}$$

for the action functional, form a system of characteristic equations corresponding to the following first-order pde's with respect to the field $S(x, t)$ and its gradient $v(x, t) = \nabla S(x, t)$:

$$\frac{\partial S}{\partial t} + \frac{1}{2}(\nabla S)^2 = U(x, t), \tag{24}$$

$$\frac{\partial v}{\partial t} + (v \cdot \nabla)v = f(x, t). \tag{25}$$

The latter have a clear-cut physical meaning as they describe the action and the velocity fields for a gas of noninteracting particles in the hydrodynamic limit.

If the external force $f(x, t)$ is a sufficiently smooth function of its arguments, then there exists a

$$t_1 > 0,$$

such that, for $0 < t < t_1$, the solutions of equations (24) and (25) exist, are unique, and continuous for any $x \in \mathbf{R}^n$. At this initial stage, until the formation of discontinuities in the profile of generalized solution (20), it coincides with the solution of equation (25).

For $t > t_1$, the boundary-value problem (21–23) may, for some x, have $N > 1$ solutions

$$\{X_m(\tau), V_m(\tau), S_m(\tau), m = 1, 2, \ldots, N\}. \tag{26}$$

Its values for $\tau = t$ and given m,

$$v_m(x, t) = V_m(\tau = t), \qquad S_m(x, t) = S_m(\tau = t),$$

can be conveniently thought of as values of a multistream solution of equations (24–25) in the mth stream. Let us enumerate the streams in the increasing order

$$S_1(x, t) < S_2(x, t) < \cdots < S_N(x, t). \tag{27}$$

Then the generalized solution (20), taking into account the appearance of discontinuities, can be written in the form

$$v(x, t) = v_1(x, t), \tag{28}$$

(see Fig. 2.1).

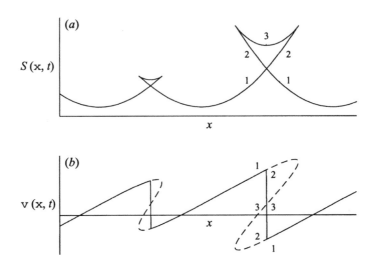

FIGURE 19.2.1.
A 1-D example of multistream fields of action $S(x,t)$ (a), and of velocity $v(x,t)$ (b). The solid line in (b) indicates the stream that corresponds to the generalized solution of the inviscid Burgers' equation. The numerals point out the stream-numbers.

The above discussion can be summarized by the following statement which forms the basis of the remainder of this chapter: *The physically significant inviscid limit solutions of the non-homogeneous Burgers' equation are fully determined by multistream properties of the gas of noninteracting particles.*

19.3 Forced inviscid Burgers turbulence and the multistream regimes

In this section we carry out a statistical analysis of solutions of the randomly forced Burgers' equation. In what follows we shall assume that the potential field $U(\boldsymbol{x},t)$ of the random force $\boldsymbol{f}(\boldsymbol{x},t)$ is a statistically homogeneous and isotropic in space, and delta-correlated in time, Gaussian random field with zero mean and correlation function

$$\langle U(\boldsymbol{x},t)U(\boldsymbol{x}+\boldsymbol{y},t+\theta)\rangle = 2a(y)\delta(\theta). \tag{1}$$

Therefore, the random force $\boldsymbol{f}(\boldsymbol{x},t)$ is also a statistically isotropic in space Gaussian field with correlation tensor

$$\left\langle f_l(\boldsymbol{x},t)f_m(\boldsymbol{x}+\boldsymbol{y},t+\theta)\right\rangle = 2\delta(\theta)\left[b(y)\delta_{lm} + \frac{y_l y_m}{y}\frac{db(y)}{dy}\right], \tag{2}$$

where

$$b(y) = -\frac{1}{y}\frac{da(y)}{dy}, \quad l, m = 1, 2, \ldots, d.$$

19.3.1. Statistical description of the auxiliary multistreams.

As we observed before, the statistical analysis of inviscid Burgers turbulence reduces to the statistical analysis of the stochastic boundary-value problem (19.2.21–23). The presence of boundary conditions, even for the delta-correlated in time random force f, does not permit a direct use of the Markov processes apparatus in the analysis of statistics of solutions (19.2.21–23). To make those powerful tools available, one has to formulate initially an auxiliary Cauchy problem, and the statistical properties thereof will determine the desired statistical properties of the boundary-value problem (19.2.21–23). It turns out that it is natural to take as such an auxiliary problem the following Cauchy problem:

$$\frac{d\boldsymbol{X}}{dt} = \boldsymbol{V}, \quad \frac{dS}{dt} = U(\boldsymbol{X}, t) + \frac{1}{2}V^2, \quad \frac{d\boldsymbol{V}}{dt} = \boldsymbol{f}(\boldsymbol{X}, t), \tag{3}$$

$$\boldsymbol{X}(\boldsymbol{y}, t = 0) = \boldsymbol{y}, \quad S(\boldsymbol{y}, t = 0) = \boldsymbol{V}(\boldsymbol{y}, t = 0) = 0,$$

$$\frac{d\hat{J}}{dt} = \hat{K}, \quad \frac{d\hat{K}}{dt} = \hat{g}(\hat{X}, t)\hat{J}, \tag{4}$$

$$\hat{J}(\boldsymbol{y}, t = 0) = \hat{I}, \quad \hat{K}(\boldsymbol{y}, t = 0) = 0,$$

for the scalar field $S(\boldsymbol{y}, t)$, vector fields $\boldsymbol{X}(\boldsymbol{y}, t)$ and $\boldsymbol{V}(\boldsymbol{y}, t)$, and also tensor fields $\hat{J}(\boldsymbol{y}, t)$ and $\hat{K}(\boldsymbol{y}, t)$ with components

$$J_{lm}(\boldsymbol{y}, t) = \frac{\partial X_l}{\partial y_m}, \quad K_{lm} = \frac{\partial V_l}{\partial y_m}.$$

In the above formulas $\hat{I}$ stands for the diagonal unit matrix and $\hat{g}$ is a random tensor with components

$$g_{lm}(\boldsymbol{x}, t) = \frac{\partial^2 U(\boldsymbol{x}, t)}{\partial x_l \partial x_m}. \tag{5}$$

The Cauchy problem (3) has a clear-cut intuitive physical interpretation. It describes the evolution of coordinates $\boldsymbol{X}$, action S, and velocity $\boldsymbol{V}$ of particles forced by $\boldsymbol{f}(\boldsymbol{x}, t)$. The notation clearly displays the dependence on the initial coordinates $\boldsymbol{y}$ of the particle. This dependence plays a fundamental role in further analysis. The Cauchy problems (3–4), together with arbitrarily distributed initial positions $\boldsymbol{y}$ can be naturally interpreted as a gas of noninteracting particles. The tensors $\hat{J}$ and $\hat{K}$ describe the deformation of an infinitesimal volume "frozen" in the gas. Recall that $\boldsymbol{y}$ is Lagrangian coordinates of this gas. Their connection with the Eulerian coordinates $\boldsymbol{x}$ is given by a vector equality

$$\boldsymbol{x} = \boldsymbol{X}(\boldsymbol{y}, t). \tag{6}$$

For given $\boldsymbol{x}$ and t it is an equation with respect to $\boldsymbol{y}$. Solving it, we obtain the Lagrangian coordinates of particles,

$$\boldsymbol{y} = \boldsymbol{Y}(\boldsymbol{x}, t), \tag{7}$$

which at time t arrive at a point with Eulerian coordinates $\boldsymbol{x}$. We should emphasize that in the general case, the gas of noninteracting particles has several, say $N(\boldsymbol{x}, t) \geq 1$, streams. It means that equation (6) may have several roots. In this case, equation (7) defines a multivalued function assuming N values

$$\boldsymbol{Y}_1(\boldsymbol{x}, t), \ \ \boldsymbol{Y}_2(\boldsymbol{x}, t), \ \ \dots \ , \boldsymbol{Y}_N(\boldsymbol{x}, t). \tag{8}$$

Consider the joint probability density of the solutions of the auxiliary Cauchy problem (3–4):

$$\mathcal{P}(\boldsymbol{x}, s, \boldsymbol{v}, \hat{\jmath}, \hat{\kappa}; \boldsymbol{y}, t) = \tag{9}$$

$$\Big\langle \delta(\boldsymbol{X}(\boldsymbol{y}, t) - \boldsymbol{x}) \delta(S(\boldsymbol{y}, t) - s) \delta(\boldsymbol{V}(\boldsymbol{y}, t) - \boldsymbol{v}) \delta(\hat{J}(\boldsymbol{y}, t) - \hat{\jmath}) \delta(\hat{K}(\boldsymbol{y}, t) - \hat{\kappa}) \Big\rangle.$$

Let us transform the right-hand side of equality (9), using the well-known identity

$$\delta(\boldsymbol{x} - \boldsymbol{X}(\boldsymbol{y}, t)) = \sum_{n=1}^{N(\boldsymbol{x}, t)} \frac{\delta(\boldsymbol{Y}_n(\boldsymbol{x}, t) - \boldsymbol{y})}{|J(\boldsymbol{Y}_n, t)|}, \tag{10}$$

for the delta function (see, e.g. Volume 1 of this monograph), where

$$J(\boldsymbol{y}, t) = \|\hat{J}(\boldsymbol{y}, t)\| = \left\| \frac{\partial X_l}{\partial y_m} \right\|, \tag{11}$$

is the Jacobian of the Eulerian-to-Lagrangian coordinate transformation. Substituting (10) into (9) and taking into account the probing property of the delta function, we have

$$|j| \mathcal{P}(\boldsymbol{x}, s, \boldsymbol{v}, \hat{\jmath}, \hat{l}; \boldsymbol{y}, t) = \tag{12}$$

$$\Bigg\langle \sum_{n=1}^{N(\boldsymbol{x}, t)} \delta(\boldsymbol{Y}_n(\boldsymbol{x}, t) - \boldsymbol{y}) \delta(s_n(\boldsymbol{x}, t) - s) \delta(\boldsymbol{v}_n(\boldsymbol{x}, t) - \boldsymbol{v}) \delta(\hat{\jmath}_n(\boldsymbol{x}, t) - \hat{\jmath})$$

$$\times \delta(\hat{\kappa}_n(\boldsymbol{x}, t) - \hat{\kappa}) \Bigg\rangle,$$

where

$$s_n(\boldsymbol{x}, t) = S(\boldsymbol{Y}_n, t), \quad \boldsymbol{v}_n(\boldsymbol{x}, t) = \boldsymbol{V}(\boldsymbol{Y}_n, t), \tag{13a}$$

$$\hat{\jmath}_n(\boldsymbol{x}, t) = \hat{J}(\boldsymbol{Y}_n, t), \quad \hat{\kappa}_n(\boldsymbol{x}, t) = \hat{K}(\boldsymbol{Y}_n, t), \tag{13b}$$

are fields that describe state of the gas in the nth of N streams which occur at point $\boldsymbol{x}$ at time t, and where j is the determinant of the matrix $\hat{\jmath}$ ($j = \|\hat{\jmath}\|$.) By the total probability formula, in view of (12),

$$|j|P(\boldsymbol{x},s,\boldsymbol{v},\hat{\jmath},\hat{\kappa};\boldsymbol{y},t) = \sum_{N=1}^{\infty} P(N;\boldsymbol{x},t) \sum_{n=1}^{N} W_n(\boldsymbol{y},s,\boldsymbol{v},\hat{\jmath},\hat{\kappa};\boldsymbol{x},t|N), \quad (14)$$

where $P(N;\boldsymbol{x},t)$ is the probability of the event that at a given point $\boldsymbol{x}$ at time t we have N streams present, and where $W_n(\boldsymbol{y},s,\boldsymbol{v},\hat{\jmath},\hat{\kappa};\boldsymbol{x},t|N)$ is the conditional joint probability density of random fields (8) and (13a,b) in the nth stream, given that the total number of streams is N.

19.3.2. Approximations for the Burgers turbulence statistics.
In view of (19.2.22–24), the sought joint probability density of the least-action functional, corresponding Lagrangian coordinates $\boldsymbol{Y}(\boldsymbol{x},t)$, the generalized solution $\boldsymbol{v}(\boldsymbol{x},t)$ of the nonhomogeneous Burgers equation in the inviscid limit, and the auxiliary fields $\hat{\jmath}, \hat{\kappa}$, are expressed in the following fashion through the components of sum (14):

$$W(\boldsymbol{y},s,\boldsymbol{v},\hat{\jmath},\hat{\kappa};\boldsymbol{x},t) = \sum_{N=1}^{\infty} P(N;\boldsymbol{x},t)W_1(\boldsymbol{y},s,\boldsymbol{v},\hat{\jmath},\hat{\kappa};\boldsymbol{x},t|N). \quad (15)$$

In the case of statistically homogeneous fields—in what follows we will restrict our attention to such fields—the probability density of the streams' number does not depend on $\boldsymbol{x}$, and the probability density in (14–15) depends only on $\boldsymbol{x} - \boldsymbol{y}$. Hence, integrating equalities (14–15) over all $\boldsymbol{x}, \hat{\jmath}, \hat{\kappa}$, we arrive at the following relations,

$$\langle |J| \rangle_{sv} P(s,v;t) = \sum_{N=1}^{\infty} P(N;t) \sum_{n=1}^{N} W_n(s,v;t|N), \quad (16)$$

$$W(s,\boldsymbol{v};t) = \sum_{N=1}^{\infty} P(N;t)W_1(s,v;t|N), \quad (17)$$

which are more convenient for further analysis. Here $\langle ... \rangle_{sv}$ denotes the average under the condition that $S(\boldsymbol{y},t) = s$, $V(\boldsymbol{y},t) = \boldsymbol{v}$ are given.

Unfortunately we cannot extract the partial sum (17), which is of interest to us, from the total sum (16). Such an operation is possible in principle, but to find (17) one has to have knowledge of all the joint probability densities for the Cauchy problem (3–4) under different initial conditions. These joint probability densities satisfy complex Kolmogorov equations whose solutions are not known. For that reason we will utilize a semi-qualitative method of finding probability densities of the forced Burgers turbulence.

Our main assumption is as follows: *there exists a number $\bar{S}(t)$— the largest value of the least action—such that*

$$\int_{-\infty}^{\bar{S}(t)} W_1(s;t|N)ds \approx 1, \tag{18}$$

and

$$\int_{-\infty}^{\bar{S}(t)} W_n(s;t|N)ds \approx 0, \quad n = 2, 3, \ldots N,$$

where

$$W_n(s;t|N) = \int_{-\infty}^{\infty} W_n(s, \boldsymbol{v}; t|N)d^d v, \quad n = 1, 2, \ldots, N.$$

If this assumption is satisfied, then the desired probability density

$$W(\boldsymbol{v};t) = \sum_{N=1}^{\infty} P(N;t)W_1(\boldsymbol{v};t|N)$$

of Burgers turbulence can be approximated by integration of equality (16) over all the values of s in the interval $(-\infty, \bar{S}(t))$, that is

$$W(\boldsymbol{v};t) = \int_{-\infty}^{\bar{S}(t)} \langle|J|\rangle_{sv} \mathcal{P}(s, \boldsymbol{v};t)\, ds. \tag{19}$$

In addition, the value of $\bar{S}(t)$ can be determined from the normalization condition

$$1 = \int_{-\infty}^{\bar{S}(t)} \langle|J|\rangle_s \mathcal{P}(s;t)\, ds, \tag{20}$$

for probability density (19), where $\mathcal{P}(s;t)$ is the probability density of random action $S(\boldsymbol{y},t)$ satisfying the auxiliary Cauchy problem (3).

Closing this subsection we will make an additional assumption that the *random Jacobian field J (11) is statistically independent from the random fields $S(\boldsymbol{y},t)$ and $\boldsymbol{V}(\boldsymbol{y},t)$.* In such a case, the expressions (19) for the solutions of the nonhomogeneous Burgers equation and equation for the maximal value of absolute minima $\bar{\boldsymbol{S}}(t)$ (20) take a particularly simple form

$$W(\boldsymbol{v};t) = \langle N(t) \rangle \int_{-\infty}^{\bar{S}(t)} \mathcal{P}(s, \boldsymbol{v};t)\, ds, \tag{21}$$

$$\langle N(t) \rangle \int_{-\infty}^{\bar{S}(t)} \mathcal{P}(s;t)\, ds = 1.$$

Note that the last assumption is not really essential and has only a technical nature. If it is not satisfied then the following calculations do not change qualitatively, but they do get more complicated. In the test case considered in the next example we will verify that the statistical dependence between J and $(S, \boldsymbol{V})$ does not significantly affect the final outcome. For that reason, in the remainder of this section we will always assume J to be statistically independent from the values of the vector $(S, \boldsymbol{V})$ and use expression (21) instead of a more correct, but much more complex formulas (19–20).

Example: Random values of actions for different streams: Let us test the conjecture underlying formulas (19–20) on the following simple model which, however, is relatively close to the problem we are considering. Let, for a given number of streams N, the values of actions of different streams $\{S_1, \ldots, S_N\}$ form a family of statistically independent random variables with identical cumulative distribution functions

$$F(s) = P(S_n < s).$$

In each realization we will form an order statistic

$$S^1 \leq S^2 \leq \cdots \leq S^N,$$

and denote the cumulative distribution function of the nth ordered variable S^n by

$$F_n^N(s) = P(S^n < s).$$

It is well known that

$$F_n^N(s) = 1 - \sum_{l=0}^{n-1} \binom{N}{l} \left(\frac{z}{N}\right)^l \left(1 - \frac{z}{N}\right)^{N-l}, \tag{22}$$

and in particular, that the cumulative distribution of the smallest $S_{min} = S^1$ is equal to

$$F_1^N(s) = P(S^1 < s) = 1 - \left(1 - \frac{z}{N}\right)^N, \tag{23}$$

where $z = z(s) = NF(s)$. Besides, it is clear that

$$\sum_{n=1}^{N} F_n^N = NF = z. \tag{24}$$

Within the framework of this example, the conditional normalizations (20) defining values of $\bar{S}$ reduce to the equality

$$NF = z = 1.$$

In addition, according to our assumption, conditions

$$F_1^N \Big|_{z=1} \approx 1, \qquad \sum_{n=2}^{N} F_n^N \Big|_{z=1} \approx 0, \tag{25}$$

analogous to (18) have to be fulfilled. Let us verify to what extent they are valid, substituting here the corresponding expressions from (22–24). This gives

$$F_1^N(\bar{s}) = F_1^N \Big|_{z=1} = 1 - \left(1 - \frac{1}{N}\right)^N, \tag{26}$$

$$R^N(\bar{S}) = \sum_{n=2}^{N} F_n^N \Big|_{z=1} = \left(1 - \frac{1}{N}\right)^N. \tag{27}$$

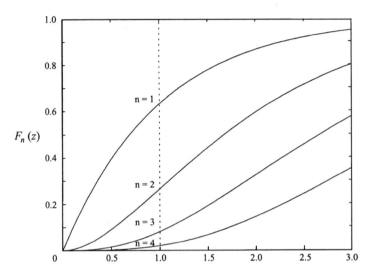

FIGURE 19.3.1.
Graphs of cumulative distributions of order statistics $F_n^\infty(z)$. The curves have a universal form and do not depend on the distributions of original random variables $S_1, S_2, \ldots$ One can see that, for $z = 1$, the value $F_1^\infty(1)$ is rather close to 1, while other cumulative distributions are still pretty small.

The values of $\{F_1^N(\bar{S})\}$ form a monotonically decreasing (as N increases) sequence; the limit thereof is

$$F_1^\infty(z = 1) = \lim_{N \to \infty} F_1^N(\bar{s}) = 1 - e^{-1}, \tag{27}$$

a quantity one can think of, in this context, as being "close" to 1. Correspondingly, $\{R^N(\bar{s})\}$ is a monotonically increasing sequence which, for $N \to \infty$, converges to a "small" number

$$R^\infty(z = 1) = \lim_{N \to \infty} R^N(\bar{s}) = e^{-1}. \tag{28}$$

In this fashion, in the model example under consideration, the relations (25) are satisfied with a "good" accuracy, expressed by the limit equalities (27–28).

Furthermore, notice that for arbitrary z and large $N \to \infty$, the probability densities of the order statistics $\{S^1, S^2, ...\}$ are described (see Fig. 19.3.1) by the main asymptotics of expressions (22–24):

$$F_1^\infty(z) = 1 - e^{-z}, \quad F_n^\infty(z) = 1 - e^{-z} \sum_{l=0}^{n-1} \frac{z^l}{l!}, \tag{29}$$

$$R^\infty(z) = \sum_{m=2}^\infty F_n^\infty(z) = z - 1 + e^{-z}.$$

We should emphasize here that the original problem of finding statistical properties of solutions of the nonhomogeneous Burgers equation is related to the situation discussed above at very large times, when the average number

$$\langle N(t) \rangle = \sum_{m=1}^\infty N P(N; t) = \langle |J| \rangle \tag{30}$$

of streams is much larger than 1. Indeed, for $\langle N \rangle \gg 1$ the law of large numbers effects take over, the random number $N(t)$ of streams is not much different from the mean number

$$\sqrt{\langle N - \langle N \rangle \rangle^2} \ll \langle N \rangle,$$

and one can assume that, for $\langle N \rangle \gg 1$, the number of streams in each realization is the same and equal to $\langle N(t) \rangle$.

In addition, it is natural to assume that in the multistream regime $\langle N \rangle \gg 1$, the particles which arrive at a given time at point $\boldsymbol{x}$, move along strongly dispersed paths, so that the forces acting on different particles $\boldsymbol{f}(\boldsymbol{X}_m(\tau), \tau); \tau \in [0, t]$, actually are statistically independent. Therefore, the values $\{S_1(\boldsymbol{x}, t), S_2(\boldsymbol{x}, t), \ldots, S_{\langle N \rangle}(\boldsymbol{x}, t)\}$ of their actions can be treated as independent parameters of the particles.

19.4 The case of 2-D unforced Burgers turbulence

We shall illustrate the above general statistical approach in the relatively well-understood case of the homogeneous Burgers turbulence and restrict ourselves to the 2-D case, $\boldsymbol{x} \in \mathbf{R}^2$. Then, the potential

$$U(\boldsymbol{x}, t) = S_0(\boldsymbol{x})\delta(t),$$

where $S_0(\boldsymbol{x})$ is the initial velocity field potential, $\boldsymbol{v}_0(\boldsymbol{x}) = \boldsymbol{\nabla} S_0(\boldsymbol{x})$. Taking this into account, the auxiliary Cauchy problem (19.3.3–4) takes the following form:

$$\frac{d\boldsymbol{X}}{dt} = \boldsymbol{V}, \quad \frac{dS}{dt} = \frac{1}{2}V^2, \quad \frac{d\boldsymbol{V}}{dt} = 0,$$

$$\boldsymbol{X}(\boldsymbol{y}, t = 0) = \boldsymbol{y}, \quad S(\boldsymbol{y}, t = 0) = S_0(\boldsymbol{y}), \quad \boldsymbol{V}(\boldsymbol{y}, t = 0) = \boldsymbol{v}_0(\boldsymbol{y}),$$

$$\frac{d\hat{J}}{dt} = \hat{K}, \qquad \frac{d\hat{K}}{dt} = 0, \tag{1}$$

$$\hat{J}(\boldsymbol{y}, t = 0) = \hat{I}, \qquad \hat{K}(\boldsymbol{y}, t = 0) = \hat{K}_0(\boldsymbol{y}),$$

where $\hat{K}_0(\boldsymbol{x})$ is a tensor with components

$$K_{0lm}(\boldsymbol{x}) = \frac{\partial^2 S_0(\boldsymbol{x})}{\partial x_l \partial x_m}.$$

Now, assume that $S_0(\boldsymbol{x})$ be a Gaussian, statistically isotropic field with zero mean, and correlation function

$$\langle S_0(\boldsymbol{x}) S_0(\boldsymbol{x} + \boldsymbol{y}) \rangle = \frac{\sigma_0^2}{\kappa^2} \exp\left(-\frac{1}{2}\kappa^2 y^2\right).$$

Then the fields $S_0(\boldsymbol{x})$, and $\boldsymbol{v}_0(\boldsymbol{x})$, are statistically independent at the same spatial point, and the joint probability density of the solutions S, and $\boldsymbol{V}$, of the Cauchy problem (1) takes the form

$$\mathcal{P}(s, \boldsymbol{v}; t) = w_v(\boldsymbol{v}) w_s(s - v^2 t/2), \tag{2}$$

where $w_v(\boldsymbol{v})$, and $w_s(s)$, are, respectively, the probability densities of the fields, $\boldsymbol{v}_0(\boldsymbol{x}), S_0(\boldsymbol{x})$ which, in the 2-D case, are

$$w_v(\boldsymbol{v}) = \frac{1}{2\pi\sigma_0^2} \exp\left(-\frac{v^2}{2\sigma_0^2}\right), \tag{3}$$

$$w_s(s) = \frac{\kappa}{\sqrt{2\pi}\sigma_0} \exp\left(-\frac{s^2\kappa^2}{2\sigma_0^2}\right). \tag{4}$$

For the sake of convenience, let us introduce a dimensionless scalar field

$$u(\boldsymbol{x}, t) = \boldsymbol{v}^2(\boldsymbol{x}, t)/2\sigma_0^2. \tag{5}$$

It follows from (19.3.21), and (2–4), that its probability density is given by the formula,

$$W(u; t) = \frac{1}{2}\langle N(t)\rangle e^{-u}\mathrm{erfc}\,(u\tau - \rho), \tag{6}$$

where the quantity ρ is determined from the normalization condition

$$\langle N(t)\rangle \int_0^\infty e^{-u}\mathrm{erfc}\,(u\tau - \rho)du = 2,$$

which is not difficult to transform into the following, more convenient for our analysis, form:

$$\langle N(t)\rangle \left[\mathrm{erfc}\,(-\rho) - \exp\left(-\rho^2 + \left(\rho - \frac{1}{2\tau}\right)^2\right)\mathrm{erfc}\left(\frac{1}{2\tau} - \rho\right)\right] = 2. \tag{7}$$

Note that in (6–7) we have introduced the following dimensionless variables,

$$\rho = \kappa \bar{S} / \sqrt{2}\sigma_0, \qquad \tau = \kappa \sigma_0 t / \sqrt{2}, \tag{8}$$

and the notation

$$\operatorname{erfc}(z) = 1 - \operatorname{erf}(z), \qquad \operatorname{erf}(z) = \frac{2}{\sqrt{\pi}} \int_0^z e^{-y^2} dy, \tag{9}$$

was used for the special error function.

Expressions (6–7) contain the mean value $\langle N(t) \rangle$ of the streams' number, which will be calculated below. For now, assuming that $\langle N(t) \rangle$ is known, observe that it is not very difficult to solve equation (7) numerically with respect to $\rho(\tau)$, and define the probability density (6), and corresponding moment functions, for any τ. Here, we will restrict ourselves to the derivation of the asymptotic formulas for the late stage when multiple discontinuities coalesce ($\tau \gg 1, \langle N(t) \rangle \gg 1$) in the Burgers turbulence. At that stage, equation (7) can be replaced, with help of the asymptotic formula,

$$\operatorname{erfc}(z) \sim \frac{1}{\sqrt{\pi} z} e^{-z^2}, \qquad z \to \infty, \tag{10}$$

by the asymptotic relation

$$\rho^2 e^{\rho^2} = \frac{\langle N(t) \rangle}{4 \tau \sqrt{\pi}}. \tag{11}$$

If the right-hand side of this equality is much larger than 1, then we get the following asymptotic formula

$$|\rho| \sim \sqrt{\ln\left(\frac{\langle N(t) \rangle}{4\tau\sqrt{\pi}}\right)}, \qquad \rho < 0, \quad |\rho| \gg 1. \tag{12}$$

Let us substitute expression (12) into (6). Using (10), we arrive at the following result:

(i) For $\tau \gg 1$, $\langle N(t) \rangle \gg 1$, and $|\rho| \gg 1$, the dimensionless kinetic energy $u = v^2 / 2\sigma_0^2$ in unforced Burgers turbulence has the probability density

$$W(u; \tau) = 2|\rho|\tau \exp(-2|\rho|\tau u), \tag{13}$$

where ρ and τ are given by (8).

In particular, it follows that the average dimensionless kinetic energy $\langle u(x,t) \rangle$ in Burgers turbulence in the late stage of multiple shock coalescence satisfies the asymptotic law

$$\langle u(\boldsymbol{x}, t) \rangle \sim 1/2|\rho|\tau. \tag{14}$$

In relations (12–14), the principal role was played by the average number $\langle N(t) \rangle$ of streams in the gas of noninteracting particles. Now we will calculate that number in the 2-D case under consideration. For that purpose recall that this average is connected by formula (19.3.30) with the statistical characteristics of the Jacobian $J(\boldsymbol{y}, t)$ (19.3.11):

$$\langle N(t) \rangle = \langle |J| \rangle.$$

It is known (see, e.g., Gurbatov, Malakhov, Saichev [19]) that in the 2-D case the Jacobian is statistically equivalent with the following random quantity

$$J = (1 + 2\alpha)^2 - 2\beta,$$

where α, $-\infty < \alpha < \infty$, and $\beta \geq 0$, are statistically independent random quantities with probability densities

$$P(\alpha; \tau) = \frac{1}{\sqrt{2\pi}\tau} \exp\left(-\frac{\alpha^2}{2\tau^2}\right), \qquad Q(\beta; \tau) = \frac{1}{\sqrt{2\tau^2}} \exp\left(-\frac{\beta}{2\tau^2}\right).$$

The above two formulas permit us to obtain an exact expression for the probability density of the Jacobian:

$$P(j; \tau) = \frac{1}{8\sqrt{3}\tau^2} \exp\left(\frac{j}{4\tau^2} - \frac{1}{12\tau^2}\right) \times \tag{15}$$

$$\times \begin{cases} 2, & \text{if } j < 0; \\ 2 - \operatorname{erf}\left(\sqrt{\frac{3j}{8\tau^2}} - \frac{1}{4\sqrt{3}\tau}\right) - \operatorname{erf}\left(\sqrt{\frac{3j}{8\tau^2}} + \frac{1}{4\sqrt{3}\tau}\right), & \text{if } j > 0. \end{cases}$$

where from, after simple calculations, we obtain that

$$\langle N(t) \rangle = 1 + \frac{8}{\sqrt{3}}\tau^2 \exp\left(-\frac{1}{12\tau^2}\right). \tag{16}$$

In particular, for $\tau \to \infty$, the average number of streams satisfies the following asymptotic power law:

$$\langle N(t) \rangle \sim \frac{8}{\sqrt{3}}\tau^2. \tag{17}$$

Substituting it into (12), we find that

$$|\rho| \sim \sqrt{\ln\left(\frac{\tau}{\sqrt{3\pi}}\right)}, \qquad \tau \gg 1, \tag{18}$$

which gives the following result:

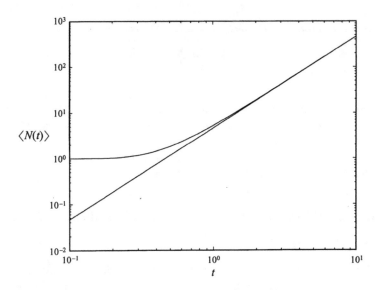

FIGURE 19.4.1.
Time evolution of the exact (top line; see (4.16)) and asymptotic (bottom line; see (4.17)) average number $\langle N(t)\rangle$ of streams in 2-D homogeneous Burgers turbulence. Initially, when the dimensionless time $\tau < 1$, the number of streams is close to 1, and the discontinuities of $v(x,t)$ are practically absent. In the late stages $\tau \gg 1$ of the multiple shock coalescence, the number of shocks is well described by the asymptotic formula (4.17).

(ii) The average kinetic energy (14) of the unforced 2-D inviscid Burgers turbulence decays, at sufficiently large times, as

$$\langle u(\boldsymbol{x},t)\rangle \sim \left(2\tau \sqrt{\ln\left(\frac{\tau}{\sqrt{3\pi}}\right)} \right)^{-1},$$

where $\tau = \kappa\sigma_0 t/\sqrt{2}$.

Remark 1. For $\tau \to \infty$, the probability density of the Jacobian (15) has the following *self-similar* property:

$$\mathcal{P}(j;\tau) \sim \frac{1}{c\langle N\rangle}\mathcal{P}_\infty\left(\frac{j}{c\langle N\rangle}\right), \qquad (19)$$

where

$$\mathcal{P}_\infty\left(\frac{z}{c}\right) = \frac{2}{3}\exp\left(\frac{z}{\sqrt{12}}\right) \times \begin{cases} 1, & \text{if } z < 0; \\ 1 - \mathrm{erf}\left(\sqrt{z\sqrt{3}}\right), & \text{if } z > 0, \end{cases}$$

and c is a normalizing constant, which in this case is

$$c = \frac{4}{\sqrt{15}} \left(3\sqrt{6} - 2\sqrt{5} \right).$$

The self-similarity (19) of the Jacobian's probability density, which is clear in the homogeneous case, can be used in the multidimensional and forced case as an assumption under which we can find the rate of growth for the time evolution of $\langle N(t) \rangle$, and of the average kinetic energy. In the 1-D case, we were able in Section 19.1 to use a more precise approach to study the convergence of Burgers turbulence to a stationary regime.

19.5 Statistical properties of forced Burgers turbulence

19.5.1. Statistics of noninteracting particles' action. In this section we shall apply the algorithm proposed above to the calculation of statistical properties of forced Burgers turbulence. Let us begin with a study of the probability density of solutions of the auxiliary Cauchy problem (19.3.3). It follows from (19.3.1–3) that action $S(\boldsymbol{y}, t)$ can be represented as the sum of two statistically independent components,

$$S(\boldsymbol{y}, t) = S_1(\boldsymbol{y}, t) + S_2(\boldsymbol{y}, t). \tag{1}$$

Moreover, the first summand is also independent of the random velocity $\boldsymbol{V}(\boldsymbol{y}, t)$ and has the probability density

$$\mathcal{P}_1(s; t) = \frac{1}{2\sqrt{\pi a t}} \exp\left(-\frac{s^2}{4at} \right), \qquad a = a(0). \tag{2}$$

Furthermore, the joint probability density of the second summand in (1) and the velocity field $\boldsymbol{V}(\boldsymbol{y}, t)$ satisfies the following Kolmogorov equation:

$$\frac{\partial \mathcal{P}_2}{\partial t} + \frac{1}{2} v^2 \frac{\partial \mathcal{P}_2}{\partial s} = b \Delta_v \mathcal{P}_2, \qquad b = b(0), \tag{3}$$

$$\mathcal{P}_2(s, \boldsymbol{v}; t = 0) = \delta(s)\delta(\boldsymbol{v}).$$

Respectively, the joint probability density of the full action $S(\boldsymbol{y}, t)$, and the velocity $\boldsymbol{V}(\boldsymbol{y}, t)$, is equal to

$$P(s, \boldsymbol{v}; t) = \mathcal{P}_1(s; t) * \mathcal{P}_2(s, \boldsymbol{v}; t), \tag{4}$$

where the symbol $*$ stands for the convolution operation, here with respect to variable s.

Equation (3) implies that the function,

$$\theta(\mu, \boldsymbol{\nu}; t) := \int_0^\infty ds \int_{-\infty}^\infty \ldots \int_{-\infty}^\infty \mathcal{P}_2(s, \boldsymbol{v}; t) \exp[-\mu s + i(\boldsymbol{\nu} \cdot \boldsymbol{v})] d^d v, \tag{5}$$

satisfies the following equation,

$$\frac{\partial \theta}{\partial t} = \frac{\mu}{2} \Delta_\nu \theta - b\nu^2 \theta, \qquad \theta(\mu, \nu; t = 0) = 1. \tag{6}$$

which has a solution of the exponential form,

$$\theta(\mu, \nu; t) = \exp[q(\mu, t) - \frac{1}{2} p(\mu, t)\nu^2]. \tag{7}$$

Substituting (7) into (6), we arrive at the following pair of equations for q and p,

$$\frac{dq}{dt} + \frac{\mu d}{2} p = 0, \qquad q(\mu, 0) = 0,$$

$$\frac{dp}{dt} + \mu p^2 = 2b, \qquad p(\mu, 0) = 0,$$

the solutions thereof, under the initial conditions indicated above, are

$$p(\mu, t) = \frac{\tau}{\sqrt{\delta}} \tanh \sqrt{\delta}, \quad q(\mu, t) = -\frac{d}{2} \ln(\cosh \sqrt{\delta}),$$

where the new variables,

$$\delta = 2\mu b t^2, \quad \text{and} \quad \tau = 2bt, \tag{8}$$

have been introduced. Substituting the above expressions for p and q into (7), we obtain that

$$\theta(\mu, \nu; t) = \left(\frac{1}{\cosh \sqrt{\delta}}\right)^{d/2} \exp\left(-\frac{\tau}{2}\nu^2 \frac{\tanh \sqrt{\delta}}{\sqrt{\delta}}\right). \tag{9}$$

In particular, for $\nu = 0$, we have the expression

$$\theta_2(\mu; t) = \left(\frac{1}{\cosh \sqrt{\delta}}\right)^{d/2} \tag{10}$$

for the Laplace transform

$$\theta_2(\mu; t) = \int_0^\infty e^{-\mu s} P_2(s; t) \, ds \tag{11}$$

of the probability density of the second action component S_2.

Finally, calculating the inverse Fourier transform with respect to ν, we pass from (9) to the following expression

$$\Phi(\mu, v; t) = \left(\frac{\sqrt{\delta}}{2\pi\tau \sinh \sqrt{\delta}}\right)^{d/2} \exp\left(-\frac{v^2\sqrt{\delta}}{2\tau \tanh \sqrt{\delta}}\right) \tag{12}$$

for the Laplace transform

$$\Phi(\mu, \boldsymbol{v}; t) = \int_0^\infty e^{-\mu s} \mathcal{P}_2(s, \boldsymbol{v}; t) \, ds \tag{13}$$

of the probability density $\mathcal{P}_2(s, \boldsymbol{v}; t)$ with respect to variable s.

Now, let us introduce an auxiliary dimensionless random variable,

$$G_2 = S_2(\boldsymbol{y}, t)/2bt^2. \tag{14}$$

It follows from (10) that the probability density $\tilde{\mathcal{P}}_2(g)$ is independent of time and has the Laplace transform,

$$\tilde{\theta}_2(\delta) = \int_0^\infty \tilde{\mathcal{P}}_2(g) e^{-\delta g} dg = \frac{1}{\cosh \sqrt{\delta}}. \tag{15}$$

Here, as in the previous section, we have taken $d = 2$. Using the inverse Laplace transform of (15) we get

$$\tilde{\mathcal{P}}_2(g) = \langle \delta(g - G_2) \rangle = \sum_{k=0}^\infty (-1)^k \frac{2k+1}{\sqrt{\pi g g}} \exp\left(-\frac{(2k+1)^2}{4g} \right). \tag{16}$$

Probability density of the full, normed with respect to (14), action is equal to the convolution

$$\tilde{\mathcal{P}}(g; \tau) = \tilde{\mathcal{P}}_2(g) \otimes \tilde{\mathcal{P}}_1(g; \tau) \tag{17}$$

of the probability density (16), and the Gaussian probability density,

$$\tilde{\mathcal{P}}_1(g; \tau) = \frac{1}{\sqrt{2\pi\epsilon^2}} \exp\left(-\frac{g^2}{2\epsilon^2} \right), \tag{18}$$

obtained from (2) by passing to dimensionless variables τ, and $g = s/2bt^2$. In (18), the dimensionless parameter

$$\epsilon = 2\sqrt{ab/\tau^3}. \tag{19}$$

For sufficiently large times, when $\epsilon \ll 1$, the probability density $\mathcal{P}_1$ (18) plays the role of a delta function in convolution (17), and we are justified to use an approximate formula

$$\tilde{\mathcal{P}}(g) \approx \tilde{\mathcal{P}}_2(g). \tag{20}$$

19.5.2. Asymptotics of the largest value of least action. The above discussion of statistical properties of action of noninteracting particles will help us find the largest value of least action $\bar{S}$ which, in turn, will determine the statistical properties of the Burgers turbulence in the inviscid limit. Let us introduce, similarly to (14), the dimensionless value,

$$\rho = \bar{S}/2bt, \tag{21}$$

For very large times, when $\epsilon \ll 1$ (see (2)) and, additionally, $\langle N(t) \rangle \gg 1$, it is sufficient to know the behavior of function (16) for small $g \ll 1$. For such g, the sum (16) is approximately equal to its first summand. As a result, we arrive at the asymptotic formula

$$\tilde{\mathcal{P}}(g) \sim \frac{1}{\sqrt{\pi g}g} \exp\left(-\frac{1}{4g}\right), \qquad \epsilon \ll 1, \; g \ll 1. \tag{22}$$

Similarly, it is not difficult to show that in the space of arbitrary dimension d, the probability density of action is described by the asymptotic expression,

$$\tilde{\mathcal{P}}(g) \sim \sqrt{\frac{2^d}{\pi g}\frac{d}{4g}} \exp\left(-\frac{d^2}{16g}\right), \qquad \epsilon \ll 1, \; g \ll 1. \tag{23}$$

Consequently, equation for ρ,

$$\langle N(t) \rangle \int_0^\rho \tilde{\mathcal{P}}(g)\,dg = 1 \tag{24}$$

assumes the form

$$\langle N(t) \rangle \sqrt{2^d}\, \mathrm{erfc}\left(-d/4\sqrt{\rho}\right) = 1. \tag{25}$$

Utilizing the asymptotic formula (10) we can reduce (25) to the transcendental equation,

$$\langle N(t) \rangle \frac{4}{d}\sqrt{\frac{2^d\rho}{\pi}} \exp\left(-\frac{d^2}{16\rho}\right) = 1, \tag{26}$$

the asymptotic solution thereof can be written in the form

$$\rho = d^2 \Big/ 16\ln\left(\langle N(t) \rangle\sqrt{\frac{2^{d+1}}{\pi}}\right). \tag{27}$$

19.5.3.   Average energy of Burgers turbulence. Now, we can return to analysis of the desired statistical characteristics of the forced Burgers turbulence. First of all, let us take a look at the behavior of the average kinetic energy

$$\langle u(\boldsymbol{x}, t) \rangle = \frac{1}{2}\langle v^2(\boldsymbol{x}, t) \rangle.$$

Let's multiply (12) by $\langle N(t) \rangle v^2/2$ and then integrate it over all the values of v. As a result, we obtain the following auxiliary function

$$T(\nu, \tau) = \langle N(t) \rangle \tau d \left(\frac{1}{\cosh\sqrt{\nu}}\right)^{d/2} \frac{\tanh\sqrt{\nu}}{\sqrt{\nu}}. \tag{28}$$

To calculate the average kinetic energy, it is necessary to find the inverse Laplace transform of that function with respect to ν, and then to integrate the obtained expression with respect to g, over the interval $(0, \rho)$. To implement these steps note that the behavior of the desired original function for small values of g, which are of interest to us, is determined by the behavior of its Laplace transform (28) for large values of ν. For that reason, we will pass in (28) to the corresponding asymptotic expression,

$$T(\nu, \tau) \sim \langle N(t) \rangle d\tau \sqrt{\frac{2^d}{\nu}} e^{-(d/2)\sqrt{\nu}} = -4\tau \langle N(t) \rangle \sqrt{2^d} \frac{d}{d\nu} e^{-(d/2)\sqrt{\nu}}, \qquad \nu \gg 1.$$

Finding the inverse Laplace transform of this function, integrating it over g in the interval $(0, g)$, we arrive at the following asymptotic formula for a kinetic energy of the Burgers turbulence:

$$\langle u(\boldsymbol{x}, t) \rangle \sim d\tau \langle N \rangle \sqrt{\frac{2^d}{\pi}} \int_0^\rho \frac{dg}{\sqrt{g}} \exp\left(-\frac{d^2}{16g} \right).$$

Replacing the integral by its main asymptotics for $\rho \ll 1$, we have

$$\langle u(\boldsymbol{x}, t) \rangle \sim \tau \rho \langle N(t) \rangle \frac{16}{d} \sqrt{\frac{2^d \rho}{\pi}} \exp\left(-\frac{d^2}{16\rho} \right).$$

Comparing this expression with equation (26) we finally obtain the following result:

(i) Let $\epsilon = 2\sqrt{ab/\tau^3}$, $\tau = 2bt$, and $\langle N(t) \rangle$ be the average number of streams of the auxiliary gas of noninteracting particles. Then, for $\epsilon \ll 1$, $\langle N \rangle \gg 1$, the average dimensionless kinetic energy in forced Burgers turbulence has the following asymptotic behavior:

$$\langle u(\boldsymbol{x}, t) \rangle \approx 4\tau\rho, \tag{29}$$

where ρ (27) is the largest possible value of least action.

The above conclusion and formula (27) give us an opportunity to formulate a necessary condition for existence of a stationary regime in forced Burgers turbulence:

(ii) A necessary condition for the existence of a stationary regime in forced d-dimensional Burgers turbulence is the exponential growth

$$\langle N(t) \rangle \sim Ce^{\gamma\tau} \tag{30}$$

of the average stream-number in the auxiliary gas of noninteracting particles. The exponent γ determines the limit average energy via the formula

$$u_\infty = \lim_{t \to \infty} \langle u(\boldsymbol{x}, t) \rangle = d^2/4\gamma. \tag{31}$$

In the multidimensional case, the verification of the exponential growth law (30) requires knowledge of the joint $2d$-dimensional probability density $P(\hat{j}, \hat{k}; t)$ for components of tensors $\hat{J}$, and $\hat{K}$. This is a formidable problem, both analytically and numerically. Nevertheless, linearity of the corresponding stochastic equations for $\hat{J}$, and $\hat{K}$, enables us to reach some conclusions about the behavior of forced Burgers turbulence for $t \to \infty$.

First of all, notice that in the 2-D case it is rather easy to derive the following exact equation for the second moment $\langle J^2 \rangle$ of the Jacobian:

$$\frac{d^6 \langle J^2 \rangle}{d\theta^6} - 14 \frac{d^3 \langle J^2 \rangle}{d\theta^3} - 20 \theta \langle J^2 \rangle = 0, \tag{32}$$

where $\theta = c^{1/3} t$ is the dimensionless time, and c is the third coefficient in the power series expansion

$$a(y) = a - \frac{b}{2} y^2 + \frac{c}{8} y^4 - \dots$$

of function $a(y)$ from (19.3.1). A suitable solution of equation (32) has the form

$$\langle J^2 \rangle = \left(1 - \sqrt{3/23}\right) \left[\exp(\beta_1 \theta) + 2 \exp(-\beta_1 \theta/2) \cos(\sqrt{3}\beta_1 \theta/2)\right]$$

$$+ \left(1 + \sqrt{3/23}\right) \left[\exp(-\beta_2 \theta) + 2 \exp(\beta_2 \theta/2) \cos(\sqrt{3}\beta_2 \theta/2)\right],$$

where $\beta_{1,2} = \sqrt{\sqrt{69} \pm 7}$, and it grows monotonically with θ. As $\theta \to \infty$, we have the exponential asymptotics

$$\langle J^2 \rangle \sim (1 - \sqrt{3/23}) e^{\beta_1 \theta}.$$

In addition, it is also clear that

$$\langle N(t) \rangle = \langle |J| \rangle < \sqrt{\langle J^2 \rangle} \sim \exp(\beta_1 \theta/2). \tag{33}$$

This means that the *average energy of the forced Burgers turbulence is bounded from below and satisfies the following asymptotic inequality:*

$$\langle u(\boldsymbol{x}, t) \rangle \geq d^2 b / \beta_1 c^{1/3}. \tag{34}$$

Remark 5.1. If the self-similarity property (19.4.19) is taken as a working hypothesis (it has been established in the previous section for the unforced Burgers turbulence), then the exponential law (30) follows, with

$$\gamma = \beta_1 c^{1/3} / 4b,$$

and the kinetic energy converges to the stationary value

$$u_\infty = d^2 b / \beta_1 c^{1/3}$$

which coincides with the right-hand side of bound (34). The 1-D case, where the crucial exponential law (30) can be derived by more precise methods, will be discussed in the next section.

19.6   Stream-number statistics for 1-D gas

In this section we discuss statistical properties of the Jacobian (19.3.11) and
find an asymptotic rate of growth of the average number of streams $\langle N(t)\rangle$
(19.3.30). We will restrict our attention to the 1-D case which was briefly
discussed in Section 19.1. Then, equations for the Jacobian (19.3.4) have a
particularly simple form,

$$\frac{dJ}{dt} = K, \qquad \frac{dK}{dt} = g(X,t)J. \tag{1}$$

In the delta-correlated approximation used in this chapter, the random field
$g(x,t)$ can be replaced by a statistically equivalent Gaussian process $g(t)$ with
zero mean and correlation function

$$\langle g(t)g(t+\theta)\rangle = 2c\delta(\theta). \tag{2}$$

We need to solve equations (1) with the initial conditions,

$$J(t=0) = 1, \qquad K(t=0) = 0. \tag{3}$$

Let us introduce an ordered sequence

$$0 < t_1 < t_2 < \cdots < t_m < \ldots \tag{4}$$

of times $\{t_m\}$ which are roots of the equation

$$J(t) = 0. \tag{5}$$

Take one of these times t_m as the initial time. Then, the sought solution of
equation (1), for $t > t_m$, can be written in the form,

$$J(t) = \tilde{K}(t_m)\tilde{J}(t|t_m), \qquad K(t) = \tilde{K}(t_m)\tilde{K}(t|t_m), \tag{6}$$

where $\tilde{J}(t|t_m)$, and $\tilde{K}(t|t_m)$, are solutions of equation (1) with the initial
conditions

$$\tilde{J}(t=t_m|t_m) = 0, \qquad \tilde{K}(t=t_m|t_m) = 1. \tag{7}$$

Expressing, in turn, $\tilde{K}(t_m)$ by $\tilde{K}(t_{m-1})$, and so on, we arrive at the equality,

$$\tilde{K}(t_m) = \prod_{p=1}^{m} K_p, \tag{8}$$

where
$$K_1 = K(t_1), \qquad K_p = \tilde{K}(t_p|t_{p-1}), \quad p > 1.$$

Additionally, observe that—according to (6)—the product of random variables (8) defines the value $J(t)$ of the solution of the initial-value problem
(1–3) at time $t > t_m$:

$$J(t) = \tilde{J}(t|t_m) \prod_{p=1}^{m} K_p. \tag{9}$$

We emphasize that, for a given value of m, all the factors in the products (8–9) are statistically mutually independent, since they are functionals of the white noise $g(t)$ on the nonoverlapping time intervals (t_{p-1}, t_p). It is not difficult to show that even a more general statement is true: elements of the sequence of random quantities $\{K_p, \tau_p\}$, where

$$\tau_p = t_p - t_{p-1},$$

with different indices p and p' are statistically independent, and the joint probability density with identical indices

$$w(\kappa, \tau) = \langle \delta(K_p - \kappa) \delta(\tau_p - \tau) \rangle, \qquad p > 1,$$

does not depend on the index p.

Recall that, in the final count, we are interested in the average stream-number $\langle N(t) \rangle$ (19.3.30)

$$\langle N(t) \rangle = \langle |J(t)| \rangle. \tag{10}$$

For sufficiently large times, when $\langle N(t) \rangle \gg 1$, using the law of large numbers one can assume that

$$m = t / \langle \tau_1 \rangle, \tag{11}$$

where $\langle \tau_1 \rangle$ is the mean length of the time interval between adjacent zeros of the process $J(t)$. In this fashion, taking into account (9), we arrive at the following conclusion:

In the forced 1-D Burgers turbulence, the average stream-number

$$\langle N(t) \rangle \sim C e^{\nu t}, \qquad t \gg \langle \tau_1 \rangle,$$

where the exponent

$$\nu = \frac{1}{\langle \tau_1 \rangle} \ln \left(\langle K \rangle \right), \tag{12}$$

and $\langle K \rangle$ is the statistical average of any of the random factors in the product (8) for $p > 1$.

Hence, the calculation of the exponent ν reduces to finding the averages $\langle \tau_1 \rangle$ and $\langle K \rangle$. These averages can be computed numerically. For that purpose we introduce a new dimensionless time

$$\theta = c^{1/3} t,$$

and transform (1) into the dimensionless equations,

$$\frac{dR}{d\theta} = K, \quad \text{and} \quad \frac{dK}{d\theta} = \alpha(\theta) R, \tag{13}$$

where $\alpha(\theta)$ is a Gaussian, delta-correlated process with autocorrelation function

$$\langle \alpha(\theta) \alpha(\theta + \eta) \rangle = 2\delta(\eta).$$

The suggested scheme of numerical calculations of $\langle \tau_1 \rangle$ and $\langle K \rangle$ requires repeated numerical solutions of equations (13) with initial conditions

$$R(0) = 0, \quad K(0) = 1,$$

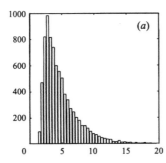 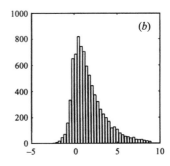

FIGURE 19.6.1.
The histograms of data $\{\theta_1^m\}$ (a) and $\{\log K^m\}$ (b) for $M \approx 8,000$.
The logarithmic scale was needed in case (b) because of the huge variance of the data

for a large number $M \gg 1$ of statistically independent realizations of $\alpha(\theta)$. Stopping the calculations at the first moment $\theta = \theta_1 > 0$ when $R_1(\theta_1) = 0$, we obtain two data arrays $\{\theta_1^m\}$ and $\{K^m\}$, $m = 1, 2, \ldots, M$, $K^m = K^m(\theta_1^m)$, the means thereof give us approximate values of statistical averages of θ_1, and K. Notice, that $\langle \theta_1 \rangle$ is related to the above-mentioned average $\langle \tau \rangle$ via an obvious equality

$$\langle \tau_1 \rangle = \langle \theta_1 \rangle c^{-1/3}.$$

The histograms on Fig. 19.6.1 illustrate the results of $M \approx 8,000$ such numerical calculations. In particular, they provide the following estimates: $\langle \theta_1 \rangle \approx 4.83$, $\langle K \rangle \approx 81.26$, and as a result

$$\nu \approx 0.91 c^{1/3}.$$

19.7 Density fields in Burgers turbulence: the basic model

Another phenomenon, adequately modeled by the chaotic solutions of the 3-D Burgers' equation, is a nonlinear evolution of the gravitational instability of matter in the Universe, and related characteristic large-scale cellular structures (see, e.g. Gurbatov et al. [19], Weinberg and Gunn [20], Vergassola et al. [21], Woyczynski [22]). In this example, it is not just the velocity field satisfying the 3-D Burgers' equation that is of physical importance, since the evolution of the matter density field in the Universe, driven by the Burgersian velocity field, is of principal interest.

In this section we consider a somewhat simplified model by augmenting the Burgers' equation by a complementary, but compatible model of the density field which makes the computations feasible while preserving all the characteristic features of the real density field. The model satisfies the mass conservation law, and in the inviscid limit, of interest to, e.g., astrophysicists, it weakly converges to a generalized solution of the continuity equation. In that case it gives a vivid analytic description of the mass distribution in the elementary cellular structures and, in particular, in their edges and vertices. A stochastic interpretation of the model density field is given in the following section.

Later on in this chapter we provide an explicit and detailed analysis of our model in the 2-D case. We deliberately discuss it before discussing the 1-D case in later sections , as the latter leads us in a natural fashion to additional topics of compressible gas dynamics and the KdV-Burgers equation, not directly related to the original problem of density fields in Burgers turbulence.

The first goal of this section is to give an analytically convenient model for a *density field* $\rho(\boldsymbol{x}, t)$, governed by the *velocity field* $\boldsymbol{v}(\boldsymbol{x}, t)$ in $\mathbf{R}^d$ satisfying the multidimensional *Burgers' equation:*

$$\frac{\partial \boldsymbol{v}}{\partial t} + (\boldsymbol{v} \cdot \nabla)\boldsymbol{v} = \mu \Delta \boldsymbol{v}, \tag{1}$$

$$\boldsymbol{v}(\boldsymbol{x}, t = 0) = \boldsymbol{v}_0(\boldsymbol{x}).$$

We deliberately avoid writing at this point the usual continuity equation for the density field $\rho(\boldsymbol{x}, t)$ since our model will subtly but significantly differ from it.

For a scalar field $S_0(\boldsymbol{y})$, and

$$\phi(\boldsymbol{y}, \boldsymbol{x}, t) := \frac{(\boldsymbol{y} - \boldsymbol{x})^2}{2t} + S_0(\boldsymbol{y}), \tag{2}$$

let us introduce the *spatial probability distribution function*[5]

$$f_\mu(\boldsymbol{y}; \boldsymbol{x}, t) := \frac{\exp[-\frac{1}{2\mu}\phi(\boldsymbol{y}, \boldsymbol{x}, t)]}{\int \exp[-\frac{1}{2\mu}\phi(\boldsymbol{y}, \boldsymbol{x}, t)]d\boldsymbol{y}}, \tag{3}$$

in $\boldsymbol{y}$ variable, which obviously satisfies the normalization condition over $\boldsymbol{y}$:

$$\int f_\mu(\boldsymbol{y}; \boldsymbol{x}, t)d\boldsymbol{y} \equiv 1.$$

---

[5]We will use the term "probability distribution function" instead of the "probability density function", to avoid confusion with the density fields that play the central role in this paper.

The *spatial averaging* of an arbitrary function $g(\boldsymbol{y})$ by means of spatial distribution function $f_\mu(\boldsymbol{y}; \boldsymbol{x}, t)$ will be denoted by the double square brackets

$$[\![g(\boldsymbol{y})]\!] = [\![g]\!](\boldsymbol{x}, t) = \int g(\boldsymbol{y}) f_\mu(\boldsymbol{y}; \boldsymbol{x}, t) d\boldsymbol{y}. \tag{4}$$

Whenever the initial field $\boldsymbol{v}_0(\boldsymbol{x})$ is of potential type, the multidimensional Burgers' equation itself has an explicit solution. More precisely, in terms of the above spatial averaging, that fact is stated as follows:

If

$$\boldsymbol{v}_0(\boldsymbol{x}) = \nabla S_0(\boldsymbol{x}),$$

then the Burgers' equation (1) *has a solution of the form*

$$\boldsymbol{v}(\boldsymbol{x}, t) = \frac{\boldsymbol{x} - [\![\boldsymbol{y}]\!](\boldsymbol{x}, t)}{t}, \tag{5}$$

where, in accordance with definition (4),

$$[\![\boldsymbol{y}]\!](\boldsymbol{x}, t) = \int \boldsymbol{y} f_\mu(\boldsymbol{y}; \boldsymbol{x}, t) d\boldsymbol{y}. \tag{5a}$$

Indeed, formula (5) is a convenient for our purposes reformulation of the well-known *Hopf–Cole formula*

$$\boldsymbol{v}(\boldsymbol{x}, t) = \frac{1}{t} \frac{\int (\boldsymbol{x} - \boldsymbol{y}) \exp[-\frac{1}{2\mu}\phi(\boldsymbol{y}, \boldsymbol{x}, t)] d\boldsymbol{y}}{\int \exp[-\frac{1}{2\mu}\phi(\boldsymbol{y}, \boldsymbol{x}, t)] d\boldsymbol{y}}, \tag{6}$$

for irrotational solutions of the Burgers equation, where ϕ is defined by (2). Its proof is immediate after the transformation,

$$\boldsymbol{v}(\boldsymbol{x}, t) = -2\mu \nabla \log U(\boldsymbol{x}, t),$$

reduces Burgers' equation for potential fields to the multidimensional heat equation[6]

$$\frac{\partial U}{\partial t} = \mu \Delta U, \qquad U(\boldsymbol{x}, t = 0) = \exp[-S_0(\boldsymbol{x})/2\mu].$$

Remark 7.1. It is important to notice that in the inviscid limit $(\mu \to 0+)$, for each $\boldsymbol{x}$, and $t > 0$, the spatial distribution function (3) weakly converges to the Dirac delta:

$$\lim_{\mu \to 0} f_\mu(\boldsymbol{y}; \boldsymbol{x}, t) = f_0(\boldsymbol{y}; \boldsymbol{x}, t) = \delta[\boldsymbol{y} - \boldsymbol{y}(\boldsymbol{x}, t)], \tag{7}$$

---

[6]See, e.g., W.A Woyczyński, *Burgers-KPZ Turbulence, Göttingen Lectures*, Springer (1998).

where $y(x, t)$ is a y-coordinate of the absolute minimum of function $\phi(y, x, t)$ defined in (2), and the velocity field (5) is transformed into the multidimensional Hopf's form of a generalized solution

$$v(x, t) = \frac{x - y(x, t)}{t} \tag{8}$$

of the Riemann equation

$$\frac{\partial v}{\partial t} + (v \cdot \nabla)v = 0. \tag{9}$$

The crucial observation which permits formulation of a rather simple approximate model of the density field ρ is that the vector function,

$$[\![y]\!](x, t) = x - v(x, t)t \tag{10}$$

(see (5a)), gives quasi-Lagrangian coordinates for particles in Burgers turbulence. Indeed, we can easily demonstrate that:

The vector function $[\![y]\!](x, t)$ satisfies equation

$$\frac{\partial [\![y]\!]}{\partial t} + (v \cdot \nabla)[\![y]\!] = \mu \Delta [\![y]\!], \qquad [\![y]\!](x, t = 0) = x. \tag{11}$$

Moreover, the mapping

$$x \longmapsto y = [\![y]\!](x, t) \tag{12}$$

establishes a one-to-one correspondence between coordinate frames y and x.

The validity of (11) is obtained by a direct verification. The second statement follows from the fact that the determinant $\|a_{ij}\|$ of the matrix

$$[a_{ij}] = [a_{ij}(x, t)] := \left[\frac{\partial [\![y_i]\!](x, t)}{\partial x_j} \right] = \left[\frac{1}{2\mu t}\Big([\![y_i y_j]\!] - [\![y_i]\!][\![y_j]\!] \Big) \right], \tag{13}$$

is positive for all $x \in \mathbf{R}^d$ since the last expression in (13) is the covariance matrix of the random vector $y/\sqrt{2\mu t}$ with the d-dimensional probability distribution function $f_\mu(y, x, t)$ defined in (3) which is nondegenerate for $\mu > 0$.

Remark 7.2. It follows from (13), that the matrix $[a_{ij}]$ is symmetric, i.e., $a_{ij} \equiv a_{ji}$. This implies that the vector field $[\![y]\!](x, t)$ is a potential field. Indeed, $[\![y]\!](x, t) = \nabla \psi(x, t)$, with

$$\psi(x, t) = \frac{x^2}{2} - S(x, t)t,$$

where

$$S(x, t) = -2\mu \log U(x, t)$$

is the potential of velocity field v, i.e., $v = \nabla S$.

Now, we are in a position to define our *analytical model* of the density field ρ. Suppose, that the initial density field,

$$\rho(\boldsymbol{x}, t = 0) = \rho_0(\boldsymbol{x}),$$

is known. Consequently, in the quasi-Lagrangian coordinates, evolution of the density field is described by the formula

$$\rho(\boldsymbol{x}, t) = \rho_0\Big([\![\boldsymbol{y}]\!](\boldsymbol{x}, t)\Big)\|a_{ij}(\boldsymbol{x}, t)\|. \tag{14}$$

Notice, that model density field (14) satisfies the mass conservation law. Indeed, in view of (11–12), the mass integral

$$m = \int \rho_0([\![\boldsymbol{y}]\!](\boldsymbol{x}, t))\|a_{ij}(\boldsymbol{x}, t)\| d\boldsymbol{x}$$

does not depend on t; one can change variables of integration from $\boldsymbol{x}$ to $\boldsymbol{y} = [\![\boldsymbol{y}]\!]$ to get that

$$m = \int \rho_0(\boldsymbol{y}) d\boldsymbol{y} = \text{ const.}$$

The above model is able to *analytically* predict the cellular intermittent structure of the density field associated with the Burgers flow. This program is carried out in the remainder of this chapter. Also, observe that numerical computations related to the model density field (14), which just involve evaluations of integrals of type (4), are much simpler than standard procedures (Shandarin, Zeldovich [16], and Weinberg, Gunn [20]) which compute the density evolution for the large-scale structure of the Universe by numerically solving a huge number of nonlinear differential equations

$$\frac{d\boldsymbol{X}_i}{dt} = \boldsymbol{v}(\boldsymbol{X}_i, t), i = 1, 2, \ldots, N,$$

where $\boldsymbol{v}(\boldsymbol{x}, t)$ is a known solution of the Burgers' equation.

Remark 7.3. Notice that in our approach equation (11) was *not* introduced as a softening, via addition of an extra Laplacian term, of the usual equation

$$\frac{\partial \boldsymbol{y}}{\partial t} + (\boldsymbol{v} \cdot \nabla)\boldsymbol{y} = 0, \qquad \boldsymbol{y}(\boldsymbol{x}, t = 0) = \boldsymbol{x}, \tag{15}$$

for exact Lagrangian coordinates (with which (11) coincides in the case $\mu = 0$), but it appeared as a natural and unavoidable description of the quasi-Lagrangian vector field $[\![\boldsymbol{y}]\!](\boldsymbol{x}, t)$ given in (10). Physical arguments justifying the replacement of the Lagrangian coordinate $\boldsymbol{y}(\boldsymbol{x}, t)$ by the quasi-Lagrangian $[\![\boldsymbol{y}]\!](\boldsymbol{x}, t)$ are provided in the next section.

19.8   Stochastic interpretation of the model density field

In this section we provide an elegant and physically convincing interpretation of the introduced above model density (19.7.14) field which takes into account the Brownian motion of passive tracer particles resulting from their collisions with the molecules of the surrounding medium.

In the theory of turbulent diffusion one often uses the following stochastic equation

$$\frac{d\boldsymbol{X}}{dt} = \boldsymbol{v}(\boldsymbol{X}, t) + \boldsymbol{\xi}(t), \tag{1}$$

$$\boldsymbol{X}(\boldsymbol{y}, t = 0) = \boldsymbol{y},$$

to describe the evolution of the passive tracer (see, e.g., Csanady (1980)). The vector field $\boldsymbol{X}(\boldsymbol{y}, t)$ stands for the coordinates of a passive tracer particle, and $\boldsymbol{v}(\boldsymbol{x}, t)$—the hydrodynamic velocity field of the continuous medium in which the particle is carried. In this section we will assume that the field, $\boldsymbol{v}(\boldsymbol{x}, t)$, is deterministic, twice continuously differentiable everywhere, with bounded first spatial derivatives:

$$\left| \frac{\partial v_i}{\partial x_j} \right| < M < \infty, \qquad \boldsymbol{x} \in \mathbf{R}^d.$$

The white noise stochastic vector process $\boldsymbol{\xi}(t)$ in (1), which reflects the influence of random collisions of passive tracer particles with molecules of the surrounding medium, will be assumed to be zero-mean, Gaussian, delta-correlated, with the correlation matrix

$$\langle \xi_i(t)\xi_j(t + \tau) \rangle = 2\mu \delta_{ij}\delta(\tau), \tag{2}$$

where $\langle . \rangle$ denotes the averaging over the statistical ensemble of the realization of the process $\boldsymbol{\xi}(t)$. In the case of a continuous medium at rest ($\boldsymbol{v} \equiv 0$), equation (1) implies that $\boldsymbol{X}$ is the usual Brownian motion, with

$$\langle \boldsymbol{X} \rangle = \boldsymbol{y}, \qquad \langle (\boldsymbol{X} - \langle \boldsymbol{X} \rangle)^2 \rangle = 6\mu t. \tag{2a}$$

Now, consider the probability distribution function

$$\mathcal{P}(\boldsymbol{x}; t | \boldsymbol{y}) = \langle \delta(\boldsymbol{X}(\boldsymbol{y}, t) - \boldsymbol{x}) \rangle \tag{3}$$

of the coordinates of the passive tracer particle. It is a well-known fact in the theory of Markov processes that the above distribution function satisfies the forward Kolmogorov equation

$$\frac{\partial \mathcal{P}}{\partial t} + \nabla(\boldsymbol{v}\mathcal{P}) = \mu \Delta \mathcal{P}, \tag{4}$$

$$\mathcal{P}(\boldsymbol{x}; t = 0 | \boldsymbol{y}) = \delta(\boldsymbol{x} - \boldsymbol{y}).$$

It is also well known that if the initial $(t = 0)$ density field is deterministic, and equal to $\rho_0(\boldsymbol{x})$, then the mean density $\rho(\boldsymbol{x}, t)$ of the passive tracer at time t is given by the formula

$$\rho(\boldsymbol{x}, t) = \int \rho_0(\boldsymbol{y}) \mathcal{P}(\boldsymbol{x}; t | \boldsymbol{y}) \, d\boldsymbol{y}. \tag{5}$$

Multiplying equation (4) by $\rho_0(\boldsymbol{y})$, and integrating it over all $\boldsymbol{y}$'s, we obtain the equation,

$$\frac{\partial \rho}{\partial t} + \nabla(\boldsymbol{v}\rho) = \mu \Delta \rho, \tag{6}$$

$$\rho(\boldsymbol{x}; t = 0 | \boldsymbol{y}) = \rho_0(\boldsymbol{x}),$$

for the density field averaged over the ensemble of realizations of the process $\boldsymbol{\xi}(t)$.

Notice that the above equation differs from the classical hydrodynamic continuity equation by the "superfluous" diffusion term, which takes into account the Brownian motion of the particle due to random molecular collisions. It is useful to recall that the diffusive term in (6) takes into account the discrete, molecular structure of the medium, that is completely ignored in the hydrodynamic derivation of the continuity equation.

At this point it may be worthwhile to provide another, equivalent to (5), form of the density of the passive tracer, by performing averaging over the statistical ensemble of random molecular diffusions. For that purpose observe that if the conditions imposed above on the velocity field $\boldsymbol{v}(\boldsymbol{x}, t)$ are satisfied, then for each separate realization of the process $\boldsymbol{\xi}(t)$, equality

$$\boldsymbol{x} = \boldsymbol{X}(\boldsymbol{y}, t) \tag{7}$$

defines a continuously differentiable and one-to-one mapping of $\boldsymbol{y} \in \mathbf{R}^d$ onto $\boldsymbol{x} \in \mathbf{R}^d$. Consequently, there exists an inverse mapping

$$\boldsymbol{y} = \boldsymbol{y}(\boldsymbol{x}, t), \tag{8}$$

with the same properties, and the Jacobian

$$\|\tilde{a}_{ij}\|, \qquad \tilde{a}_{ij} = \frac{\partial y_i(\boldsymbol{x}, t)}{\partial x_j}, \tag{9}$$

is continuous, bounded everywhere, and strictly positive:

$$0 < \|\tilde{a}_{ij}\| < \infty.$$

Under the above conditions, the following distribution-theoretic formula is valid:

$$\delta(\boldsymbol{X}(\boldsymbol{y}, t) - \boldsymbol{x}) = \|\tilde{a}_{ij}\| \delta(\boldsymbol{y}(\boldsymbol{x}, t) - \boldsymbol{y}). \tag{10}$$

Averaging this equality over the ensemble of realization of $\boldsymbol{\xi}(t)$ we arrive at another useful expression for the probability distribution (3):

$$\mathcal{P}(\boldsymbol{x}; t | \boldsymbol{y}) = \langle \|\tilde{a}_{ij}\| \delta(\boldsymbol{y} - \boldsymbol{y}(\boldsymbol{x}, t)) \rangle. \tag{10a}$$

Substituting it into (5) and using the basic property of the Dirac delta, we arrive at the promised expression for the density of the passive tracer averaged over the ensemble of $\boldsymbol{\xi}(t)$:

$$\rho(\boldsymbol{x}, t) = \langle \|\tilde{a}_{ij}\| \rho_0(\boldsymbol{y}(\boldsymbol{x}, t)) \rangle \tag{11}$$

which is an alternative to (5).

Using the *mean field approach* let us replace this exact equation by an approximate equation. The approach permits the drastic replacement of the averages of functions of random arguments by functions of their averages. For example,

$$\left\langle f\left(\boldsymbol{y}(\boldsymbol{x}, t), \frac{\partial y_i}{\partial x_j}\right) \right\rangle \Longrightarrow f\left(\langle \boldsymbol{y} \rangle, \frac{\partial \langle y_i \rangle}{\partial x_j}\right).$$

Applying the mean field approach to the right-hand side (11) we arrive at the equality

$$\rho(\boldsymbol{x}, t) = \|a_{ij}\| \rho_0(\langle \boldsymbol{y}(\boldsymbol{x}, t) \rangle), \tag{12}$$

where

$$\|a_{ij}\| = \|\tilde{a}_{ij}\| \big|_{\boldsymbol{y} = \langle \boldsymbol{y} \rangle} = \left\| \frac{\partial \langle y_i \rangle}{\partial x_j} \right\|.$$

Leaving aside the question of the degree of validity of the mean field approach, let us turn our attention to the fact that the equality (12) formally coincides, up to the replacement of $\langle \boldsymbol{y} \rangle$ by $[\![\boldsymbol{y}]\!]$, with equality (19.7.14). This suggests that the proposed stochastic interpretation of the model density field is as follows: *Model density field* (19.7.14) *coincides with the average (with respect to the Brownian motion) of the passive tracer density, computed by the mean field approach.* To convince ourselves about the validity of this statement it suffices to observe that:

The statistical average $\langle \boldsymbol{y}(\vec{x}, t) \rangle$ *of the random mapping* (8) *satisfies equation* (11).

Indeed, consider an auxiliary function (Wronskian)

$$J(\boldsymbol{y}, t) = \left\| \frac{\partial X_i(\boldsymbol{y}, t)}{\partial y_j} \right\|, \tag{13}$$

satisfying equation

$$\frac{dJ}{dt} = J(\nabla \cdot \boldsymbol{v}), \qquad J(\boldsymbol{y}, t = 0) = 1, \tag{14}$$

where

$$(\nabla \cdot \boldsymbol{v}) = \frac{\partial v_i(\boldsymbol{X}(\boldsymbol{y}, t), t)}{\partial X_i}.$$

Clearly, the vector $(\boldsymbol{X}(\boldsymbol{y},t), J(\boldsymbol{y},t))$ forms a $(d+1)$-dimensional Markov process with joint probability distribution function

$$\mathcal{P}(\boldsymbol{x}, j; t|\boldsymbol{y}) = \langle \delta(\boldsymbol{X}(\boldsymbol{y},t) - \boldsymbol{x})\delta(J(\boldsymbol{y},t) - j)\rangle, \tag{15}$$

satisfying the forward Kolmogorov equation

$$\frac{\partial \mathcal{P}}{\partial t} + \nabla(\boldsymbol{v}(\boldsymbol{x},t)\mathcal{P}) + (\nabla \cdot \boldsymbol{v}(\boldsymbol{x},t))\frac{\partial}{\partial j}(j\mathcal{P}) = \mu\Delta\mathcal{P}, \tag{16}$$

$$\mathcal{P}(\boldsymbol{x}, j; t = 0|\boldsymbol{y}) = \delta(\boldsymbol{x} - \boldsymbol{y})\delta(j - 1).$$

Using the obvious identity

$$J(\boldsymbol{y},t)\|\tilde{a}_{ij}\| \Big|_{\boldsymbol{x} = \boldsymbol{X}(\boldsymbol{y},t)} \equiv 1,$$

we can rewrite the relation (10) in the form,

$$\delta(\boldsymbol{y}(\boldsymbol{x},t) - \boldsymbol{y}) = J(\boldsymbol{y},t)\delta(\boldsymbol{X}(\boldsymbol{y},t) - \boldsymbol{x}).$$

Therefore, it follows that the probability distribution function,

$$\mathcal{Q}(\boldsymbol{y}; t|\boldsymbol{x}) = \langle \delta(\boldsymbol{y}(\boldsymbol{x},t) - \boldsymbol{y})\rangle,$$

of the vector stochastic process $\boldsymbol{y}(\boldsymbol{x},t)$, is related to the probability distribution (15) by the equality

$$\mathcal{Q}(\boldsymbol{y}; t|\boldsymbol{x}) = \int j\mathcal{P}(\boldsymbol{x}, j; t|\boldsymbol{y})\, dj.$$

Multiplying equation (16) by j, and integrating it over j, we arrive at the conclusion that $\mathcal{Q}(\boldsymbol{y}; t|\boldsymbol{x})$ satisfies the following *backward Kolmogorov equation*:

$$\frac{\partial \mathcal{Q}}{\partial t} + (\boldsymbol{v}(\boldsymbol{x},t) \cdot \nabla)\mathcal{Q} = \mu\Delta\mathcal{Q}, \tag{17}$$

$$\mathcal{Q}(\boldsymbol{y}; t|\boldsymbol{x}) = \delta(\boldsymbol{y} - \boldsymbol{x}).$$

The above equation implies, in particular, that the statistical average

$$\langle \boldsymbol{y}(\boldsymbol{x},t)\rangle = \int \boldsymbol{y}\mathcal{Q}(\boldsymbol{y}; t|\boldsymbol{x})\, d\boldsymbol{y}, \tag{18}$$

which is of principal interest to us, is a solution of the Cauchy problem

$$\frac{\partial \langle \boldsymbol{y}\rangle}{\partial t} + (\boldsymbol{v} \cdot \nabla)\langle \boldsymbol{y}\rangle = \mu\Delta\langle \boldsymbol{y}\rangle, \tag{19}$$

$$\langle \boldsymbol{y}(\boldsymbol{x}, t = 0)\rangle = \boldsymbol{x}.$$

This concludes the verification of the statement italicized above. Note that its formulation (replacing $\langle \boldsymbol{y}\rangle$ by $[\![\boldsymbol{y}]\!]$) is similar to the formulation of the first part of statement (19.7.11–12). The essential difference is that the above field $\vec{v}(\boldsymbol{x},t)$ need not be a solution of the Burgers' equation (19.7.1).

Remark 8.1. In the case velocity field $\boldsymbol{v}(\boldsymbol{x}, t)$ satisfies the Burgers' equation with the coefficient μ identical to the coefficient of molecular diffusion in (2), then the average field $\langle \boldsymbol{y}(\boldsymbol{x}, t) \rangle$ is expressed by the velocity field with the help of a simple formula (19.7.10)). This serendipitous coincidence provided us with analytical advantages of the suggested model density field.

Remark 8.2. We have derived equation (19) utilizing equality (18), where Q satisfied equation (17). In the case of the Burgers' velocity field $\boldsymbol{v}(\boldsymbol{x}, t)$, where $[\![\boldsymbol{y}]\!](\boldsymbol{x}, t)$ (see (19.7.5a)) satisfies equation (19.7.11), it is natural to assume that f_μ, like Q, also satisfies equation (17). A direct substitution shows that, indeed, this is the case. Hence, the spatial probability distribution function f_μ, like the density field itself, has a clear-cut stochastic interpretation: $f_\mu(\boldsymbol{y}; \boldsymbol{x}, t)$ *is the probability distribution of the Lagrangian coordinates of passive tracer particle driven by the Burgers' velocity field and subjected to Brownian motion.*

Remark 8.3. Another important quantity in the theory of turbulent diffusion is the *mean concentration*

$$C(\boldsymbol{x}, t) = \int C_0(\boldsymbol{y}) Q(\boldsymbol{y}; t | \boldsymbol{x}) \, d\boldsymbol{y},$$

of the passive tracer which satisfies equation

$$\frac{\partial C}{\partial t} + (\boldsymbol{v}(\boldsymbol{x}, t) \nabla) C = \mu \Delta C,$$

$$C(\boldsymbol{x}, t = 0) = C_0(\boldsymbol{x}).$$

Recall that the density ρ is proportional to the number of passive tracer particles in the unit volume, whereas the concentration C is proportional to the *ratio* of the number of passive tracer particles in the unit volume to the number of particles of the surrounding fluid in the same volume. In view of the preceding remark, in the case of Burgers' velocity field $\boldsymbol{v}(\boldsymbol{x}, t)$, we also obtain an *exact* solution

$$C(\boldsymbol{x}, t) = \int C_0(\boldsymbol{y}) f_\mu(\boldsymbol{y}; \boldsymbol{x}, t) \, d\boldsymbol{y}$$

of the equation for the concentration field.

Remark 8.4. The 1-D case occupies a special place. Then the determinant $\|\tilde{a}\|$ is tied to $y(x, t)$ via a simple linear equation

$$\|\tilde{a}\| = \frac{\partial y(x, t)}{\partial x}, \tag{20}$$

which implies that in the 1-D case we have the following identity:

$$\langle \|\tilde{a}\| \rangle \equiv \|a\|.$$

If, in addition, the initial density is the same everywhere ($\rho_0 = const$), then the "real" density

$$\rho(x,t) = \rho_0 \int \mathcal{P}(x;t|y)\, dy$$

is *exactly* equal to the density calculated via the mean field approach:

$$\rho(x,t) = \rho_0 \|a\| = \rho_0 \frac{\partial \langle y\rangle (x,t)}{\partial x}. \tag{20a}$$

19.9 2-D cellular structures

In general, model density field $\rho(\boldsymbol{x},t)$ described in (19.7.14) may be found only by numerical calculations of integrals similar to (19.7.4). However, in the most interesting for astrophysical applications case of small viscosity μ (Shandarin and Zeldovich [16], Weinberg and Gunn [20], Vergassola et al. [21]), which corresponds to the strongly nonlinear regime of evolution of the velocity and density fields, it is possible to obtain a physically acceptable analytical approximation for $\rho(\boldsymbol{x},t)$. To escape unwieldy formulas which arise in the 3-D case, we shall demonstrate this fact on the 2-D case. Hence, for the remainder of this section, $\boldsymbol{x} \in \mathbf{R}^2$.

Assume that the initial velocity field potential $S_0(\boldsymbol{x})$ has second partial derivatives which are bounded from below. This implies that there exists a $t_1 > 0$ such that, if $0 < t < t_1$, then for any given $\boldsymbol{x}$ function $\phi(\boldsymbol{y},\boldsymbol{x},t)$ defined in (19.7.2) has only one local minimum. Let's denote coordinates of this minimum by $\boldsymbol{y}(\boldsymbol{x},t)$. As far as the probability distribution function $f_\mu(\boldsymbol{y};\boldsymbol{x},t)$ is concerned, the existence of a unique minimum of $\phi(\boldsymbol{y},\boldsymbol{x},t)$ means that, at any $\boldsymbol{x}$, the Gaussian approximation is asymptotically valid for small μ. It corresponds to an application for $f_\mu(\boldsymbol{y};\boldsymbol{x},t)$ of the steepest descent method for calculation of integrals similar to (19.7.4). The method justifies an approximation of $\phi(\boldsymbol{y},\boldsymbol{x},t)$ by the first three terms of its Taylor expansion:

$$\phi(\boldsymbol{y},\boldsymbol{x},t) \approx \phi\Big(\boldsymbol{y}(\boldsymbol{x},t),\boldsymbol{x},t\Big) + \vec{\kappa}\cdot\left(\boldsymbol{v}_0\Big(\boldsymbol{y}(\boldsymbol{x},t)\Big) + \frac{\boldsymbol{y}(\boldsymbol{x},t)-\boldsymbol{x}}{t}\right) + \frac{1}{2t}A_{ij}(\boldsymbol{x},t)\kappa_i\kappa_j, \tag{1}$$

where

$$A_{ij}(\boldsymbol{x},t) = \left(\delta_{ij} + t\frac{\partial^2 S_0(\boldsymbol{y})}{\partial y_i \partial y_j}\right)\Bigg|_{\boldsymbol{y}=\boldsymbol{y}(\boldsymbol{x},t)},$$

and $\vec{\kappa} = \boldsymbol{y} - \boldsymbol{y}(\boldsymbol{x},t)$. Since $\boldsymbol{y}(\boldsymbol{x},t)$ are coordinates of a minimum of a smooth function $\phi(\boldsymbol{y},\boldsymbol{x},t)$ over $\boldsymbol{y}$'s, we have that

$$\boldsymbol{v}_0\Big(\boldsymbol{y}(\boldsymbol{x},t)\Big) + \frac{\boldsymbol{y}(\boldsymbol{x},t)-\boldsymbol{x}}{t} = 0. \tag{2}$$

Consequently, in this approximation, the spatial distribution function (19.7.3) is described by the asymptotic formula

$$f_\mu(\boldsymbol{y};\boldsymbol{x},t) \approx \frac{1}{4\pi\mu t}\sqrt{j(\boldsymbol{x},t)}\exp\left(-\frac{1}{4\mu t}A_{ij}(\boldsymbol{x},t)\kappa_i\kappa_j\right). \tag{3}$$

Here, again $\vec{\kappa} = \boldsymbol{y} - \boldsymbol{y}(\boldsymbol{x},t)$, where $\boldsymbol{y}(\boldsymbol{x},t)$ is a root of equation (2), and

$$j(\boldsymbol{x},t) = J(\boldsymbol{y},t)\Big|_{\boldsymbol{y}=\boldsymbol{y}(\boldsymbol{x},t)} = \|A_{ij}\|\Big|_{\boldsymbol{y}=\bar{\boldsymbol{y}}(\boldsymbol{x},t)}$$

$$= \left[\left(1+t\frac{\partial^2 S_0(\boldsymbol{y})}{\partial y_1^2}\right)\left(1+t\frac{\partial^2 S_0(\boldsymbol{y})}{\partial y_2^2}\right) - t^2\left(\frac{\partial^2 S_0(\boldsymbol{y})}{\partial y_1 \partial y_2}\right)^2\right]\Bigg|_{\boldsymbol{y}=\boldsymbol{y}(\boldsymbol{x},t)}. \tag{4}$$

It is worth recalling that, as $\mu \to 0$, function f_μ (3) weakly converges to the Dirac delta (19.7.7). Nevertheless, for the density field itself, it follows from (19.7.13–14) that

$$\rho(\boldsymbol{x},t) = \frac{\rho_0\left(\llbracket\boldsymbol{y}\rrbracket(\boldsymbol{x},t)\right)}{(2\mu t)^2}\left(\llbracket(y_1 - \llbracket y_1\rrbracket)^2\rrbracket\llbracket(y_2 - \llbracket y_2\rrbracket)^2\rrbracket - \llbracket(y_1 - \llbracket y_1\rrbracket)(y_2 - \llbracket y_2\rrbracket)\rrbracket^2\right), \tag{5}$$

where, as above, the square brackets signify the "spatial average" with respect to the probability distribution function (19.7.3), and a calculation of the "thickness" of distribution (3) is of principal importance. In particular, we have the following result:

If $t < t_1$ then the limit

$$\lim_{\mu\to 0}\rho(\boldsymbol{x},t) = \frac{\rho_0(\boldsymbol{y}(\boldsymbol{x},t))}{j(\boldsymbol{x},t)}, \tag{6}$$

of the model density field (5) exists, where $\boldsymbol{y}(\boldsymbol{x},t)$ are the Lagrangian coordinates and $j(\boldsymbol{x},t)$ is the Jacobian of the transformation from Lagrangian to Eulerian coordinates of the continuous medium, the velocity field thereof satisfies the Riemann equation (19.7.9) with the initial condition (19.7.1).

Indeed, direct calculation shows that the desired limit expression for the model density field is described by the right-hand side of equation (6), where $\boldsymbol{y}(\boldsymbol{x},t)$ are the coordinates of the minimum of function ϕ in formula (19.7.2) and $j(\boldsymbol{x},t)$ is given by expression (4). Now, it suffices to prove that the fields $\boldsymbol{y}(\boldsymbol{x},t)$ and $j(\boldsymbol{x},t)$ defined in this fashion are indeed, respectively, the fields of Lagrangian coordinates and of the Jacobian. For that purpose, observe that the unique (for $t < t_1$) solution of the Riemann equation (19.7.9) can be written in the form

$$\boldsymbol{v}(\boldsymbol{x},t) = \boldsymbol{v}_0(\boldsymbol{y}(\boldsymbol{x},t)),$$

where $\boldsymbol{y}(\boldsymbol{x},t)$ is the field of Lagrangian coordinates. Comparing the last equation with (2), and (2) with equality (19.7.8), we realize that the coordinates of the minimum of function (19.7.2) are indeed simultaneously the Lagrangian coordinates.

Furthermore, note that the Riemann equation describes the field of velocities in the continuous medium all the particles thereof move with uniform speed on straight lines. This means that the Eulerian and Lagrangian coordinates of such a medium are connected by equality

$$\boldsymbol{x} = \boldsymbol{y} + \boldsymbol{v}_0(\boldsymbol{y})t, \qquad \boldsymbol{v}_0(\boldsymbol{y}) = \nabla S_0(\boldsymbol{y}).$$

The Jacobian

$$J(\boldsymbol{y}, t) = \left\| \frac{\partial x_i}{\partial y_j} \right\|$$

coincides with the expression (4) in the 2-D case under consideration. Thus the above statement about the model density field (6) has been verified.

Remark 9.1. The above proof can be also easily carried out in the case of $\boldsymbol{x} \in \mathbf{R}^d$ for arbitrary dimension d.

Remark 9.2. The time instant t_1, which appeared in the above considerations, is the infimum of positive roots of the equation $J(\boldsymbol{y}, t) = 0$ over all $\boldsymbol{y}$.

Remark 9.3. The right-hand side of equation (6) is familiar for the physicists and has a transparent physical meaning: The density at a given point $\boldsymbol{x}$ is equal to the initial density at this point of the continuous medium divided by the Jacobian $j(\boldsymbol{x}, t)$ which describes the influence of squeezing and stretching of the fluid.

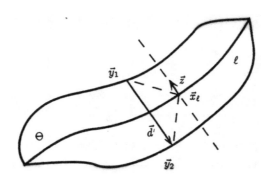

FIGURE 19.9.1.
A typical example of curve ℓ—a 2-D analog of 3-D "pancake structures," and corresponding region Θ bounded by curves $\boldsymbol{y}_1(\boldsymbol{x}_\ell, t)$ and $\boldsymbol{y}_2(\boldsymbol{x}_\ell, t)$. In the inviscid limit, the region Θ has a vivid physical interpretation: All particles inside it at the initial time $t = 0$, and driven by the Burgers' velocity field, end up on the "pancake" curve ℓ at time t.

To make the subsequent analysis more transparent we will assume now that the initial density is the same at all points $\boldsymbol{x}$, i.e., that

$$\rho_0(\boldsymbol{x}) \equiv \rho_0 = \text{ const}, \qquad (7)$$

and, under these circumstances, we will analyze evolution of the density fields. This case corresponds, for instance, to the initial stage of development of the gravitation instability of matter in the Universe (cf. e.g., Shandarin and Zeldovich [16], Weinberg and Gunn [20], Vergassola et al. [21].

For $0 < t < t_1$, the auxiliary field $J(\boldsymbol{y}, t)$ defined in (4) is positive for any $\boldsymbol{y}$. For $t > t_1$, there appear regions in the $\boldsymbol{y}$ plane, where $J(\boldsymbol{y}, t)$ is negative. At this stage, there appear on $\boldsymbol{x}$-plane curves ℓ, at all points $\boldsymbol{x}_\ell$ thereof function $\phi(\boldsymbol{y}, \boldsymbol{x}, t)$ defined in (19.7.2) has two minima of equal value. Let's mark the coordinates of these two minima as $\boldsymbol{y}_1(\boldsymbol{x}_\ell, t)$ and $\boldsymbol{y}_2(\boldsymbol{x}_\ell, t)$. As point $\boldsymbol{x}$ travels over curve ℓ, points $\boldsymbol{y}_1, \boldsymbol{y}_2$ draw a closed curve in the $\boldsymbol{y}$-plane which bounds a certain region Θ (see Fig. 19.9.1).

Curve ℓ shown on Fig. 19.9.1 represents a 2-D analog of three-dimensional "pancake" structures which appear in the large-scale distribution of matter in the Universe (see Gurbatov et al. (1991) and the astrophysical papers quoted above).

If μ is sufficiently small, the density in a small vicinity of these curves is very large and we can calculate the density field there via a bimodal approximation. In this approximation, function $f_\mu(\boldsymbol{y}; \boldsymbol{x}, t)$ from (19.7.3) is represented by a superposition of two unimodal distributions (3)

$$f_\mu(\boldsymbol{y}; \boldsymbol{x}, t) = \frac{f_1(\boldsymbol{y}; \boldsymbol{x}_\ell, t) + R f_2(\boldsymbol{y}; \boldsymbol{x}_\ell, t)}{1 + R}, \tag{8}$$

where

$$R = \sqrt{\frac{J_1}{J_2}} \exp\left(\frac{d z}{2\mu t}\right), \quad d(\boldsymbol{x}_\ell, t) = \boldsymbol{y}_2(\boldsymbol{x}_\ell, t) - \boldsymbol{y}_1(\boldsymbol{x}_\ell, t), \tag{9}$$

and where $z = \boldsymbol{x} - \boldsymbol{x}_\ell$ is a vector located on the dashed straight line r perpendicular to $\boldsymbol{x}_\ell$ (see Fig. 19.9.1). Functions f_1, J_1, f_2, J_2 are now obtained from (3) and (4), after substituting for $\boldsymbol{y}(\boldsymbol{x}, t)$ vectors $\boldsymbol{y}_1(\boldsymbol{x}_\ell, t)$ and $\boldsymbol{y}_2(\boldsymbol{x}_\ell, t)$, respectively.

Let's calculate, with the help of bimodal distribution (8), the density field

$$\rho(\boldsymbol{x}, t) = \rho_0 \|a_{ij}\| \tag{10}$$

in the vicinity of curve ℓ . The elements of matrix $[a_{ij}]$, after some calculations, can be expressed in the form

$$a_{ij} = \frac{1}{2\mu t} \left(\frac{b_{ij}^1 + R b_{ij}^2}{1 + R} + \frac{R}{(1 + R)^2} d_i d_j \right), \tag{11}$$

where

$$b_{11}(\boldsymbol{y}, t) = \frac{2\mu t}{J(\boldsymbol{y}, t)} \left(1 + t \frac{\partial^2 S_0(\boldsymbol{y})}{\partial y_2^2} \right),$$

$$b_{22}(\boldsymbol{y}, t) = \frac{2\mu t}{J(\boldsymbol{y}, t)} \left(1 + t \frac{\partial^2 S_0(\boldsymbol{y})}{\partial y_1^2} \right),$$

$$b_{12}(\boldsymbol{y}, t) = -\frac{2\mu t^2}{J(\boldsymbol{y}, t)} \frac{\partial^2 S_0(\boldsymbol{y})}{\partial y_1 \partial y_2}.$$

The superscripts in (11) indicate that in the formulas for b_{ij}, components of vector $\boldsymbol{y}$ must be replaced by components of vectors $\boldsymbol{y}_1(\boldsymbol{x}_\ell, t)$ and $\boldsymbol{y}_2(\boldsymbol{x}_\ell, t)$, respectively.

Substituting (11) into (10) we obtain that

$$\rho(\boldsymbol{x}, t) = \frac{\rho_1}{(1 + R^2)} + \frac{\rho_2}{(1 + 1/R)^2} + \frac{\rho_{12}}{(\sqrt{R} + 1/\sqrt{R})^2}$$

$$+ \frac{\rho_0}{2\mu} \frac{R}{(1 + R)^3} \left((\boldsymbol{d} \cdot \nabla_1)^2 \phi(\boldsymbol{y}_1, t) + R(\boldsymbol{d} \cdot \nabla_2)^2 \phi(\boldsymbol{y}_2, t) \right), \quad (12)$$

with

$$\phi(\boldsymbol{y}, t) = \phi(\boldsymbol{y}, 0, t) = \frac{y^2}{2t} + S_0(\boldsymbol{y}).$$

Here ρ_1, and ρ_2, are values of the density for different sides of pancake ℓ in the immediate neighborhood of the pancake, and ρ_{12} represents a certain additional "mixed" density:

$$\rho_{12} = \frac{\rho_0}{4\mu t^2} \left(b_{11}^1 b_{22}^2 + b_{11}^2 b_{22}^1 - 2b_{12}^1 b_{21}^2 \right).$$

The last summand in (12) is the "main" component of the density, which describes the high density of matter getting stuck in the vicinity of the "pancake rim." It can also be conveniently estimated by the "double delta function" approximation

$$f_\mu(\boldsymbol{y}; \boldsymbol{x}, t) = \frac{\delta(\boldsymbol{y} - \boldsymbol{y}_1(\boldsymbol{x}_\ell, t)) + R\delta(\boldsymbol{y} - \boldsymbol{y}_2(\boldsymbol{x}_\ell, t))}{1 + R}, \quad (13)$$

which is simpler than (8). It corresponds to neglecting lower-order densities $\rho_1, \rho_2, \rho_{12}$.

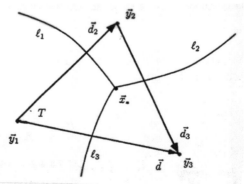

FIGURE 19.9.2.
Pancakes intersect at points $\vec{x}_*$, and nodes of very high density are created.

The next stage of the density field evolution is associated with appearance of points $\boldsymbol{x}_*$, where pancakes intersect and nodes of very high density are created (see Fig. 19.9.2). For a rough calculation of the density in a vicinity of these nodes we may use a "triple delta function" approximation:

$$f_\mu = \frac{\delta(\boldsymbol{y} - \boldsymbol{y}_1) + R_2\delta(\boldsymbol{y} - \boldsymbol{y}_2) + R_3\delta(\boldsymbol{y} - \boldsymbol{y}_3)}{1 + R_2 + R_3}, \tag{14}$$

where

$$R_2 = \sqrt{\frac{\rho_2}{\rho_1}} \exp\left(\frac{\boldsymbol{d}_2\boldsymbol{z}}{2\mu t}\right),$$

$$R_3 = \sqrt{\frac{\rho_3}{\rho_1}} \exp\left(\frac{\boldsymbol{d}_3\boldsymbol{z}}{2\mu t}\right),$$

with

$$\boldsymbol{d}_2 = \boldsymbol{y}_2 - \boldsymbol{y}_1, \quad \boldsymbol{d}_3 = \boldsymbol{y}_3 - \boldsymbol{y}_1, \quad \boldsymbol{d} = \boldsymbol{y}_3 - \boldsymbol{y}_2, \quad \boldsymbol{z} = \boldsymbol{x} - \boldsymbol{x}_*.$$

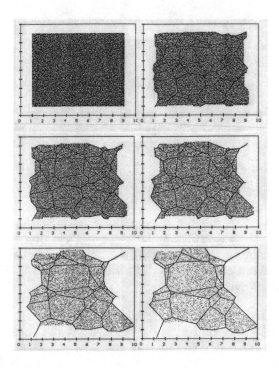

FIGURE 19.9.3.
Evolution of 2-D Burgers turbulence in the inviscid limit. The consecutive frames show the location of shock fronts at t = 0.0, 0.3, 0.6, 1.0, 2.0, 3.0 (from Janicki, Surgailis and Woyczyński [23].)

It follows from (14) that

$$a_{ij} = \frac{R_2 d_{2i} d_{2j} + R_3 d_{3i} d_{3j} + R_2 R_3 d_i d_j}{2\mu t (1 + R_2 + R_3)^2}. \tag{15}$$

Substituting this expression into (10) we get, after simple calculations, that

$$\rho(\boldsymbol{x}, t) = \frac{\rho_0 |T|^2 R_2 R_3}{\mu^2 t^2 (1 + R_2 + R_3)^3}. \tag{16}$$

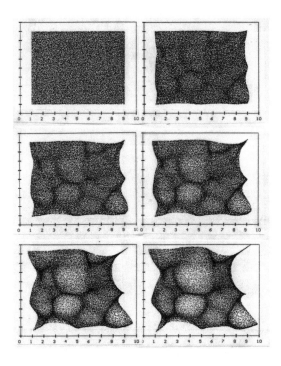

FIGURE 19.9.4.
Evolution of 2-D Burgers turbulence in the viscous case with shot-noise initial velocity data. The consecutive frames show the location of shock fronts at t = 0.0, 0.3, 0.6, 1.0, 2.0, 3.0 (from Janicki, Surgailis, and Woyczyński [23]).

Here $|T|$ is the area of triangle T with vertices at points $\boldsymbol{y}_1, \boldsymbol{y}_2$, and $\boldsymbol{y}_3$ (see Fig. 19.9.2). The total mass concentrated at the node is obtained by integration of the density field (16) over $\boldsymbol{x}$:

$$m = \int \rho(\boldsymbol{x}, t) d\boldsymbol{x} = \rho_0 |T|. \tag{17}$$

This indicates that the passive tracer mass, which at time $t = 0$ was uniformly distributed over triangle T, eventually concentrates at such a node.

We illustrate this phenomenon in Fig. 19.9.3 which presents results of simulations from Janicki, Surgailis, and Woyczyński [23] for 100,000 particles with the initial uniform mass distribution. The figure shows the evolution of cellular structure of the true density field in Burgers velocity field. The lines actually indicate the location of shock fronts in the velocity field at different epochs. For comparison we also present in Fig. 19.9.4 a similar evolution of cellular structure in the viscous case.

19.10 Exercises

1. Verify that if the random force potential $U(\boldsymbol{x}, t)$ is statistically homogeneous in space then the necessary condition (19.1.16) for the existence of stationary regime in forced Burgers turbulence is satisfied.

2. Check that the Hopf–Cole substitution (19.1.22) transforms the non-linear Burgers' equation (19.1.3) into a linear Schrōdinger-type equation (19.1.23).

3. Prove that if the velocity field $\boldsymbol{v}(\boldsymbol{x}, t)$ satisfying the Burger's equation with external forces $\boldsymbol{f}(\boldsymbol{x}, t)$, then the velocity potential $S(\boldsymbol{x}, t)$ satisfies the Hamilton–Jacobi equation (19.1.3) with the potential of external forces $\boldsymbol{f}$.

4. End the proof of the formula (19.9.6) to the general d-dimensional case.

Chapter 20

Probability Distributions of Passive Tracers in Randomly Moving Media

The main goal of the present chapter is to investigate probabilistic and spatiotemporal characteristics of the passive tracer in a randomly moving medium with random velocity field $v(x, t)$. Both, the case of compressible, and of incompressible ($\nabla \cdot v \equiv 0$) media will be considered. The material is based on the authors' paper [1]. In the previous chapter we have investigated the passive tracer transport in randomly moving media governed by a simplified hydrodynamic model of Burgers turbulence. The full program of such investigation should include prior recovery of the statistical properties of the velocity field $v(x, t)$ satisfying fully nonlinear equations of the hydrodynamic type such as the Navier–Stokes equation for an incompressible fluid, which however, is beyond the scope of this book.[1]

20.1 The basic model

A study of evolution of statistical characteristics of random fields of passive tracers in chaotically moving media has been of a continuing interest for a long time (see e.g., Batchelor [4], Kreichnan [5], Csanady [6], Avellaneda, Majda [2], Majda [3]). The interest was motivated by the fact that the knowledge of those characteristics is important for solution of a number of applied problems, including the environmental problems of pollution of atmosphere and the oceans. Most of the publications devoted to the study of transport processes and diffusion of the passive tracer restrict themselves

---

[1]See e.g., Avellaneda, and Majda [2], Majda [3].

© Springer International Publishing AG, part of Springer Nature 2018
A. I. Saichev and W. A. Woyczynski, *Distributions in the Physical and Engineering Sciences, Volume 3*, Applied and Numerical Harmonic Analysis, https://doi.org/10.1007/978-3-319-92586-8_20

to efforts to derive and analyze equations for the mean values of the passive tracer density, or to an analysis of the diffusion of a single particle of the tracer (see e.g., Davis [7], Lipscomb et al. [8], Careta et al. [9]). However, these characteristics are not able to describe various subtle peculiarities of the density fields of passive tracers, such as creation of complex, layered structures of the passive tracer field, density thereof rapidly changes with a slight displacement of the point of observation. These, and other phenomena are more adequately described by probability distributions, and their analysis occupies the significant portion of this chapter.

In experimental studies of behavior of passive tracers in turbulent flows two, basically different, types of methods of measurement are utilized. The first type uses fixed, immobile (or moving along a determined in advance path) sensors which measure Eulerian characteristics of the chaotically moving medium.[2] The second type uses atmospheric probes, meteorological balloons, buoys and floating devices that move with the medium, " frozen" into the flow, which measure Lagrangian characteristics of such fields such as velocity and density fields. For that reason, in the present chapter we pay great attention to a comparison of statistical properties of random fields in Eulerian and Lagrangian representations, and also to finding connections between analogous Lagrangian and Eulerian statistical characteristics of the randomly moving medium itself. The recovery of such connections is important not only for the sake of comparison of results of the above-mentioned two types of experiments. These connecting formulas are also useful in the theoretical analysis of a randomly moving medium, when it may be more convenient to conduct analysis in one (say Lagrangian) representation, but the final results are needed in the other (say, Eulerian) representation.

We will assume that the statistical properties of the random velocity field $\boldsymbol{v}(\boldsymbol{x}, t)$ are given in advance. More precisely, we will work under the assumption that $\boldsymbol{v}(\boldsymbol{x}, t)$ *is a zero-mean, smooth isotropic and homogeneous in space, and stationary in time, Gaussian random field with a known covariance matrix*

$$\langle v_i(\boldsymbol{x}, t) v_j(\boldsymbol{x} + \boldsymbol{s}, t + \tau) \rangle = b_{ij}(\boldsymbol{s}, \tau). \tag{1a}$$

Recall, that the covariance matrix of a statistically isotropic and homogeneous three-dimensional velocity field can be represented in the form

$$b_{ij}(\boldsymbol{s}, \tau) = \kappa \left[F_\perp(s, \tau) \delta_{ij} + \frac{s_i s_j}{s} \frac{d}{ds} F_\perp(s, \tau) \right] \tag{1b}$$

$$+ (1 - \kappa) \left[G_\parallel(s, \tau) \delta_{ij} + \left(\delta_{ij} - \frac{s_i s_j}{s^2} \right) \frac{s}{2} \frac{d}{ds} G_\parallel(s, \tau) \right],$$

where $F_\perp(s, \tau)$ corresponds to the potential component of the velocity field, and $G_\parallel(s, \tau)$—to the rotational component. Coefficient $0 \leq \kappa \leq 1$ takes

---

[2]See, for example. T. Day, W. Hickey, B. Parakkal, and W.A. Woyczynski [10], where transport of the oil particles caused by the Deepwater Horizon 2010 disastrous Gulf of Mexico spill was examined.

into account the relative contribution of the potential component, and of the rotational component, to the dispersion of velocity fluctuations. For $\kappa = 1$, the velocity field is purely potential, and for $\kappa = 0$, it is purely rotational.

Functions $F_\perp$ and $G_\parallel$ have a transparent physical meaning: $F_\perp$ is the transversal correlation of the potential component of the velocity field, and $G_\parallel$ is the longitudinal correlation of its rotational component. The longitudinal correlation of the potential component is expressed in terms of the transversal correlation by the formula

$$F_\parallel = \frac{d}{ds}(sF_\perp),$$

and the transversal correlation of the rotational component is computable from its longitudinal correlation through the Karman relation

$$G_\perp = \frac{1}{2s}\frac{d}{ds}(s^2 G_\parallel).$$

In what follows, we will often use the simplest "Gaussian" model for functions $F_\perp$ and $G_\parallel$:

$$F_\perp(s,\tau) = G_\parallel(s,\tau) = \frac{1}{3}\sigma_v^2 \exp\left(-\frac{s^2}{2l_v^2} - \frac{\tau^2}{2\tau_v^2}\right). \tag{1c}$$

There are reasons to expect that even such a crude model of fluctuations of the velocity field permits adequate description of numerous qualitative and quantitative features of evolution of the passive tracer in flows of randomly moving media.

As is well known, in the Eulerian coordinate system, the density field $\rho(\boldsymbol{x},t)$ of the passive tracer satisfies the continuity equation

$$\frac{\partial\rho}{\partial t} + \boldsymbol{\nabla}\cdot\boldsymbol{v}\rho = \kappa\Delta\rho, \tag{2}$$

$$\rho(\boldsymbol{x},t=0) = \rho_0(\boldsymbol{x}), \tag{3}$$

where $\rho_0(\boldsymbol{x})$ is the initial density field of the passive tracer and κ is the coefficient of molecular diffusion. Throughout this chapter we will restrict ourselves to the case of the "massive enough" tracer particles which, practically, are not affected by collisions with the particles of the surrounding medium, and at the same time are "light enough" to be carried without resistance by the flow of the medium. In such a case, one can neglect the molecular diffusion term in (2) and use a more simple "standard" continuity equation

$$\frac{\partial\rho}{\partial t} + \boldsymbol{\nabla}\cdot\boldsymbol{v}\rho = 0. \tag{4}$$

The density of the passive tracer is directly connected with the laws of motion of individual particles. To recall this connection, let us write a random vector differential equation

$$\frac{d\boldsymbol{X}}{dt} = \boldsymbol{v}(\boldsymbol{X},t), \quad \boldsymbol{X}(\boldsymbol{y},t=0) = \boldsymbol{y}, \tag{5}$$

describing the random position $\boldsymbol{X}(\boldsymbol{y},t)$ of a particle with initial (Lagrangian) coordinates $\boldsymbol{y}$. It is equivalent to (4) (but not to (2)). Whereas, continuity equations (2) and (4) are the basic relationships describing evolution of the passive tracer density in the Eulerian system of coordinates, equation (5) constitutes the basic equation of motion in the Lagrangian coordinate system. If one knows the laws of motion of a tracer particle with arbitrary Lagrangian coordinates $\boldsymbol{y}$, one can write the solution of the continuity equation (4) in the form

$$\rho(\boldsymbol{x},t) = \int \rho_0(\boldsymbol{y})\delta\Big(\boldsymbol{x} - \boldsymbol{X}(\boldsymbol{y},t)\Big)d\boldsymbol{y}. \tag{6}$$

Notice, that under relatively weak conditions imposed on the velocity field $\boldsymbol{v}(\boldsymbol{x},t)$, the vector function

$$\boldsymbol{x} = \boldsymbol{X}(\boldsymbol{y},t) \tag{7}$$

gives a one-to-one smooth mapping $\boldsymbol{x} \leftrightarrow \boldsymbol{y}$. In addition, the formula

$$\delta\Big(\boldsymbol{x} - \boldsymbol{X}(\boldsymbol{y},t)\Big) = \frac{1}{J(\boldsymbol{y},t)}\delta\Big(\boldsymbol{y} - \boldsymbol{Y}(\boldsymbol{x},t)\Big). \tag{8}$$

holds true, where the vector function

$$\boldsymbol{y} = \boldsymbol{Y}(\boldsymbol{x},t), \tag{9}$$

is the inverse function to function (7), and

$$J(\boldsymbol{y},t) = \left|\frac{\partial \boldsymbol{X}(\boldsymbol{y},t)}{\partial \boldsymbol{y}}\right| \tag{10}$$

is the Jacobian of the transformation from Lagrangian coordinates $\boldsymbol{y}$ to Eulerian coordinates $\boldsymbol{x}$. The Jacobian satisfies equation

$$\frac{dJ}{dt} = u(\boldsymbol{X},t)J, \quad J(\boldsymbol{y},t=0) = 1, \tag{11}$$

where the auxiliary scalar field

$$u(\boldsymbol{x},t) = \boldsymbol{\nabla} \cdot \boldsymbol{v}(\boldsymbol{x},t), \tag{12}$$

is identically equal to 0 in an incompressible fluid.

Substituting (8) into (6), we obtain another useful expression for the density of passive tracer:

$$\rho(\boldsymbol{x},t) = \frac{\rho_0(\boldsymbol{Y}(\boldsymbol{x},t))}{J(\boldsymbol{Y}(\boldsymbol{x},t))}. \tag{13}$$

20.2 Statistical description of Lagrangian fields in a randomly moving compressible flow

20.2.1. General framework. We begin with a survey of some key problems of the statistical description of random fields in the Lagrangian coordinate system and, in particular, of the fluctuations of the Lagrangian field of a passive tracer in a randomly moving compressible medium.

First of all, observe that in physical situations passive tracer can be subject to compression and rarefaction even in the case when the fluid carrying the tracer is itself incompressible. It is true for instance for floating tracers (such as oil slicks and other pollutants) which are carried on the water surface in a quasi-2-D motions of the ocean surface. Indeed, the divergence of the velocity field

$$\nabla \cdot \boldsymbol{v} = \frac{\partial v_1}{\partial x_1} + \frac{\partial v_2}{\partial x_2} + \frac{\partial v_3}{\partial x_3}$$

in the depth of the ocean is practically equal to 0. Let x_3 be the vertical coordinate and (x_1, x_2) be horizontal coordinates in the plane in which the passive tracer is constrained. The divergence of the horizontal component of the velocity field, which determines the behavior of a floating passive tracer is not equal to zero and, in view of the previous observation,

$$\text{div}_h \, \boldsymbol{v} = \frac{\partial v_1}{\partial x_1} + \frac{\partial v_2}{\partial x_2} = -\frac{\partial v_3}{\partial x_3} \neq 0,$$

so that a floating passive tracer is subject to the laws of motion of a passive tracer in a 2-D *compressible* flow. The situation is similar in the case of a channel flow where water, despite vertical mixing, has a horizontal motion only in the direction of channel walls. Motion of a floating tracer on its surface can then be described as the motion of passive tracer in a 1-D compressible medium. The above comments also apply to the analysis of heavy, bottom hugging tracer particles which are mixed by the fluid motions but remain constrained to the bottom layer. A simple example of such a situation can be observed in a stirred teapot with tea leaves floating just above its bottom.

An analytic description of statistical fluctuations of tracer density is simplest in the Lagrangian coordinate system, in which its behavior is controlled by a system of ordinary differential equations (20.1.5) and (20.1.11). Statistical properties of a fixed, physically infinitesimal volume of the passive tracer are determined by the joint probability distribution,

$$f_L(\boldsymbol{x}, j; \boldsymbol{y}, t) = \Big\langle \delta(\boldsymbol{x} - \boldsymbol{X}(\boldsymbol{y}, t)) \delta(j - J(\boldsymbol{y}, t)) \Big\rangle, \qquad (1)$$

of its coordinate $\boldsymbol{X}$ and Jacobian J, which quantitatively describes the effects of compression and expansion of the volume of the passive tracer. The subscript L signifies that the probability distribution of the random field is considered in the Lagrangian representation. In what follows we shall simply call similar statistical characteristics, and related probability distributions, Lagrangian.

Our problem has been already significantly simplified by an assumption that the velocity field $v(x, t)$ is Gaussian, with a known correlation tensor (20.1.1). However, even in this case, for arbitrary spatial and temporal characteristic scales of the velocity field, the problem of statistical analysis of evolution of a passive tracer field remains quite intractable. For that reason, we will make one more simplifying assumption in our model, and suppose that

$$\sigma_v \tau_v \ll l_v, \tag{2}$$

where $\sigma_v = \sqrt{\langle v^2 \rangle}$ is the standard deviation of fluctuations of the velocity, τ_v is the characteristic time scale, and l_v is the characteristic length scale of the velocity field $v(x, t)$. Inequality (2) has a clear physical interpretation. It requires that the displacement $\Delta X \sim \sigma_v \tau_v$ during the characteristic time τ_v of changes in its velocity, be much smaller than the length scale l_v of the velocity field.

Inequality (2) is a condition of applicability of the diffusion approximation which permits a passage from the stochastic equations (20.1.5) and (20.1.11) to a closed equation for the probability distribution (1). A detailed and quite complex derivation of that equation will appear later in this section. Another approach, using a powerful apparatus of the Furutsu–Novikov–Donsker formula will be discussed later in this chapter.

The diffusion approximation takes into account, fundamental for the description of the processes of turbulent diffusion, differences between Eulerian and Lagrangian statistics of the velocity field. We will illustrate this by comparing statistical means of the divergence of the velocity field (20.1.12). The Eulerian mean

$$\langle u(x, t) \rangle = 0,$$

(see (20.1.12)) in view of the general assumptions imposed in Section 20.1 on the velocity field v. Next, let us calculate the mean of field u in the diffusion approximation and in the Lagrangian coordinate system, where it has the form,

$$U(y, t) = u(X(y, t), t). \tag{3}$$

For this purpose, we shall split the solution of the stochastic equation (20.1.5) into a sum of two vector fields

$$X(y, t) = X_0(y, t) + Z(t_0, t), \quad (t_0 > 0) \tag{4}$$

where the first field satisfies a stochastic differential equation

$$\frac{dX_0}{dt} = v(X_0, t)\chi(t_0 - t), \tag{5}$$

$$\boldsymbol{X}_0(\boldsymbol{y}, t = 0) = \boldsymbol{y},$$

and where $\chi(t_0 - t)$ is the Heaviside function equal 1 to the left of t_0, and equal 0 to the right of t_0. The inclusion of the latter eliminates the influence on $\boldsymbol{X}_0(\boldsymbol{y}, t)$ of fluctuations of the velocity field in the time interval $[t_0, t]$. That influence on coordinates of a fixed tracer particle is taken into account by the second term in (4), which satisfies equation

$$\frac{d\boldsymbol{Z}_0}{dt} = \boldsymbol{v}\Big(\boldsymbol{X}_0 + \boldsymbol{Z}(t_0, t), t\Big), \tag{6}$$

with the initial condition $\boldsymbol{Z}_0(t_0, t = t_0) = 0$ and notation $\boldsymbol{X}_0(\boldsymbol{y}, t) = \boldsymbol{X}_0$.

In the first stage, the diffusion approximation replaces the exact solution of equation (6) by what is often called the Born approximation,

$$\boldsymbol{Z}_1 = \int_{t_0}^{t} \boldsymbol{v}(\boldsymbol{X}_0, t') dt'. \tag{7}$$

Roughly speaking, the replacement of $\boldsymbol{Z}$ by $\boldsymbol{Z}_1$ is justified for

$$|\boldsymbol{Z}| \ll l_v. \quad (|\boldsymbol{Z}_1| \ll l_v) \tag{8}$$

Replacing $\boldsymbol{Z}$ in (4) by $\boldsymbol{Z}_1$, and then substituting (4) into (3), we arrive at an approximate equality

$$U(\boldsymbol{y}, t) \approx u(\boldsymbol{X}_0 + \boldsymbol{Z}_1, t).$$

Assumption (8) also justifies the second stage of the diffusion approximation, where the field u appearing on the right-hand side of the above approximate equality is expanded into the Taylor series in powers of $\boldsymbol{Z}_1$, and only terms of order ≤ 1 are retained. This gives

$$U(\boldsymbol{y}, t) \approx u(\boldsymbol{X}_0, t) + (\boldsymbol{Z}_1 \cdot \boldsymbol{\nabla}_{\boldsymbol{X}_0}) u(\boldsymbol{X}_0, t). \tag{9}$$

Suppose that the velocity field $\boldsymbol{v}(\boldsymbol{x}, t)$ and, consequently, field $u(\boldsymbol{x}, t)$ have a finite temporal range τ_v of statistical dependence. This means that, for arbitrary $\boldsymbol{x}_1$ and $\boldsymbol{x}_2$, random vectors $\boldsymbol{v}(\boldsymbol{x}_1, t_1)$ and $\boldsymbol{v}(\boldsymbol{x}_2, t_2)$ are statistically independent whenever $|t_1 - t_2| \geq \tau_v$. Let us select the length of the time interval $[t_0, t]$ considered above to be

$$t - t_0 = 2\tau_v. \tag{10}$$

Then, the statistical mean of the first term on the right-hand side of (9) is equal to 0 in view of the statistical independence of $u(\boldsymbol{x}, t)$ and $\boldsymbol{X}_0(\boldsymbol{y}, t)$. Hence, in view of (9), (7), and the statistical homogeneity of the velocity field, the mean of the Lagrangian field $U(\boldsymbol{y}, t)$ is approximately equal to

$$\langle U(\boldsymbol{y}, t) \rangle \approx - \int_{t-2\tau_v}^{t} \langle u(\boldsymbol{X}_0, t') u(\boldsymbol{X}_0, t) \rangle dt'. \tag{11}$$

The approximation also reflects the fact that, in our case, $X_0(y, t') = X_0(y, t) = X_0$.

Next, consider the conditional mean,

$$\langle u(X_0, t')u(X_0, t) \rangle_x, \qquad (12)$$

under the condition that the random vector X_0 takes a given value x. Then the means under the integral sign in (11) will be obtained by an addition averaging over X_0. Notice, that in the right half of the interval of integration $(t - \tau_v, t)$, in view of the statistical independence of X_0 from $u(x, t)$, and $u(x, t')$, the conditional mean (12) is equal to the unconditional mean

$$\langle u(x, t')u(x, t) \rangle = b_u(0, t - t'), \qquad (13)$$

where

$$b_u(s, \tau) = -\sum_{i,j} \frac{\partial^2 b_{ij}(s, \tau)}{\partial s_i \partial s_j}, \qquad (14a)$$

is the covariance function of the Gaussian random field $u(x, t)$, and $b_{ij}(s, \tau)$ are defined in (20.1.1). In particular, substituting expression (20.1.14b) into (14), we get that

$$b_u(s, \tau) = -\kappa \left(\frac{8}{s} \frac{dF_\perp}{ds} + 7 \frac{d^2 F_\perp}{ds^2} + s \frac{d^3 F_\perp}{ds^3} \right). \qquad (14b)$$

In the left half of the interval of integration, $t' \in (t - 2\tau_v, t - \tau_v)$, the field $u(x, t)$ is statistically independent from both X_0, and $u(x, t')$, and the conditional mean (12) is split into a product of means,

$$\langle u(X_0, t')u(X_0, t) \rangle_x = \langle u(X_0, t') \rangle_x \langle u(X_0, t) \rangle = 0.$$

For $t' < t - \tau_v$, the covariance function (14) of the field $u(x, t)$ vanishes as well. Therefore, in the whole domain of integration $[t - 2\tau_v, t]$ one can replace the conditional mean (12), and with it the means under the integral sign in (11), by the covariance function (13). Respectively, the lower limit of integration in (11) can be replaced by $-\infty$. As a result, the diffusion approximation gives that

$$\langle U(y, t) \rangle = -B, \qquad (15)$$

where $B = B(0)$, and

$$B(s) = \int_{-\infty}^{0} b_u(s, \tau)d\tau. \qquad (16)$$

Notice that, in the case of Gaussian velocity fields, inequality (2), and conditions (8), and (10), can be viewed as equivalent. It should be also mentioned that there exist approximations more sophisticated and exact than the diffusion approximation sketched above, which permit calculation of similar

Lagrangian means. However, the diffusion approximation, as simple as it is, permits an adequate description of many universal characteristic features of fluctuations of the density field in both, compressible and incompressible randomly moving media.

20.2.2. Equation for the Lagrangian probability distribution in the diffusion approximation. The above subsection provided a detailed illustration of the diffusion approximation ideas on the simple example of calculation of the statistical mean $\langle U(\boldsymbol{y}, t) \rangle$ (see (15)). Now, the same ideas will be used to derive a closed equation for the probability distribution (1) itself.

To begin with, consider a smooth function $\phi(\boldsymbol{x}, j)$. Differentiating the composition, $\phi(\boldsymbol{X}(\boldsymbol{y}, t), J(\boldsymbol{y}, t))$, with respect to time, using equalities (20.1.5), and (20.1.11), and then averaging the obtained identity, we get that

$$\frac{d}{dt} \langle \phi(\boldsymbol{X}, J) \rangle = \left\langle \left(\boldsymbol{v}(\boldsymbol{X}, t) \boldsymbol{\nabla}_X \right) \phi(\boldsymbol{X}, J) \right\rangle + \left\langle u(\boldsymbol{X}, t) J \frac{\partial}{\partial J} \phi(\boldsymbol{X}, J) \right\rangle. \quad (17)$$

Let us compute, in the diffusion approximation, the means on the right-hand side of (17). We start with representing the random field $J(\boldsymbol{y}, t)$ in the form analogous to (4),

$$J(\boldsymbol{y}, t) = J_0(\boldsymbol{y}, t) + Q(t_0, t), \quad (18)$$

where $J_0(\boldsymbol{y}, t)$ satisfies the equation,

$$\frac{dJ_0}{dt} = u(\boldsymbol{X}, t) J \chi(t_0 - t), \quad J_0(\boldsymbol{y}, t = 0) = 1, \quad (19)$$

in which, in analogy with equation (5), we have excluded the influence of the velocity field $\boldsymbol{v}(\boldsymbol{x}, t)$ on the behavior of field $J_0(\boldsymbol{y}, t)$ for $t > t_0$. The second summand in (18) satisfies an equation analogous to (6), namely

$$\frac{dQ}{dt} = u(\boldsymbol{X}_0 + \boldsymbol{Z}, t)(J_0 + Q), \quad Q(t_0, t = t_0) = 0. \quad (20)$$

Suppose, that all the assumptions imposed on field $\boldsymbol{v}(\boldsymbol{x}, t)$ in derivation of relation (15) are still in force. Then, as in (7), one can restrict oneself to the Born approximation of the solution of equation (20); that is,

$$Q_1(t_0, t) = J_0(\boldsymbol{y}, t) \int_{t_0}^{t} u(\boldsymbol{X}_0, t') \, dt'. \quad (21)$$

Clearly, we have taken into account the fact that, for $t > t_0$, neither $\boldsymbol{X}_0$ nor J_0 depend on t'.

Now, let us proceed directly to the calculation of the last mean in equation (17). For that purpose, let us rewrite it in the following form:

$$\left\langle u(\boldsymbol{X},t)J\frac{\partial}{\partial J}\phi(\boldsymbol{X},J)\right\rangle = \left\langle u(\boldsymbol{X}_0+\boldsymbol{Z},t)(J_0+Q)\frac{\partial}{\partial J_0}\phi(\boldsymbol{X}_0+\boldsymbol{Z},J_0+Q)\right\rangle. \tag{22}$$

Assume that $\phi(\boldsymbol{x},j)$ is a sufficiently smooth function of its arguments so that the expression inside the mean brackets on the right-hand side of equality (22) can be expanded in a Taylor series in powers of $\boldsymbol{Z}$, and Q, and then sufficiently accurately represented by the part of that expansion that contains terms of degree at most 1. In addition, we replace $\boldsymbol{Z}$ and Q by their Born approximations $\boldsymbol{Z}_1$ and Q_1. Hence,

$$\left\langle u(\boldsymbol{X},t)J\frac{\partial}{\partial J}\phi(\boldsymbol{X},J)\right\rangle \approx \left\langle u(\boldsymbol{X}_0,t)J_0\frac{\partial}{\partial J_0}\phi(\boldsymbol{X}_0,J_0)\right\rangle \tag{23}$$

$$+\left\langle J_0\frac{\partial}{\partial J_0}\phi(\boldsymbol{X}_0,J_0)(\boldsymbol{Z}_1\nabla_{\boldsymbol{X}_0})u(\boldsymbol{X}_0,t)\right\rangle$$

$$+\left\langle u(\boldsymbol{X}_0,t)Q_1(t_0,t)\frac{\partial}{\partial J_0}J_0\frac{\partial}{\partial J_0}\phi(\boldsymbol{X}_0,J_0)\right\rangle$$

$$+\left\langle u(\boldsymbol{X}_0,t)(\boldsymbol{Z}_1(t_0,t)\nabla_{\boldsymbol{X}_0})J_0\frac{\partial}{\partial J_0}\phi(\boldsymbol{X}_0,J_0)\right\rangle.$$

If equation (10) is satisfied, then the first mean on the right-hand side of the above equation splits into the product of means, and

$$\left\langle u(\boldsymbol{X}_0,t)J_0\frac{\partial}{\partial J_0}\phi(\boldsymbol{X}_0,J_0)\right\rangle = \langle u(\boldsymbol{X}_0,t)\rangle\left\langle J_0\frac{\partial}{\partial J_0}\phi(\boldsymbol{X}_0,J_0)\right\rangle = 0, \tag{24}$$

because of the statistical independence of the field $u(\boldsymbol{x},t)$ from the values of fields $J_0(\boldsymbol{y},t)$, and $\boldsymbol{X}_0(\boldsymbol{y},t)$, and the fact that $\langle u(\boldsymbol{x},t)\rangle = 0$.

Now, let us turn to calculation of the second mean on the right-hand side of equality (23) which we will write in the form,

$$\left\langle J_0\frac{\partial}{\partial J_0}\phi(\boldsymbol{X}_0,J_0)(\boldsymbol{Z}_1\nabla_{\boldsymbol{X}_0})u(\boldsymbol{X}_0,t)\right\rangle \tag{25}$$

$$= \left\langle J_0\frac{\partial}{\partial J_0}\phi(\boldsymbol{X}_0,J_0)\left\langle(\boldsymbol{Z}_1\nabla_{\boldsymbol{X}_0})u(\boldsymbol{X}_0,t)\right\rangle_{X_0,J_0}\right\rangle,$$

where $\langle\,.\,\rangle_{X_0,j_0}$ signifies the averaging under condition that values of $\boldsymbol{X}_0$, and J_0, are given, and the outside brackets signify the averaging with respect to random quantities $\boldsymbol{X}_0$, and J_0. Substituting the explicit expression (7) for $\boldsymbol{Z}_1$ into the conditional mean, that average can now be written in the form,

$$\left\langle (Z_1 \nabla_{X_0}) u(X_0, t) \right\rangle_{X_0, J_0} = \int_{t-2\tau_v}^{t} \left\langle \left(v(X_0, t') \nabla_{X_0} \right) u(X_0, t) \right\rangle_{X_0, J_0} dt'.$$

Repeating considerations that permitted us to pass from equality (11) to equality (15), we obtain that the above conditional mean is independent of the conditions and, in the diffusion approximation, is equal to

$$\left\langle (Z_1 \nabla_{X_0}) u(X_0, t) \right\rangle_{X_0, J_0} = -B.$$

As a result, equality (25) assumes the form,

$$\left\langle J_0 \frac{\partial}{\partial J_0} \phi(X_0, J_0)(Z_1 \nabla_{X_0}) u(X_0, t) \right\rangle = -B \left\langle J_0 \frac{\partial}{\partial J_0} \phi(X_0, J_0) \right\rangle \qquad (26)$$

In an analogous fashion one can also show that the two remaining means on the right-hand side of (23) are, in the diffusion approximation, as follows:

$$\left\langle u(X_0, t) Q_1(t_0, t) \frac{\partial}{\partial J_0} J_0 \frac{\partial}{\partial J_0} \phi(X_0, J_0) \right\rangle = B \left\langle \frac{\partial}{\partial J_0} J_0 \frac{\partial}{\partial J_0} \phi(X_0, J_0) \right\rangle,$$

$$\left\langle u(X_0, t)(Z_1(t_0, t) \nabla_{X_0}) J_0 \frac{\partial}{\partial J_0} \phi(X_0, J_0) \right\rangle = 0.$$

Consequently, equality (23) takes the form

$$\left\langle u(X, t) J \frac{\partial}{\partial J} \phi(X, J) \right\rangle = -B \left\langle J_0 \frac{\partial}{\partial J_0} \phi(X_0, J_0) \right\rangle + B \left\langle \frac{\partial}{\partial J_0} J_0 \frac{\partial}{\partial J_0} \phi(X_0, J_0) \right\rangle,$$

or, after simplifications,

$$\left\langle u(X, t) J \frac{\partial}{\partial J} \phi(X, J) \right\rangle = B \left\langle J_0^2 \frac{\partial^2 \phi(X_0, J_0)}{\partial J_0^2} \right\rangle. \qquad (27)$$

With the help of similar arguments, in the diffusion approximation, we arrive at the following expression for the first mean on the right-hand side of equality (17):

$$\left\langle v(X, t) \nabla_X \phi(X, J) \right\rangle = D \left\langle \Delta_{X_0} \phi(X_0, J_0) \right\rangle, \qquad (28)$$

where the diffusion coefficient $D = D(0)$, with

$$D(s) = \frac{1}{N} \int_{-\infty}^{0} b_{ii}(s, \tau) \, d\tau, \qquad (29)$$

and N is the dimension of the space. Notice, that in the three-dimensional case, in view of (20.1.14b), the integrand in (29) is equal to

$$b_{ii} = \frac{1}{s^2} \frac{d}{ds} s^3 [\kappa F + (1 - \kappa) G].$$

Utilizing relations (27–29), we can rewrite equality (17) as follows:

$$\frac{d}{dt}\langle\phi(\boldsymbol{X},J)\rangle = D\Big\langle\Delta_{\boldsymbol{X}_0}\phi(\boldsymbol{X}_0,J_0)\Big\rangle + B\left\langle J_0^2\frac{\partial^2\phi(\boldsymbol{X}_0,J_0)}{\partial J_0^2}\right\rangle. \qquad (30)$$

Next, we will utilize once more the smoothness assumption of function $\phi(\boldsymbol{X},J)$. More exactly, we will assume the function to be sufficiently nice so that, without going outside the boundaries of accuracy of the approximate equality (30), we can replace on its right-hand side, $\boldsymbol{X}_0$, and J_0, by $\boldsymbol{X}$, and J, respectively. In this situation, equation (30) takes the following form:

$$\frac{d}{dt}\langle\phi(\boldsymbol{X},J)\rangle - D\Big\langle\Delta_{\boldsymbol{X}}\phi(\boldsymbol{X},J)\Big\rangle - B\left\langle J^2\frac{\partial^2\phi(\boldsymbol{X},J)}{\partial J^2}\right\rangle = 0. \qquad (31)$$

Observe, that to calculate the statistical means entering in this equation only knowledge of density $f_L(\boldsymbol{x},j;\boldsymbol{y},t)$ (see (20.2.1)) is needed. Putting together all the summands in (31), and writing the statistical averages explicitly, we obtain that

$$\int d\boldsymbol{x}\int dj\left[\phi(\boldsymbol{x},j)\frac{\partial f_L}{\partial t} - Df_L\Delta_{\boldsymbol{x}}\phi(\boldsymbol{x},j) - Bf_Lj^2\frac{\partial^2\phi(\boldsymbol{x},j)}{\partial j^2}\right] = 0. \qquad (32)$$

If we additionally ssume that $\phi(\boldsymbol{x},j)$, together with its partial derivatives, converges to zero as $|x|\to\infty$, and $|j|\to\infty$, then, after integration by parts, we can pass from (32) to an equivalent equation,

$$\int d\boldsymbol{x}\int dj\,\phi(\boldsymbol{x},j)\left[\frac{\partial f_L}{\partial t} - D\Delta_{\boldsymbol{x}}f_L - B\frac{\partial^2}{\partial j^2}(j^2f_L)\right] = 0. \qquad (32a)$$

The above equation, given that the supply of functions $\phi(\boldsymbol{x},j)$ satisfying conditions imposed above is sufficient, implies that the probability density $f_L(\boldsymbol{x},j;\boldsymbol{y},t)$ fulfills a closed, Fokker–Planck–Kolmogorov type equation

$$\frac{\partial f_L}{\partial t} = D\Delta_{\boldsymbol{x}}f_L + B\frac{\partial^2}{\partial j^2}(j^2f_L) \qquad (33)$$

that is well known in the theory of Markov processes. It should be considered together with the initial condition

$$f_L(\boldsymbol{x},j;\boldsymbol{y},t=0) = \delta(\boldsymbol{x}-\boldsymbol{y})\delta(j-1). \qquad (34)$$

Remark 1. Without getting bogged down by mathematical details, we would like to comment that the way we used the smoothness conditions on function $\phi(\boldsymbol{x},j)$ means that the smoother probability densities $f_L(\boldsymbol{x},j;\boldsymbol{y},t)$ vary as functions of variables $\boldsymbol{x}$ and j (or, alternatively, the slower f_L varies as a function of time), the more accurate the approximate equation (33) is. From the physical viewpoint, the above comment means that equation (33) is valid in the "large time scale," with the characteristic scale of variability τ_f, much larger than the correlation time τ_v of the random velocity field $v(x, t)$.

Finally, it is useful to observe that the Lagrangian mean (15) which is part of the structure of the last term in equation (33), determines numerous characteristic features of the evolution of the Jacobian field $J(y, t)$, which will be studied in detail in the following sections.

20.3 Probabilistic properties of realizations of Lagrangian Jacobian and density fields

20.3.1. Auxiliary Markov processes. In this section we will provide a detailed analysis of implications of the Fokker–Planck–Kolmogorov equation (20.2.33) for Lagrangian probability distributions of fields $X(y, t)$, and $J(y, t)$. Its solution, in view of the initial conditions, (20.2.33), splits into a product of two factors

$$f_L(x, j; y, t) = f_X(x; y, t) f_J(j; t),$$ (1)

each of which has a clear cut physical meaning. The first factor represents the probability density for the coordinates of a fixed particle at time t, that is

$$f_X(x; y, t) = \langle \delta(x - X(y, t)) \rangle,$$ (2)

and it satisfies the standard diffusion equation

$$\frac{\partial f_X}{\partial t} = D \Delta f_X, \qquad f_X(x; y, t = 0) = \delta(x - y).$$ (3)

In particular, the standard law of diffusion of the passive tracer,

$$\langle (X(y, t) - y)^2 \rangle = ND = \int_{-\infty}^{0} b_{ii}(0, \tau) \, d\tau$$

follows form (3).

The second factor on the right-hand side of (1) gives a quantitative description of probabilistic properties of compressions and expansions of physically infinitesimal volumes of the passive tracers and satisfies the equation,

$$\frac{\partial f_J}{\partial t} = B \frac{\partial^2}{\partial j^2} (j^2 f_J), \qquad f_J(j; t = 0) = \delta(j - 1).$$ (4)

From now on, our analysis of evolution of physical fields will actively utilize the notion of statistically equivalent random processes, that is processes with identical finite-dimensional distributions, and in particular with identical equations for their joint probability densities. For Markov processes, which are determined by their infinitesimal generators, the above two conditions are equivalent.

More precisely, we will discuss, at a physical level of rigorousness, properties of simpler, but statistically equivalent auxiliary processes, and then attribute these properties to corresponding random fields which are of principal interest in this paper. This will be done for fixed space points, and only time dependence results will be discussed. Although such an approach has a different degree of validity in different concrete situations, it is always heuristically useful and helps a deeper understanding of the fine structure of realizations of a random field.

Equation (4) can be interpreted as a Fokker–Planck–Kolmogorov equation for transition probabilities of an auxiliary Markov process $J(t)$, which satisfies a (Stratonovich-type) stochastic differential equation

$$\frac{dJ}{dt} + BJ = \eta(t)J, \qquad J(t=0) = 1, \tag{5}$$

where $\eta(t)$ is a Gaussian white noise with covariance function

$$\langle \eta(t)\eta(t+\tau) \rangle = 2B\delta(\tau). \tag{6}$$

The solution of stochastic equation (5) is of the form

$$J(t) = \exp[-Bt + w(t)], \tag{7}$$

where $w(t)$ is the Brownian motion process such that

$$\langle w(t) \rangle = 0, \quad \text{and} \quad \langle w^2(t) \rangle = 2Bt. \tag{8}$$

Formulas (7), and (4), imply that the probability density for process $J(t)$, and thus also field $J(\boldsymbol{y}, t)$ in a neighborhood of a fixed tracer particle with Lagrangian coordinate $\boldsymbol{y}$, is lognormal, and

$$f_J(j; t) = \frac{1}{2j\sqrt{\pi Bt}} \exp\left[-\frac{\ln^2(je^{Bt})}{4Bt} \right], \quad j > 0, \tag{9}$$

with the cumulative probability distribution function

$$F_J(j; t) = \int_0^j f_J(j; t)dj = \Phi\left(\frac{\ln(je^{Bt})}{2\sqrt{Bt}} \right), \tag{10}$$

where $\Phi(z) = (1/\sqrt{\pi}) \int_{-\infty}^z \exp(-y^2)dy$ is the standard error function.

Observe, that the realizations of process (7) (and thus also those of $J(\boldsymbol{y}, t)$, for a fixed $\boldsymbol{y}$) have a complex fine structure which apparently contradicts their statistical properties. For example, consider the time behavior of statistical moments of process $J(t)$. It follows from (4) that the nth moment, $\langle J^n(t) \rangle$, satisfies the closed equation,

$$\frac{d\langle J^n \rangle}{dt} = Bn(n-1)\langle J^n \rangle, \qquad \langle J^n(0) \rangle = 1,$$

with the solution

$$\langle J^n \rangle = \exp[n(n-1)Bt]. \tag{11}$$

Hence, it follows that the mean Jacobian is conserved, that is,

$$\langle J \rangle = \langle J(\boldsymbol{y}, t) \rangle = 1. \tag{12}$$

This result is not difficult to justify as a consequence of the physically obvious law of conservation of the fluid volume. According to such conservation law, even in compressible medium, statistically homogeneous flows of the medium behave in such a way that the mean volume occupied by an arbitrary fixed portion of the medium remains constant. For an incompressible medium, this conservation law becomes a trivial dynamical identity $J(\boldsymbol{y}, t) \equiv 1$.

Formula (11) also implies that all the higher moments of J, beginning with $n = 2$, grow exponentially with the passage of time. For example, for $n = 2$,

$$\langle J^2(t) \rangle = e^{2Bt}. \tag{13}$$

This property seems to imply appearance, as t increases, of very high peaks in the realizations of $J(t)$, which would explain the exponential growth of higher moments, and of the variance

$$\sigma_J(t) = \langle J^2 \rangle - \langle J \rangle^2 = e^{2Bt} - 1. \tag{14}$$

Geometrically, appearance of peaks with $J(\boldsymbol{y}, t) \gg 1$ signifies the sojourn of a fixed particle in a strongly rarefied region. However, the picture of a particle finding itself at large times in an increasingly rarefied region is contradicted by the behavior of probabilistic characteristics (9–10) of the process $J(t)$ and the field $J(\boldsymbol{y}, t)$. In particular, it follows from (10) that the probability that the random field $J(\boldsymbol{y}, t)$ exceeds the mean level $\langle J \rangle \equiv 1$ is equal to

$$P\Big(J(\boldsymbol{y}, t) > 1\Big) = \Phi\left(-\frac{\sqrt{Bt}}{2}\right) \sim \frac{2}{\sqrt{\pi Bt}} \exp\left(-\frac{Bt}{4}\right), \qquad (t \to \infty) \tag{15}$$

and for $Bt \gg 1$, when the moments of J grow exponentially, it converges exponentially to 0.

The geometric meaning of (15) becomes clear if we recall that the probability (15) can be written as a statistical mean

$$P(J > 1) = \langle \chi(J(\boldsymbol{y}, t) - 1) \rangle, \tag{16}$$

where $\chi(z)$ is the usual Heaviside function, equal to 1, for $z \geq 0$, and equal to 0, for $z < 0$. Integrating equality (16) over a fixed domain Ω in the Lagrangian coordinate system, and taking into account the statistical homogeneity of the random field $J(\boldsymbol{y}, t)$, we arrive at the equality,

$$P(J > 1) = \frac{\langle V(J > 1)\rangle}{V},$$

where V is the volume of the domain of integration Ω, and $V(J > 1)$—volume of the region where $J(\boldsymbol{y}, t) > 1$. Equation (15) shows that, on the average, the ratio of volumes of these domains exponentially decays as t increases, which means that, as time grows, the overwhelming majority of particles find themselves in the strongly compressed regions where $J \ll 1$.

20.3.2. Extremal properties of realizations of the Jacobian and the density fields. The above argument about concentration of particles in the areas of high compression can be even more convincingly illustrated by the extremal properties of realizations of the process $J(t)$, which is statistically equivalent to the random field $J(\boldsymbol{y}, t)$. Recall, that in view of the well known extremal property of the process $w(t) - Bt$ (Brownian motion with a regular drift), it is easy to show that, with probability

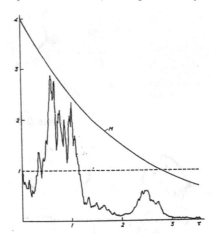

FIGURE 20.3.1.
Dominating curve for one-half of the realizations of the Jacobian random field $J(\boldsymbol{y}, t)$.

$$Q = 1 - A^{(r-1)}, \tag{17}$$

realizations of process $J(t)$, for all t, are dominated by an exponentially decreasing (for $r > 0$) curve

$$M(t) = Ae^{-rBt}, \tag{18}$$

where $A > 1$, and $0 < r < 1$. In particular, one-half of all realizations of $J(t)$ are located beneath the curve (see Fig. 20.3.1)

$$M(t) = 4\exp(-Bt/2).$$

In other words, beginning with time

$$t_0 = \frac{4}{B}\ln 2,$$

one-half of the particles of the medium find themselves in the compressed region of the space where

$$J(t) < 4\exp\left(-\frac{Bt}{2}\right) < 1,$$

whenever $t > t_0$.

The above result on "trapping" of particles in the compressed regions is also supported by another result of the theory of Markov processes according to which the random area

$$S = \int_0^\infty J(t)dt, \tag{19}$$

under the realizations of process $J(t)$, has a nondegenerate probability distribution

$$\mathbf{P}(S < \theta) = \exp\left(-\frac{1}{B\theta}\right). \tag{20}$$

This implies that the realizations of $J(t)$ converge to 0 (as $t \to \infty$) sufficiently fast for the improper integral (19) to be finite with probability 1. More exactly, for any $p < 1$,
$$\mathbf{P}(S < -1/B\ln p) = p.$$

It is interesting to observe that although, as expected, all moments of the distribution (20) are infinite, and in particular the mean

$$\langle S \rangle = \int_0^\infty \langle J \rangle dt = \int_0^\infty dt = \infty,$$

the inverse moments, for example

$$\langle 1/S \rangle = B$$

are finite. This means that the area under realizations of $J(t)$ may not be too small.

Let us summarize the basic facts concerning statistics of the Jacobian $J(\mathbf{y}, t)$, and of the auxiliary, statistically equivalent process $J(t)$. The main conclusion is that the exponential growth of moments of Jacobian (11), and (13), does not mean that the sensor moving with the medium, which measures the degree of compression or expansion of the medium, will in course of time fall into more and more rarefied region where $J \gg 1$. Just the opposite happens, as is clear from the behavior of the probability distribution of the Jacobian (9–10), of the dominating curves (18), and of the probability distribution (20) of the area (19) under the realizations of $J(t)$. The sensor will eventually fall in compressed ($J \ll 1$) regions, where the degree of compression increases with time. Hence, the time dependence of the Jacobian's moments $\langle J^n \rangle$ does not reflect the true behavior of the realizations of the Jacobian J. The exponential growth of the Jacobian's moments is caused by the so-called ensemble effect. In our particular case it means that some (but far from all) realizations of $J(\boldsymbol{y}, t)$ display large peaks. The characteristic height level of these peaks increases with time. However, the relative portion of realizations of $J(\boldsymbol{y}, t)$ which develop such peaks $J \gg 1$, decreases. As a result, although $\langle J^2 \rangle$—the ensemble mean of the square of the Jacobian—increases exponentially in time because of high peaks in separate realizations, practically all the experiments will register a more or less monotone decay of $J(t)$, so that, beginning with a certain instant of time the inequality $J < 1$ is fulfilled.

Physical phenomena involving compression and expansion of infinitesimally small volumes of the passive tracer are usually described in terms of density. For that reason we will rephrase the above results in the language of density fluctuations.

It is clear from (13) that the density field $R(\boldsymbol{y}, t)$ of the passive tracer in the Lagrangian coordinate system (as different from $\rho(\boldsymbol{x}, t)$ —the density field in the Eulerian coordinate system) is described by a simple expression

$$R(\boldsymbol{y}, t) = \frac{\rho_0(\boldsymbol{y})}{J(\boldsymbol{y}, t)}, \tag{21}$$

where $\rho_0(\boldsymbol{y})$ is the initial given density field. For a fixed Lagrangian coordinate $\boldsymbol{y}$, statistical properties of realizations of the random Lagrangian density field are more conveniently studied with the help of an auxiliary process

$$R(t) = \rho_0 / J(t). \tag{22}$$

It follows from (6) that R satisfies a stochastic differential equation

$$\frac{dR}{dt} = BR - \eta(t)R,$$

with the initial condition $R(0) = \rho_0$, the solution thereof is of the form,

$$R(t) = \rho_0 \exp[Bt - w(t)].$$

This (or the solution of corresponding Fokker-Planck equation) gives the cumulative distribution function for the density:

$$P\Big(R(\boldsymbol{y}, t) < \rho\Big) = \Phi\left(\frac{\ln((\rho/\rho_0)e^{-Bt})}{2\sqrt{Bt}}\right).$$

The median curve

$$\bar{\rho} = \rho_0 e^{Bt}$$

corresponding to the equality,

$$\mathbf{P}\left(R(\boldsymbol{y}, t) < \rho\right) = \frac{1}{2},$$

grows exponentially with time, as do curves

$$m(t) = \frac{\rho_0}{A} e^{rBt},$$

which bound from below, with probability Q (see (17)), the realizations of the density field for $t > 0$. In other words, as time progresses, practically every particle of the tracer falls in the compressed region with an exponentially growing density. In this case, the moments of the density field

$$\langle R^n \rangle = \rho_0^n \exp[n(n+1)Bt] \tag{23}$$

and, in particular, the mean density of the passive tracer is described by the formula

$$\langle R(\boldsymbol{y}, t) \rangle = \rho_0(\boldsymbol{y}) e^{2Bt},$$

which adequately reflects evolution of the Lagrangian density field at a point with fixed Lagrangian coordinate $\boldsymbol{y}$.

20.4 Lagrangian vs. Eulerian statistical characteristics of randomly moving media

The above analysis of evolution of passive tracer density fields was carried out in the Lagrangian coordinate system. However, fixed sensors are also used to study fluctuations of the passive tracer fields, and they register behavior of these fields in the Eulerian coordinate system. Eulerian statistical characteristics are, in general, not only quantitatively but also qualitatively different from the corresponding Lagrangian statistical characteristics. Nevertheless, both types of characteristics and, in particular, characteristics of density fields, are connected by universal relationships, and with their help one can recover Eulerian statistics of a random field from known Lagrangian statistics of the same field. Some of such formulas will be established in the present section.

To begin, consider the joint Eulerian probability density

$$f_E(\boldsymbol{y}, j; \boldsymbol{x}, t) = \Big\langle \delta(\boldsymbol{y} - \boldsymbol{Y}(\boldsymbol{x}, t)) \delta(j - j(\boldsymbol{x}, t)) \Big\rangle \tag{1}$$

of the Lagrangian coordinates $\boldsymbol{Y}(\boldsymbol{x}, t)$ (see (20.1.9)) of a particle which is located at a point with Eulerian coordinates $\boldsymbol{x}$, and of the Eulerian Jacobian field

$$j(\boldsymbol{x}, t) = J(\boldsymbol{Y}(\boldsymbol{x}, t), t),$$

which is obtained from the corresponding Lagrangian field $J(\boldsymbol{y}, t)$ by replacing $\boldsymbol{y}$ with $\boldsymbol{Y}(\boldsymbol{x}, t)$. We transfom the right-hand side of (1) using formula (20.1.8) to get that

$$f_E(\boldsymbol{y}, j; \boldsymbol{x}, t) = j f_L(\boldsymbol{x}, j; \boldsymbol{y}, t), \tag{2}$$

where f_L is the Lagrangian probability distribution (20.2.1) encountered before. Relation (2) is the first in a series of promised formulas connecting Lagrangian and Eulerian characteristics.

A closer look at (2) shows that, in an incompressible medium, where

$$j(\boldsymbol{x}, t) = J(\boldsymbol{y}, t) = 1,$$

it is equivalent with the identity

$$f_Y(\boldsymbol{y}; \boldsymbol{x}, t) = f_X(\boldsymbol{x}; \boldsymbol{y}, t), \tag{3}$$

where

$$f_Y(\boldsymbol{y}; \boldsymbol{x}, t) = \langle \delta(\boldsymbol{y} - \boldsymbol{Y}(\boldsymbol{x}, t)) \rangle. \tag{4}$$

Hence, in an incompressible medium, the probability distribution (20.3.2) of the Eulerian coordinates of a fixed particle, and that of the Lagrangian coordinates of the particle which is located at point $\boldsymbol{x}$ at time t satisfies the reciprocity condition (3). In the general case of a compressible medium that relationship may fail.

In the general case of a compressible medium, if the random field $\boldsymbol{v}(\boldsymbol{x}, t)$ of velocities is statistically homogeneous, then another useful formula connecting Lagrangian and Eulerian probability distributions of the Jacobian follows from (2). Indeed, in this case, the probability distributions entering in (2) depend, apart from j, only on the difference of the spatial coordinates $\boldsymbol{x}$ and $\boldsymbol{y}$, that is

$$f_E(\boldsymbol{y} - \boldsymbol{x}, j; t) = j f_L(\boldsymbol{x} - \boldsymbol{y}, j; t). \tag{5}$$

Integrating both sides with respect to either $\boldsymbol{x}$ or $\boldsymbol{y}$, we get that

$$f_j(j; t) = j f_J(j; t). \tag{6}$$

This equality connects the probability distribution of the Eulerian Jacobian field

$$f_j(j; t) = \langle \delta(j - j(\boldsymbol{x}, t)) \rangle \tag{7}$$

with the probability distribution

$$f_J(j; t) = \langle \delta(j - J(\boldsymbol{y}, t)) \rangle \tag{8}$$

of the corresponding Lagrangian field, which satisfies, in the diffusion approximation, equation (20.3.4). Factor j in (6), by which these two distributions differ, takes into account the increase in the proportion of expanded particles of the fluid for which $j > 1$ in the statistical ensemble of Eulerian fields, as compared to the statistical ensemble of Lagrangian fields.

Since both probability distributions in (6) have to satisfy the normalization condition, we automatically get conservation laws

$$\langle J(\boldsymbol{y}, t) \rangle = 1, \quad \text{and} \quad \langle 1/j(\boldsymbol{x}, t) \rangle = 1. \tag{9}$$

Both of them have a transparent physical meaning. The first coincides with equality (20.3.12), which was obtained earlier in the diffusion approximation. At that time we noticed its universal character, and its relationship with the law of conservation of the volume occupied by the medium. The meaning of the second conservation law in (9) also becomes clear if we assume that, at the initial instant of time, $t = 0$, the density of the medium, say ρ_0, is identical at all points of space. Multiplying the second equality in (9) by ρ_0, and noticing that

$$\frac{\rho_0}{j(\boldsymbol{x}, t)} = \rho(\boldsymbol{x}, t) \tag{10}$$

is the Eulerian density field of the medium, we get that

$$\langle \rho(\boldsymbol{x}, t) \rangle = \rho_0. \tag{11}$$

So, the second equation in (9) expresses the conservation law for the mean Eulerian density of the medium. The latter, in turn, is a consequence of the dynamical mass conservation law for the medium.

Substituting in (6) the Lagrangian probability distribution (20.3.9) of the Jacobian found in the diffusion approximation, we obtain, also in the diffusion approximation, an explicit expression for the Eulerian probability distribution of the Jacobian:

$$f_j(j; t) = \frac{1}{2j\sqrt{\pi Bt}} \exp\left[-\frac{\ln^2(je^{-Bt})}{4Bt}\right], \quad j > 0. \tag{12}$$

However, in this case, the equation satisfied by the probability distribution is more enlightening than the probability distribution itself. Substituting $f_j = f_j/j$ in (20.3.4), we arrive at the equation,

$$\frac{\partial f_j}{\partial t} = B \frac{\partial}{\partial j}\left(j^2 \frac{\partial f_j}{\partial j}\right), \quad f(j; t = 0) = \delta(j - 1). \tag{13}$$

This equation can be also treated as a Fokker–Planck–Kolmogorov equation for the transition probabilities of the auxiliary Markov process $j(t)$, which is statistically equivalent with the Jacobian field $j(\boldsymbol{x}, t)$ at a fixed point $\boldsymbol{x}$. This auxiliary process satisfies a stochastic differential equation

$$\frac{dj}{dt} = Bj - \eta(t)j, \qquad j(t=0) = 1. \tag{14}$$

Notice, that this equation can be formally obtained from equation (20.3.5) by substituting $J(t) = 1/j(t)$. Therefore, in the case under consideration, a sort of inversion principle is satisfied: the Eulerian Jacobian j (the Jacobian at a fixed point $\boldsymbol{x}$ of the space) behaves as a function of time as an inverse Lagrangian Jacobian J^{-1} (where J is the Jacobian in the neighborhood of a fixed particle of the medium with the Lagrangian coordinate $\boldsymbol{y}$.)

In particular, if—with a given probability Q—one can find an exponentially decreasing dominating curves $M(t)$ (see (20.3.18)) under which lie graphs of $Q \cdot 100$ percent of realizations of $J(t)$, then the same percentage of realization of $j(t)$ will, for any $t > 0$, lie above the exponential growing minorant curves $1/M(t)$.

In other words, if a given number of particles of the medium fall (as time t grows) once and forever in the region with exponentially growing compression (where $J \ll 1$), then the same number of fixed immobile sensors, uniformly distributed in the whole space, will at the same time register a strong expansion of the medium (that is $j \gg 1$) in their neighborhood. The observed, mutually inverse behavior of the auxiliary processes $J(t)$, and $j(t)$, which are (in different representations) statistically equivalent with corresponding physical Jacobian field, is not a contradiction. It explains the nature of the spatial structure of the distribution of the passive tracer in chaotic flows of a compressible medium: As time grows, tracer particles clump up into relatively compact clusters of high density which are surrounded by large regions, where the density of passive tracer is much smaller than the average.

Formula (2) connects the Lagrangian and Eulerian probability distribution of the Jacobian field. From the view point of physical applications, it is desirable to obtain a similar formula which would connect probability distributions of the random density fields of the passive tracer. Its derivation is analogous to the derivation of formula (2) and gives the final result of the form,

$$\rho \varphi_E(\boldsymbol{y}, \rho; \boldsymbol{x}, t) = \rho_0(\boldsymbol{y}) \varphi_L(\boldsymbol{x}, \rho; \boldsymbol{y}, t), \tag{15}$$

where

$$\varphi_L(\boldsymbol{x}, \rho; \boldsymbol{y}, t) = \Big\langle \delta(\boldsymbol{x} - \boldsymbol{X}(\boldsymbol{y}, t)) \delta(\rho - R(\boldsymbol{y}, t)) \Big\rangle \tag{16}$$

is the joint Lagrangian probability distribution of random fields $\boldsymbol{X}(\boldsymbol{y}, t)$, and $R(\boldsymbol{y}, t)$, and

$$\varphi_E(\boldsymbol{y}, \rho; \boldsymbol{x}, t) = \Big\langle \delta(\boldsymbol{y} - \boldsymbol{Y}(\boldsymbol{x}, t)) \delta(\rho - \rho(\boldsymbol{x}, t)) \Big\rangle \tag{17}$$

is the Eulerian joint probability distribution of the Lagrangian coordinate $\boldsymbol{Y}(\boldsymbol{x}, t)$, and of the density $\rho(\boldsymbol{x}, t)$. In (15), as in the rest of this chapter, the initial density field $\rho_0(\boldsymbol{x})$ is assumed to be deterministic. Integrating (15) over all $\boldsymbol{y}$ we arrive at the formula

$$\rho\varphi_\rho(\rho; \boldsymbol{x}, t) = \int \rho_0(\boldsymbol{y})\varphi_L(\boldsymbol{x}, \rho; \boldsymbol{y}, t)d\boldsymbol{y}, \tag{18}$$

which expresses the one-point probability distribution of the Eulerian density through a joint Lagrangian distribution of $\boldsymbol{X}$ and R defined in (16). Now, integrating both sides of (18) with respect to ρ we arrive at the well known statement of the basic theorem of turbulent diffusion:

$$\langle \rho(\boldsymbol{x}, t) \rangle = \int \rho_0(\boldsymbol{y}) f_X(\boldsymbol{x}; \boldsymbol{y}, t)d\boldsymbol{y}. \tag{19}$$

It expresses the mean density of the passive tracer through the probability distribution (20.3.2) of the coordinates of fixed particles.

We shall also indicate another, sometimes more convenient, modification of formula (18). The problem is that the Lagrangian probability distribution entering on the right-hand side of (18) depends not only on the "objective" properties of the chaotically moving medium, but also on the "subjective," often created by the experimenter, initial density $\rho_0(\boldsymbol{y})$ of the tracer. In order to clearly separate the impact of "objective" and "subjective" factors, we can express φ_L (see (16)) through a "completely objective" Lagrangian probability distribution f_L (see (20.2.1)). As a result, after simple transformations, we get that

$$\varphi_\rho(\rho; \boldsymbol{x}, t) = \frac{1}{\rho^3} \int \rho_0^2(\boldsymbol{y}) f_L\left(\boldsymbol{x}, \rho_0(\boldsymbol{y})/\rho; \boldsymbol{y}, t\right)d\boldsymbol{y}. \tag{20}$$

20.5 Eulerian statistics of the passive tracer

In this section we will recover the one-point Eulerian probability distributions of the passive tracer density field. Recall, that they are expressed (see formula (20.4.20)) via the Lagrangian joint probability distribution of the coordinates and of the Jacobian. Given that, in the diffusion approximation the Lagrangian probability distribution splits into a product (20.3.1), we obtain that

$$\varphi_\rho(\rho; \boldsymbol{x}, t) = \Big\langle J(t)\varphi_n(\rho; \boldsymbol{x}, t|\rho_0/J(t)) \Big\rangle, \tag{1}$$

where $J(t)$ is an auxiliary process satisfying the stochastic differential equation (20.3.5), and

$$\varphi_n(\rho; \boldsymbol{x}, t|\rho_0(\boldsymbol{x})) = \frac{1}{\rho} \int \rho_0(\boldsymbol{y}) f_Y(\boldsymbol{y}; \boldsymbol{x}, t)\delta(\rho - \rho_0(\boldsymbol{y}))d\boldsymbol{y} \tag{2}$$

is the probability distribution of passive tracer density in an incompressible medium. In particular, (1–2), and (20.4.3) imply that the statistical moments of the density field

$$\langle \rho^n(\boldsymbol{x}, t) \rangle = \langle J^{1-n}(t) \rangle \int \rho_0^n(\boldsymbol{y}) f_X(\boldsymbol{x}; \boldsymbol{y}, t)d\boldsymbol{y}. \tag{3}$$

They differ only by the factor of

$$\langle J^{1-n}(t) \rangle = e^{n(n-1)Bt} \tag{4}$$

from the moments of the density in an incompressible medium.

Observe the following interesting fact. It is clear from equations (4), and (20.3.11), that

$$\langle J^{1-n}(t) \rangle = \langle J^{n}(t) \rangle.$$

This equality, together with (3), means that the probability distribution (1) of the passive tracer density field coincides with the probability distribution of the auxiliary process

$$\rho_{\mathrm{aux}}(\boldsymbol{x}, t) = \rho_0(\boldsymbol{Y}(\boldsymbol{x}, t)) J(t). \tag{5}$$

Now, let us consider properties of this process in some detail, in the case when the initial passive tracer density is identical at all points of space, that is when

$$\rho_0(\boldsymbol{x}) = \rho_0 = const. \tag{6}$$

In this situation

$$\rho_{\mathrm{aux}}(\boldsymbol{x}, t) = \rho(t) = \rho_0 J(t). \tag{7}$$

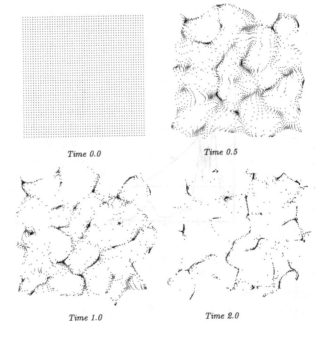

Time 0.0   Time 0.5

Time 1.0   Time 2.0

FIGURE 20.5.1.
Evolution of clusters of high tracer density in a compressible flow.

This equality looks somewhat paradoxical, since it equates the density and the Jacobian which is inverse to it in the aritmethical sense. A similar inversion phenomenon was already encountered when we considered the time evolution of Lagrangian and Eulerian Jacobians J, and j. The paradox is explained by the fact that passing from Lagrangian to Eulerian statistics of the same field (say, the density fields $\rho(\boldsymbol{x}, t)$ and $R(\boldsymbol{y}, t)$) the probability measures undergo a deformation: The compression of a region of space has a smaller statistical weight in the Eulerian ensemble than in the Lagrangian ensemble. Based on equation (7), we can repeat most of what has been said about the contradictory statistical properties of $J(t)$.

In particular, notice that although density moments

$$\langle \rho^n(\boldsymbol{x}, t) \rangle = \rho_0^n e^{n(n-1)Bt} \qquad (8)$$

grow exponentially with time in view of (7) and (20.3.11), the probability that random field $\rho(\boldsymbol{x}, t)$ exceeds the mean level $\langle \rho \rangle = \rho_0$ is given by expression (20.3.15), and converges exponentially to 0 as $t \gg 1/B$.

Extending an analogy between the time evolutions of $\rho(\boldsymbol{x}, t)$, and of $J(t)$, we can presume that their realizations behave in a similar fashion. Hence, realizations of $\rho(\boldsymbol{x}, t)$ lie underneath curves $\rho_0 M(t)$ (see (20.3.18)) and eventually decay to 0. The exponential growth of moments of the Eulerian density (8) can be explained by an appearance in some (but far from all) realizations of large peaks of density $\rho \gg \rho_0$. In the physical language it means that a majority of fixed sensors finds itself eventually in the regions of low tracer concentration (where $\rho \ll \rho_0$), and only a few of them will be lucky enough to lie inside a "macroparticle" — a compact region of high density, and register a high ($\rho \gg \rho_0$) values of the passive tracer density (see Fig. 20.5.1).

20.6   Fluctuations of the passive tracer density gradient

Although compressions and expansions, leading to density fluctuations in a compressible medium, are absent in an incompressible medium, the random motions of the latter can transform, initially smooth and simple curves of constant passive tracer concentration into more and more complex objects. This is a result of initially close particles being separated and initially far particles being brought together. Consequently, regions with vastly different tracer densities can be located next to each other, separated by clear cut boundaries in the neighborhood thereof the density changes rapidly.

In this section we give a quantitative description of the increasing spatial complexity of density fields in a randomly moving incompressible medium. For the sake of simplicity, we restrict ourselves to an analysis of a 2-D case, where components of the velocity field can be expressed by a stream function

$\Psi(\boldsymbol{x}, t)$ via formulas

$$v_1 = \frac{\partial \Psi}{\partial x_2}, \quad v_2 = -\frac{\partial \Psi}{\partial x_1}. \tag{1}$$

The statistical properties of the stream function are assumed to be known.

The spatial complexity of density field realizations can be characterized, for example, by the gradient $\boldsymbol{\nabla}_{\boldsymbol{x}} \rho(\boldsymbol{x}, t)$ of density field ρ, and we will study its statistical properties. To begin, let us recall that, in an incompressible medium, the density field is given by the expression,

$$\rho(\boldsymbol{x}, t) = \rho_0(\boldsymbol{Y}(\boldsymbol{x}, t)), \tag{2}$$

where $\rho_0(\boldsymbol{x})$ is the initial deterministic density profile, and $\boldsymbol{y} \mapsto \boldsymbol{Y}(\boldsymbol{x}, t)$ is the mapping of Eulerian coordinates into Lagrangian coordinates. As a result, we get that

$$\boldsymbol{\nabla}_{\boldsymbol{x}} \rho = \boldsymbol{e}_1 \left(\frac{\partial \rho_0}{\partial y_1} j_{11} + \frac{\partial \rho_0}{\partial y_2} j_{21} \right) + \boldsymbol{e}_2 \left(\frac{\partial \rho_0}{\partial y_1} j_{12} + \frac{\partial \rho_0}{\partial y_2} j_{22} \right), \tag{3}$$

where $\boldsymbol{e}_1, \boldsymbol{e}_2$ are the unit coordinate vectors in Eulerian coordinates, and j_{lm} are components of the transition matrix from Eulerian to Lagrangian coordinates, that is

$$j_{lm} = \frac{\partial Y_l(\boldsymbol{x}, t)}{\partial x_m}, \quad l, m = 1, 2. \tag{4}$$

The statistical properties of the gradient (3) are more conveniently studied in the Lagrangian coordinate system. For that reason, we shall express fields $j_{lm}(\boldsymbol{x}, t)$ entering into (3) through the components

$$J_{lm}(\boldsymbol{y}, t) = \frac{\partial X_l(\boldsymbol{y}, t)}{\partial y_m} \tag{5}$$

of the transition matrix from Lagrangian to Eulerian coordinates. After differentiation of the obvious vector identity

$$\boldsymbol{x} = \boldsymbol{X}(\boldsymbol{Y}(\boldsymbol{x}, t), t)$$

with respect to x_1, and x_2, and after solving obtained equations with respect to j_{lm}, we get that

$$j_{11} = J_{22}/J, \quad j_{22} = J_{11}/J, \quad j_{12} = -J_{12}/J, \quad J_{21} = -J_{21}/J. \tag{6}$$

Substituting (6) into (3), and taking into account that in the case of an incompressible medium $J \equiv 1$, we find that

$$\boldsymbol{\nabla}_{\boldsymbol{x}} \rho = |\boldsymbol{\nabla}_{\boldsymbol{y}} \rho_0|(\boldsymbol{e}_1 J_1 + \boldsymbol{e}_2 J_2), \tag{7}$$

where $|\boldsymbol{\nabla}_{\boldsymbol{y}} \rho_0|$ is the initial norm of the density gradient at point $\boldsymbol{y}$,

$$J_1(\boldsymbol{y}, t, \theta_0) = J_{22} \cos \theta_0 - J_{21} \sin \theta_0, \quad J_2(\boldsymbol{y}, t, \theta_0) = J_{11} \sin \theta_0 - J_{12} \cos \theta_0, \tag{8}$$

and θ_0 is the angle between the y_1-axis and the initial direction of the gradient vector.

It is not hard to verify that, in the Lagrangian coordinate system, random fields J_1, J_2 in (8) satisfy the following system of stochastic differential equations

$$\frac{dJ_2}{dt} = \alpha J_2 - \beta J_1, \quad \frac{dJ_1}{dt} = -\alpha J_1 + \gamma J_2, \tag{9}$$

$$J_1(\boldsymbol{y}, t = 0, \theta_0) = \cos\theta_0, \quad J_2(\boldsymbol{y}, t = 0, \theta_0) = \sin\theta_0,$$

where

$$\alpha(\boldsymbol{x}, t) = \frac{\partial^2 \Psi(\boldsymbol{x}, t)}{\partial x_1 \partial x_2}, \quad \beta(\boldsymbol{x}, t) = \frac{\partial^2 \Psi(\boldsymbol{x}, t)}{\partial x_2^2}, \quad \gamma(\boldsymbol{x}, t) = \frac{\partial^2 \Psi(\boldsymbol{x}, t)}{\partial x_1^2}.$$

Solutions of equations (9) can, in turn, be represented in the form,

$$J_1 = e^\chi \cos\theta, \quad J_2 = e^\chi \sin\theta. \tag{10}$$

Thus, expression for the gradient (7) assumes a particularly transparent form

$$\boldsymbol{\nabla}_{\boldsymbol{x}}\rho = |\boldsymbol{\nabla}_{\boldsymbol{y}}\rho_0| g \boldsymbol{n}, \tag{11}$$

where

$$g(\boldsymbol{y}, t, \theta_0) = e^\chi = |\boldsymbol{\nabla}_{\boldsymbol{x}}\rho|/|\boldsymbol{\nabla}_{\boldsymbol{y}}\rho_0| \tag{12}$$

is the relative change in the magnitude of the gradient, and

$$\boldsymbol{n} = \boldsymbol{e}_1 \cos\theta + \boldsymbol{e}_2 \sin\theta \tag{13}$$

is the unit vector describing fluctuations of the gradient's direction vector as a function of time.

It follows from (9–10) that χ and θ satisfy stochastic differential equations

$$\frac{d\chi}{dt} = -\alpha \cos 2\theta + \mu \sin 2\theta, \quad \chi(\boldsymbol{y}, t = 0, \theta_0) = 0, \tag{14a}$$

$$\frac{d\theta}{dt} = \alpha \sin 2\theta + \mu \cos 2\theta - \nu, \quad \theta(\boldsymbol{y}, t = 0, \theta_0) = \theta_0, \tag{14b}$$

where

$$\nu = (\gamma + \beta)/2, \quad \mu = (\gamma - \beta)/2. \tag{15}$$

In the subsequent statistical analysis of fields χ and θ we shall assume that the velocity field potential $\Psi(\boldsymbol{x}, t)$ is a Gaussian, statistically isotropic in space and stationary in time random field with a known covariance function. In the diffusion approximation, which we will employ again, for statistical characteristics of the above fields we will need parameter B, defined by equality

$$\int_{-\infty}^{0} \langle \Psi(\boldsymbol{x}, t)\Psi(\boldsymbol{x}, +\boldsymbol{s}, t + \tau)\rangle d\tau = \cdots + \frac{B}{8}(s_1^2 + s_2^2)^2 - \cdots,$$

where s_1, s_2 are components of vector $\boldsymbol{s}$.

Let us consider the joint probability distribution,

$$f(\chi, \theta; t, \theta_0) = \left\langle \delta(\chi(\boldsymbol{y}, t, \theta_0) - \chi)\delta(\theta(\boldsymbol{y}, t, \theta_0) - \theta) \right\rangle \qquad (16)$$

of random fields χ and θ. One can show that, in the diffusion approximation, it satisfies the equation,

$$\frac{\partial f}{\partial t} = B\left(\frac{\partial^2 f}{\partial \chi^2} - 2\frac{\partial f}{\partial \chi} + 3\frac{\partial^2 f}{\partial \theta^2}\right), \qquad (17)$$

$$f(\chi, \theta; t = 0, \theta_0) = \delta(\chi)\delta(\theta - \theta_0),$$

which implies that random fields $\chi(\boldsymbol{y}, t, \theta_0)$ and $\theta(\boldsymbol{y}, t, \theta_0)$, and the field of the relative change of the gradient of $g(\boldsymbol{y}, t, \theta_0)$ (the latter being of our main interest), are statistically equivalent with the following stochastic processes:

$$\chi(t) = \omega(t) + 2Bt, \qquad (18a)$$

$$\theta(t, \theta_0) = \theta_0 + 3\kappa(t), \qquad (18b)$$

and

$$g(t) = \exp[\omega(t) + 2Bt], \qquad (18c)$$

where $\omega(t)$ and $\kappa(t)$ are statistically independent Brownian motions with identical variances

$$\langle \omega^2(t) \rangle = \langle \kappa^2(t) \rangle = 2Bt.$$

In particular, it follows from (18) that the gradient's mean value

$$\langle g(\boldsymbol{y}, t, \theta_0) \rangle = \exp(3Bt),$$

increases exponentially, but that, at the same time, the mean value of the unit vector $\boldsymbol{n}$ decays exponentially as

$$\langle \boldsymbol{n}(\boldsymbol{y}, t, \theta_0) \rangle = \boldsymbol{n}_0 \exp(-3Bt),$$

where

$$\boldsymbol{n}_0 = \boldsymbol{e}_1^0 \cos\theta_0 + \boldsymbol{e}_2^0 \sin\theta_0, \qquad (19)$$

and $\boldsymbol{e}_1^0$, and $\boldsymbol{e}_2^0$, are the unit vectors in the Lagrangian coordinate system. Consequently, the tracer density's mean gradient, in the vicinity of a fixed tracer particle with given Lagrangian coordinate $\boldsymbol{y}$, remains unchanged so that

$$\langle \boldsymbol{\nabla}_x \rho \rangle = \boldsymbol{\nabla}_y \rho_0.$$

The corresponding Eulerian density gradient's mean, at a point with fixed coordinate x, can be obtained by averaging the above equality with respect to random Lagrangian coordinates $Y(x, t)$. This gives

$$\langle \nabla_x \rho(x, t) \rangle = \int f_Y (y; x, t) \nabla_y \rho_0(y) \, dy.$$

In contrast to the mean with fixed Lagrangian coordinates, the Eulerian mean, in general, depends on time. In the diffusion approximation, the probability distribution $f_Y (y; x, t)$ (see (20.4.4)) appearing in the above formula is of the form,

$$f_Y (y; x, t) = \left(\frac{1}{4\pi Dt} \right)^{N/2} \exp \left(-\frac{(y - x)^2}{4Dt} \right),$$

where N is the dimension of space (in our case $N = 2$).

Now, we turn to a discussion of statistical properties of the relative rate of change of the gradient's magnitude g. Statistically equivalent process $g(t)$ (see (18c)) enjoys properties that are similar to those of process $R(t)$ in (20.3.22). Thus, process $g(t)$ has a lognormal distribution with the cumulative distribution function

$$P\Big(g(t) < g\Big) = \Phi \left(\frac{\ln(ge^{-2Bt})}{2\sqrt{Bt}} \right),$$

and the probability that it assumes values smaller than the initial value $g(0) = 1$ quickly decays to 0 as time increases. More exactly,

$$P\Big(g(t) < 1\Big) = \Phi(-\sqrt{Bt}).$$

The moments of g behave in a fashion similar to those of R in (20.3.3), and grow exponentially with time:

$$\langle g^n \rangle = \exp[n(n + 2)Bt]$$

for $n > 0$.

Exponentially increasing minorant curves for process $g(t)$, which bound it from below with arbitrarily close to 1 probability $\bar{P} < 1$, can also be found.

In summary, process $R(t)$ models the density of a randomly moving compressible medium at a point with Lagrangian coordinate y, and its statistical properties are drastically different from the properties of process $\rho(t)$, modulating its density at a point with Eulerian coordinate x. In the case considered in this section, process $g(t)$ models a relative rate of change of the density gradient in both Lagrangian and Eulerian coordinates. To be more precise one has to remember that $g(t)$ models the rate of change of the gradient at an Eulerian point x, relative to the initial magnitude of the gradient in the neighborhood of a particle which, at time instant t, is located at x.

20.7   Fluctuations of parameters of contours carried by a randomly moving incompressible medium

In this section we will study the evolution of the length $l(t)$ of a contour $\mathcal{L}$ of a constant tracer density—another, important for applications, geometric characteristic of a randomly moving incompressible medium. It can be expressed by a formula

$$l(t) = \int \delta\Big(\rho_0(\boldsymbol{Y}(\boldsymbol{x},t)) - \rho\Big)|\boldsymbol{\nabla}_x\rho|d^2x. \tag{1}$$

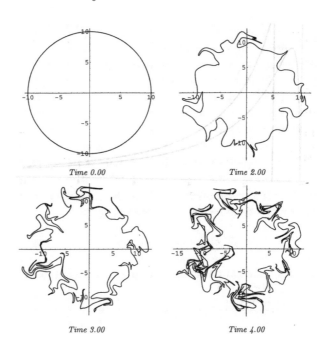

Time 0.00   *Time 2.00*

Time 3.00   *Time 4.00*

FIGURE 20.7.1.
Contour carried by an incompressible random flow.

Another related characteristic of the spatial evolution of the density field is the integral

$$A(t) = \int \delta\Big(\rho_0(\boldsymbol{Y}(\boldsymbol{x},t)) - \rho\Big)(\boldsymbol{\nabla}_x\rho)^2d^2x, \tag{2}$$

which will be called here the *total gradient of the density field*. Its physical meaning becomes clear if we observe that the double integral (2) is equal to the line integral,

$$A(t) = \oint |\boldsymbol{\nabla}_x\rho|\,dl,$$

of the magnitude of the density gradient along the curve $\mathcal{L}$ of constant density. As in the case of the density gradient studied in the previous section, the statistical properties of integrals of type (1–2) are easier to investigate by switching integrals from Eulerian to Lagrangian variables of integration. We shall show this on the example of the total gradient (2). In that case, after changing variables, we get that

$$A(t) = \int \delta(\rho_0(\boldsymbol{y}) - \rho)(\boldsymbol{\nabla}_x \rho)^2 d^2 y.$$

We shall assume that the initial density is given by a deterministic, radially symmetric function $\rho_0(\boldsymbol{y}) = \rho_0(|\boldsymbol{y}|)$, which monotonically decreases with the distance from the origin of the Lagrangian coordinate system. The initial contour in this particular case, and the computer simulation of the corresponding contours $\mathcal{L}$, for selected instants $t > 0$, are shown in Fig. 20.7.1.

Under these circumstances, the above integral takes the form

$$A(t) = \frac{r_0(\rho)}{|\boldsymbol{\nabla}_y \rho_0|} \int_{-\pi}^{\pi} (\boldsymbol{\nabla}_x \rho)^2 \Big|_{\boldsymbol{y}=r_0 \boldsymbol{n}_0} d\theta_0, \tag{3}$$

where $r_0(\rho)$ is the solution of equation $\rho_0(r) = \rho$, and $\boldsymbol{n}_0$, is determined by (20.6.19). Substituting in (3) the explicit expression (20.6.11) for the density gradient, we get that

$$A(t) = A_0 G,$$

where $A_0 = l_0 |\boldsymbol{\nabla}_y \rho_0|$ is the initial total gradient, $l_0 = 2\pi r_0$ is the initial length of the contour, and the dimensionless quantity,

$$G(t) = \frac{1}{2\pi} \int_{-\pi}^{\pi} g^2(\boldsymbol{y}, t, \theta_0) \Big|_{\boldsymbol{y}=r_0 \boldsymbol{n}_0} d\theta_0, \tag{4}$$

characterizes the relative rate of change in time of the total gradient of the selected contour.

It is also easy to see that, similarly, the relative rate of change of length of that contour, can be expressed by the integral

$$\Omega(t) = l(t)/l_0 = \frac{1}{2\pi} \int_{-\pi}^{\pi} g(\boldsymbol{y}, t, \theta_0) \Big|_{\boldsymbol{y}=r_0 \boldsymbol{n}_0} d\theta_0.$$

The statistical means of the above introduced relative quantities are thus expressed by the formulas,

$$\langle G(t) \rangle = \langle g^2(t) \rangle = e^{8Bt}, \quad \text{and} \quad \langle \Omega(t) \rangle = \langle g(t) \rangle = e^{3Bt}.$$

We should stress, however, that only the statistical means of processes G, and g^2, and Ω, and g coincide. Their other statistical characteristics differ, both quantitatively and qualitatively. We shall demonstrate this on the example of the relative rate of change of the total gradient G from (4). In our argument we will assume, for simplicity, that our contour of integration is small enough so that we can take $y = 0$ in (4).

Let us substitute in (4) $g^2 = J_1^2 + J_2^2$, utilize the formulas (20.6.8), and take into account the fact that the matrix elements, J_{lm}, do not depend on angle θ_0. Then, after integration, we get that

$$G = \frac{1}{2}(J_{11}^2 + J_{22}^2 + J_{12}^2 + J_{21}^2).$$

In view of the condition of incompressibility,

$$J_{11}J_{22} - J_{12}J_{21} \equiv 1$$

the previous equation can be rewritten in the form,

$$G = 1 + \frac{1}{2}[(J_{11} - J_{22})^2 + (J_{12} + J_{21})^2],$$

wherefrom it follows that inequality $G(t) \geq 1$ is always satisfied, even as random field g^2 appearing in (4) takes values smaller than 1 with positive probability.

Finally, we turn to the statistical properties of G. Introducing auxiliary functions

$$C = \frac{1}{2\pi} \int_{-\pi}^{\pi} g^2 \cos 2\theta d\theta_0, \quad \text{and} \quad S = \frac{1}{2\pi} \int_{-\pi}^{\pi} g^2 \sin 2\theta d\theta_0,$$

and passing from equations (20.6.14) to a system of stochastic differential equations for G, C, and S, gives

$$\frac{1}{2}\frac{dG}{dt} = \mu S - \delta C, \quad G(0) = 1,$$

$$\frac{1}{2}\frac{dC}{dt} = \nu S - \delta G, \quad C(0) = 0,$$

$$\frac{1}{2}\frac{dS}{dt} = \mu G - \nu C, \quad S(0) = 0.$$

If we substitute polar coordinates A, and φ, for C, and S, by means of equations

$$C = A\cos\varphi, \quad \text{and} \quad S = A\sin\varphi,$$

the above system yields three stochastic differential equations

$$\frac{1}{2}\frac{dG}{dt} = (\mu\sin\varphi - \delta\cos\varphi)A, \quad G(0) = 1, \tag{5a}$$

$$\frac{1}{2}\frac{dA}{dt} = (\mu\sin\varphi - \delta\cos\varphi)G, \quad A(0) = 0, \tag{5b}$$

$$A\frac{1}{2}\frac{d\varphi}{dt} = (\mu\cos\varphi + \delta\sin\varphi)G - \nu A, \quad \varphi(0) = 0 \qquad (5c)$$

for G, A, and φ. Hence, it is clear that equations (5a) and (5b) have solutions

$$G = \cosh Z(t), \quad A = \sinh Z(t), \qquad (6)$$

where auxiliary process $Z(t)$ and $\varphi(t)$ satisfy a closed system of two equations:

$$\frac{dZ}{dt} = 2(\mu\sin\varphi - \delta\cos\varphi), \quad Z(0) = 0,$$

$$\frac{d\varphi}{dt} = 2(\mu\cos\varphi + \delta\sin\varphi)\coth Z - 2\nu.$$

After rather straightforward but tedious calculations one can show that the one-point probability distribution

$$w(z;t) = \langle\delta(Z(t) - z)\rangle$$

of the random process $Z(t)$ satisfies, in the diffusion approximation, the equation,

$$\frac{\partial w}{\partial\tau} + \frac{\partial}{\partial z}(\coth zw) = \frac{\partial^2 w}{\partial z^2}, \quad w(z;0) = \delta(z), \qquad (7)$$

where $\tau = 4Bt$ denotes the dimensionless time. It means, that the statistically equivalent process $Z(\tau)$ satisfies the stochastic differential equation,

$$\frac{dZ}{d\tau} = \coth Z + \xi(\tau).$$

Consequently, according to (6), the corresponding statistically equivalent process $G(\tau)$ satisfies the Stratonovich stochastic differential equation,

$$\frac{dG}{d\tau} = G + \sqrt{G^2 - 1}\xi(\tau), \quad G(0) = 1,$$

where $\xi(\tau)$ is a Gaussian white noise with covariance function

$$\langle\xi(\tau)\xi(\tau')\rangle = 2\delta(\tau - \tau').$$

Hence, the probability distribution

$$f(u;t) = \langle\delta(G(t) - u)\rangle = \langle\delta(\cosh Z - u)\rangle,$$

of G satisfies the following Fokker–Planck–Kolmogorov equation:

$$\frac{\partial f}{\partial\tau} = \frac{\partial}{\partial u}(u^2 - 1)\frac{\partial f}{\partial u}, \quad f(u;\tau = 0) = \delta(u - 1). \qquad (8)$$

In particular, it follows from (8) that the moment functions of G satisfy a chain of equations,

$$\frac{d\langle G^n \rangle}{d\tau} = n(n+1)\langle G^n \rangle - n(n-1)\langle G^{n-2} \rangle,$$

so that, for example, the mean square of G and its third moment are equal, respectively, to

$$\langle G^2 \rangle = \frac{1}{3}(1 + 2e^{6\tau}), \quad \text{and} \quad \langle G^3 \rangle = \frac{1}{5}(2e^{12\tau} + 3e^{2\tau}).$$

Luckily, the above equation (8), can be solved explicitly[3] in terms of the Legendre function of the first kind $P_{-1/2+i\mu}(u)$ ($\mu \geq 0$), which is the solution of the Legendre equation,

$$\frac{d}{du}(u^2 - 1)\frac{d}{du}P_{-1/2+i\mu}(u) = -(\mu^2 + 1/4)P_{-1/2+i\mu}(u), \qquad (9)$$

and which has various integral representations in terms of elementary function, such as

$$P_{-1/2+i\mu}(u) = \frac{2\cosh\mu\pi}{\pi} \int_0^\infty \frac{\cos\mu\theta}{\sqrt{2(u + \cosh\theta\tau)}} d\theta, \quad \alpha \geq 0. \qquad (10)$$

Multiplying both sides of (8) by the Legendre function, and integrating them twice by parts with respect to u, in the limits from 1 to ∞, we get an easy ordinary differential equation,

$$\frac{d}{d\tau}\tilde{f}(\mu; \tau) = -(\mu^2 + 1/4)\tilde{f}(\mu; \tau) \qquad (11)$$

for the Fourier–Legendre transform,

$$\tilde{f}(\mu; \tau) = \int_1^\infty du\, f(u, \tau) P_{-1/2+i\mu}(u),$$

of function f. With the initial condition, corresponding to the initial condition in (8), its solution is

$$\tilde{f}(\mu; \tau) = \exp[-(\mu^2 + 1/4)\tau].$$

Taking its inverse Fourier–Legendre transform we get the following solution of the initial problem (8),

$$f(u; \tau) = \int_0^\infty d\mu\, \mu \tanh(\pi\mu) \exp[-(\mu^2 + 1/4)\tau] P_{-1/2+i\mu}(u),$$

---

[3]See Papanicolaou [11], Kesten and Papanicolaou [12].

which, using the integral representation (10), can be written in a more explicit form

$$f(u; \tau) = \frac{\exp[-\tau/4 + \pi^2/4\tau]}{2\sqrt{2\pi\tau}} \int_0^\infty \frac{\exp[-\theta^2/4\tau]\sinh\theta\sin(\pi\theta/2\tau)d\theta}{\sqrt{(\cosh\theta + u)^3}}. \quad (12)$$

For $\tau = 1$, the graph of the probability distribution $f(u, \tau)$ of the relative rate G of change of the total gradient is provided on Fig. 20.7.2 together with the graph of the lognormal density

$$f_{g^2}(u; \tau) = \frac{1}{2u\sqrt{\pi\tau}} \exp\left[-\frac{\ln^2(ue^{-\tau})}{4\tau}\right],$$

of the quantity g^2, where g is the gradient's magnitude.

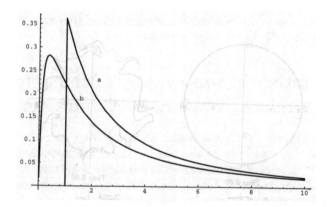

FIGURE 20.7.2.
Comparison of probability densities (a) $f(u; 1)$, and (b) $f_{g^2}(u; 1)$ of, respectively, relative rate G of change of the total gradient, and of the square of gradient's magnitude g^2.

Finally, we provide an expression for the moment generating function

$$\Theta(\lambda; \tau) = \int_1^\infty f(u; \tau)e^{-\lambda u}du$$

of the quantity G. In our case,

$$\Theta(\lambda; \tau) = e^{-t} - \sqrt{\frac{\lambda}{2\tau}}\exp\left(-\lambda - \frac{\tau}{4} + \frac{\pi^2}{4\tau}\right) \times$$

$$\int_0^\infty \exp\left(-\frac{\theta^2}{4\tau} + 2\lambda\cosh^2\frac{\theta}{2}\right)\sin\frac{\pi\theta}{2\tau}\sinh\theta\left[1 - F\left(\sqrt{2\lambda}\cosh\frac{\theta}{2}\right)\right]d\theta,$$

where

$$F(z) = \frac{2}{\sqrt{\pi}} \int_0^z e^{-t^2} dt.$$

Since statistical moments of the stochastic process $G(\tau)$ serve as coefficients in the Taylor expansion,

$$\Theta(\lambda; \tau) = \sum_{n=0}^{\infty} \frac{(-1)^n \lambda^n}{n!} \langle G^n \rangle,$$

using the formula

$$\frac{1}{\sqrt{\pi\tau}} \int_0^{\infty} \exp\left(-\frac{\theta^2}{4\tau}\right) \sin\frac{\pi\theta}{2\tau} \sinh(\beta\theta) \, d\theta = \exp\left(\beta^2\tau - \frac{\pi^2}{4\tau}\right) \sin(\beta\pi),$$

borrowed from the tables of integrals, one can also obtain the expressions for $\langle G \rangle$, $\langle G^2 \rangle$, and $\langle G^3 \rangle$, which were provided above.

20.8 Statistical topography in the delta-correlated random field approximation

20.8.1.   General observations.   In this section we return to a study of statistical characteristics of the passive tracer concentration, and of its spatial gradient, in random incompressible velocity fields from the viewpoint of statistical topography. Those include mean values, probability distributions, as well as various functionals characterizing topographical features. In contrast to the previous sections the functional approach is used. We consider the influence of the mean flow (the linear shear flow) and the molecular diffusion coefficient on the statistics of the tracer. Most of our analysis is carried out in the framework of the delta-correlated (in time) approximation and conditions for its applicability are established.

The initially smooth tracer concentration $\rho(\boldsymbol{x}, t)$, which undergoes diffusion in a random velocity field, acquires in time complex spatial structure. For example, individual realizations of 2-D fields often resemble a complex mountain landscape with randomly distributed peaks, valleys, saddles, and ridges, all of which evolve in time. The mean values such as statistical moments $\langle \rho(\boldsymbol{x}, t) \rangle$, $\langle \rho(\boldsymbol{x}_1, t) \rho(\boldsymbol{x}_2, t) \rangle$, where $\langle ., . \rangle$ denotes averaging over the ensemble of realizations of the random velocity field, smooth out fine details. Such averaging usually brings forth spatiotemporal scales of the whole tracer domain while neglecting its fine dynamics. The detailed structure of the tracer field can be described, as in standard topographic maps, in terms of level curves of the concentration field (2-D case), or level surfaces (3-D case), which are defined by the equation

$$\rho(\boldsymbol{x}, t) - \rho = const.$$

Alternatively, and often more conveniently, we shall employ the distributional (indicator) function

$$\Phi_{t,x}(q) = \delta(\rho(\boldsymbol{x},t) - \rho),$$

which has "values" concentrated on the level curve (surface). Fig. 20.7.1 shows schematically numerical simulation of time evolution of the level curve $\rho(\boldsymbol{x},t) - \rho = const$ of the 2-D concentration field. In view of the incompressibility of the fluid flow, the area bounded by the level curve is conserved, but the picture clearly becomes increasingly fragmented; we observe both the steepening of gradients and the contour dynamics at progressively smaller scales. With help of the distributional indicator function one can study dynamics of various functionals of level curves (surfaces).

For example, the integral

$$A_0(t) = \int d\boldsymbol{x}\, |\boldsymbol{p}(\boldsymbol{x},t)| \delta(\rho(\boldsymbol{x},t) - \rho) = \oint d\ell,$$

where $\boldsymbol{p}(\boldsymbol{x},t) = \nabla\rho(\boldsymbol{x},t)$ is the spatial tracer gradient, gives the level curve length in the 2-D case, and the level surface area in the 3-D case. On the other hand, the integral

$$S(t,\rho) = \pm\frac{1}{2}\int d\boldsymbol{x}\, \boldsymbol{x}\boldsymbol{p}(\boldsymbol{x},t)\delta(\rho(\boldsymbol{x},t) - \rho),$$

gives the area enclosed by the level contour, and

$$\mathcal{N}(\rho,t) \le \int_0^\infty dx\, \frac{|\boldsymbol{x}\boldsymbol{p}(\boldsymbol{x},t)|}{r}\delta(\rho(\boldsymbol{x},t) - \rho), \qquad x = |\boldsymbol{x}|,$$

estimates the number of connected level components as they evolve in time. Note, that averaging indicator functions

$$\mathbf{P}_{t,x}(q) = \langle\delta(\rho(\boldsymbol{x},t) - \rho),$$

$$\mathbf{P}_{t,x}(\rho,\boldsymbol{p}) = \Big\langle\delta(\rho(\boldsymbol{x},t) - \rho)\delta(\boldsymbol{p}(\boldsymbol{x},t) - \boldsymbol{p})\Big\rangle$$

over velocity ensemble defines, respectively, the one-point probability density of the random field ρ and the joint probability density of the tracer and its spatial gradient. In this fashion, even one-point statistical characteristics of tracer permit us to determine statistical means of various functionals of the above types and make statements about the dynamics of individual tracer realizations in a random velocity field. This is particularly useful for problems of passive tracer diffusion in the atmosphere and the ocean where, typically, one does not deal with ensembles, but rather individual realizations. The study of such problems constitutes the subject matter of *statistical topography.*

This picture is well exemplified by the tracer dynamics and can be demonstrated by very simple statistical models of the velocity field. For example, it is relatively easy to write down equations for the statistics of passive tracer in the so-called *delta-correlated* random velocity field which we will adopt later on in this section. It can be seen as an approximation to other, more realistic situations. There, a Lagrangian particle behaves like an ordinary Brownian particle.

The term *statistical topography* was widely used in the physical literature, but in mathematical community-related problems have been extensively studied within the theory of *random surfaces* or *geometry of random fields*. However, the latter emphasized static geometric properties of classical "probabilistically" defined random fields like, Brownian sheets, spatially homogeneous fields, etc, whereas the main interest in the physical community was on "dynamically" defined random fields, that is on random fields satisfying certain partial differential equations.

We have to recognize that the delta-correlated case hides many special features connected to the finiteness of the correlation radius. The latter could be approached via the diffusion approximation method which we studied in the previous sections.

20.8.2.   Another look at the evolution of passive tracer concentration.   In this subsection we return to the dynamical problem in the Lagrangian and Eulerian descriptions, establish their connection, and lay the ground for the statistical analysis of mean concentration and its correlation on the one hand, and the probability distribution of the tracer concentration and its spatial gradient on the other.

Recall that the basic equation that describes the evolution of passive tracer density $\rho(\boldsymbol{x}, t)$ has the form

$$\left(\frac{\partial}{\partial t} + \boldsymbol{V}(\boldsymbol{x}, t) \cdot \frac{\partial}{\partial \boldsymbol{x}} \right) \rho(\boldsymbol{x}, t) = \kappa \frac{\partial^2}{\partial \boldsymbol{x}^2} \rho(\boldsymbol{x}, t), \qquad \rho(\boldsymbol{x}, 0) = \rho_0(\boldsymbol{x}), \quad (1)$$

where κ denotes the "molecular" diffusion coefficient, and $\boldsymbol{V}$ stands for the random velocity field. For incompressible flows, with $\operatorname{div} \boldsymbol{V} = 0$, equation (1) has the form of a conservation law, the quantity

$$Q = \int d\boldsymbol{x} \, \rho(\boldsymbol{x}, t) = \int d\boldsymbol{x} \, \rho_0(\boldsymbol{x})$$

being conserved.

We assume the velocity field, $\boldsymbol{V}$, to be random with finite expectations and decompose it into the mean component,

$$\boldsymbol{v}(\boldsymbol{x}, t) = \langle \boldsymbol{V}(\boldsymbol{x}, t) \rangle,$$

and the random fluctuation,

$$\boldsymbol{F}(\boldsymbol{x}, t) = \boldsymbol{V}(\boldsymbol{x}, t) - \boldsymbol{v}(\boldsymbol{x}, t).$$

Although equation (1), which gives the *Eulerian description* of the system. is linear, the equations for moments of the density field, $\rho^n \equiv \rho^n(\boldsymbol{x}, t)$, are nonlinear,

$$\left(\frac{\partial}{\partial t} + \boldsymbol{V} \cdot \frac{\partial}{\partial \boldsymbol{x}}\right)\rho^n = \kappa\frac{\partial^2}{\partial \boldsymbol{x}^2}\rho^n + \kappa n(n-1)\rho^{n-2}\boldsymbol{p}^2 + n\rho^{n-1}\Delta\rho, \qquad (1')$$

and they involve the spatial gradient $\boldsymbol{p} \equiv \partial\rho(\boldsymbol{x}, t)/\partial\boldsymbol{x}$ of the tracer concentration field.

A direct study of the probability distribution of $\rho(\boldsymbol{x}, t)$ is difficult if (1) contains the second order (diffusion) term in $\boldsymbol{x}$. But the limiting behavior of solutions as $\kappa \to 0$ is feasible and yields significant insight. Here, we require the initial tracer concentration $\rho_0(\boldsymbol{x})$ and its gradient $\boldsymbol{p}_0(\boldsymbol{x})$ to be large-scale (the precise meaning will be explained later). Then one can drop terms containing κ in (1), and consider a transport problem described by equation

$$\left(\frac{\partial}{\partial t} + \boldsymbol{V}(\boldsymbol{x}, t) \cdot \frac{\partial}{\partial \boldsymbol{x}}\right)\rho(\boldsymbol{x}, t) = 0, \qquad \rho(\boldsymbol{x}, 0) = \rho_0(\boldsymbol{x}). \qquad (2)$$

However, one must recognize that dynamic equation (2) describes the physical reality only over a limited time interval.

For a more complete statistical analysis in this time interval it is necessary to include the gradient field $\boldsymbol{p}(\boldsymbol{x}, t) = \partial\rho(\boldsymbol{x}, t)/\partial\boldsymbol{x}$ that obeys, as a consequence of (2), the equations,

$$\left(\frac{\partial}{\partial t} + \boldsymbol{V}(\boldsymbol{x}, t)\frac{\partial}{\partial \boldsymbol{x}}\right)p_i(\boldsymbol{x}, t) = -\frac{\partial V_k}{\partial x_i}p_k(\boldsymbol{x}, t), \qquad \boldsymbol{p}(\boldsymbol{x}, 0) = \boldsymbol{p}_0(\boldsymbol{x}) = \frac{\partial}{\partial \boldsymbol{x}}\rho_0(\boldsymbol{x}). \qquad (3)$$

Here the repeated indices indicate, as usual, the summation over them.

Let us introduce a distributional (indicator) function

$$\Phi_{t,\boldsymbol{x}}(\rho, \boldsymbol{p}) = \delta(\rho(\boldsymbol{x}, t) - q)\delta(\boldsymbol{p}(\boldsymbol{x}, t) - \boldsymbol{p}), \qquad (4)$$

that determines the joint one-point probability distribution of fields ρ and $\boldsymbol{p}$ at a given spatial point in the Eulerian coordinates. We shall also consider a more general two-point distribution of the tracer concentration field and its gradient,

$$\Phi_{t,\boldsymbol{x}_1,\boldsymbol{x}_2}(\rho_1, \boldsymbol{p}_1; \rho_2, \boldsymbol{p}_2) = \delta(\rho(\boldsymbol{x}_1, t) - \rho_1)\delta(\boldsymbol{p}(\boldsymbol{x}_1, t) - \boldsymbol{p}_1)\delta(\rho(\boldsymbol{x}_2, t) - \rho_2)\delta(\boldsymbol{p}(\boldsymbol{x}_2, t) - \boldsymbol{p}_2)$$

$$\equiv \Phi_{t,\boldsymbol{x}_1}(\rho_1, \boldsymbol{p}_1)\Phi_{t,\boldsymbol{x}_2}(\rho_2, \boldsymbol{p}_2). \qquad (4')$$

It contains an additional information in comparison to (4). In particular, it permits the analysis of various functionals constructed from fields ρ and $\boldsymbol{p}$.

Based on equations (2), and (3), one can easily obtain a description of the dynamic evolution of functions (4), and (4'). In particular, $\Phi_{t,\boldsymbol{x}}(\rho, \boldsymbol{p})$ satisfies the *Liouville equation*,

$$\left(\frac{\partial}{\partial t} + \boldsymbol{V}(\boldsymbol{x}, t) \cdot \frac{\partial}{\partial \boldsymbol{x}}\right)\Phi_{t,\boldsymbol{x}}(\rho, \boldsymbol{p}) = \frac{\partial V_k(\boldsymbol{x}, t)}{\partial x_i}\frac{\partial}{\partial p_i}\left(p_k\Phi_{t,\boldsymbol{x}}(\rho, \boldsymbol{p})\right), \qquad (5)$$

with the initial condition

$$\Phi_{0,x}(\rho,\boldsymbol{p}) = \delta(\rho_0(\boldsymbol{x}) - \rho)\delta(\boldsymbol{p}_0(\boldsymbol{x}) - \boldsymbol{p}).$$

The first-order partial differential equations (2–3) can be solved by the method of characteristics. Indeed, if

$$\frac{d\boldsymbol{x}(t|\boldsymbol{\xi})}{dt} = \boldsymbol{V}(\boldsymbol{x},t), \qquad \boldsymbol{x}(0|\boldsymbol{\xi}) = \boldsymbol{\xi}, \tag{6}$$

then (2–3) are reduced to the initial value problem for the ordinary differential equation,

$$\frac{d}{dt}\rho(t|\boldsymbol{\xi}) = 0, \qquad \rho(0|\boldsymbol{\xi}) = \rho_0(\boldsymbol{\xi}), \tag{6'}$$

$$\frac{d}{dt}p_i(t|\boldsymbol{\xi}) = -\frac{\partial V_k(\boldsymbol{x},t)}{\partial x_i}p_k(t|\boldsymbol{\xi}), \qquad p_i(0|\boldsymbol{\xi}) = \frac{\rho_0(\boldsymbol{\xi})}{\partial \xi_i},$$

along the characteristic curves. Equations (6–6′) form a closed system in the *Lagrangian description*. Here and elsewhere $(\ldots|\boldsymbol{\xi})$ indicates conditioning by the initial marker $\boldsymbol{\xi}$ in the Lagrangian formulation. Solution ρ remains constant along the characteristics, so

$$\rho(t|\boldsymbol{\xi}) \equiv \rho_0(\boldsymbol{\xi}).$$

Introducing a new distributional (indicator) function,

$$\Phi_t(\boldsymbol{x},\rho,\boldsymbol{p}|\boldsymbol{\xi}) = \delta(\boldsymbol{x}(t|\boldsymbol{\xi}) - \boldsymbol{x})\delta(\rho(t|\boldsymbol{\xi}) - \rho)\delta(\boldsymbol{p}(t|\boldsymbol{\xi}) - \boldsymbol{p}), \tag{7}$$

that determines the joint density of particle distribution, we can write a similar Liouville equation,

$$\left(\frac{\partial}{\partial t} + \boldsymbol{V}(\boldsymbol{x},t)\cdot\frac{\partial}{\partial \boldsymbol{x}}\right)\Phi_t(\boldsymbol{x},\rho,\boldsymbol{p}|\boldsymbol{\xi}) = \frac{\partial V_k(\boldsymbol{x},t)}{\partial x_i}\frac{\partial}{\partial p_i}\Big(p_k\Phi_t(\boldsymbol{x},\rho,\boldsymbol{p}|\boldsymbol{\xi})\Big), \tag{7'}$$

with the initial condition

$$\Phi_0(\boldsymbol{x},\rho,\boldsymbol{p}|\boldsymbol{\xi}) = \delta(\boldsymbol{\xi} - \boldsymbol{x})\delta(\rho_0(\boldsymbol{\xi}) - \rho)\delta\Big(\frac{\partial \rho_0(\boldsymbol{\xi})}{\partial \boldsymbol{\xi}} - \boldsymbol{p}\Big), \tag{7''}$$

The problems (5), and (7′), describe essentially the same quantity but viewed from, respectively, the Lagrangian, and the Eulerian perspective. Indeed, taking into account the incompressibility of the flow, function (7) can be rewritten as

$$\Phi_t(\boldsymbol{x},\rho,\boldsymbol{p}|\boldsymbol{\xi}) = \delta(\boldsymbol{\xi}(\boldsymbol{x},t) - \boldsymbol{\xi})\Phi_{t,x}(\rho,\boldsymbol{p}),$$

where function $\boldsymbol{\xi}(\boldsymbol{x},t)$ inverts $\boldsymbol{x} = \boldsymbol{x}(t|\boldsymbol{\xi})$, i.e., restores the Lagrangian marker $\boldsymbol{\xi}$ from $(\boldsymbol{x},t)$. Subsequent integration over the marker $\boldsymbol{\xi}$, yields the essential relation between the Eulerian and Lagrangian densities:

$$\Phi_{t,x}(\rho,\boldsymbol{p}) = \int d\boldsymbol{\xi}\,\Phi_t(\boldsymbol{x},\rho,\boldsymbol{p}|\boldsymbol{\xi}). \tag{8}$$

Since parameter $\boldsymbol{\xi}$ enters into (7′) only through the initial condition (7″), clearly, the equations for the Eulerian and the Lagrangian densities should coincide.

Let us also observe that the variable ρ in equations (5) and (7′) enters only through the initial conditions. For that reason, multiplying those by ρ^n and integrating over ρ and $\boldsymbol{p}$ we get a dynamic evolution equation for moments ρ^n that coincides with (5) and (7′). The latter property is connected with the conservation of ρ along characteristics, and $\rho(\boldsymbol{x}, t) = \rho_0(\boldsymbol{\xi}(\boldsymbol{x}, t))$.

To recapitulate, the quantities of interest to us, the Eulerian $\rho^n(\boldsymbol{x}, t)$ and $\Phi_{t,\boldsymbol{x}}(\rho, \boldsymbol{p})$, obey the same dynamic equations as the corresponding Lagrangian probability densities $\Phi_t(\boldsymbol{x}|\boldsymbol{\xi})$ and $\Phi_t(\boldsymbol{x}, \rho, \boldsymbol{p}|\boldsymbol{\xi})$.

In a similar manner we can consider a system of two particles

$$
\begin{aligned}
\frac{d\boldsymbol{x}_1(t)}{dt} &= \boldsymbol{V}(\boldsymbol{x}_1, t), & \boldsymbol{x}_1(0) &= \boldsymbol{\xi}_1, \\
\frac{dp_{1i}(t)}{dt} &= -\frac{\partial V_k(\boldsymbol{x}_1, t)}{\partial x_{1i}} p_{1k}(t), & p_{1i}(0) &= \frac{\partial \rho_0(\boldsymbol{\xi}_1)}{\partial \xi_{1i}}, \\
\frac{d\boldsymbol{x}_2(t)}{dt} &= \boldsymbol{V}(\boldsymbol{x}_2, t), & \boldsymbol{x}_2(0) &= \boldsymbol{\xi}_2, \\
\frac{dp_{2i}(t)}{dt} &= -\frac{\partial V_k(\boldsymbol{x}_2, t)}{\partial x_{2i}} p_{2k}(t), & p_{2i}(0) &= \frac{\partial \rho_0(\boldsymbol{\xi}_2)}{\partial \xi_{2i}}.
\end{aligned}
\tag{9}
$$

The corresponding two-point Lagrangian density will then be described by the same equation as the two-point Eulerian density (4′). All (Eulerian and Lagrangian) densities above obey stochastic partial differential equations with the randomness introduced through the velocity fluctuations $\boldsymbol{F}$. Their ensemble averaging yields the evolution of the probability densities

$$
P_{t,\boldsymbol{x}}(\rho, \boldsymbol{p}) = \langle \Phi_{t,\boldsymbol{x}}(\rho, \boldsymbol{p}) \rangle_F, \qquad P_t(\boldsymbol{x}, \rho, \boldsymbol{p}|\boldsymbol{\xi}) = \langle \Phi_t(\boldsymbol{x}, \rho, \boldsymbol{p}|\boldsymbol{\xi}) \rangle_F,
\tag{10}
$$

in both the Eulerian and the Lagrangian descriptions.

20.8.3.   Statistical averaging.

Here we shall implement the averaging procedure of Subsection 20.8.2 in a number of special cases. For instance, averaging equations (1) over the $\boldsymbol{F}$-ensemble yields an evolution of the mean field, where random velocities $\boldsymbol{F}$ are coupled to random solution $\rho = \rho[\boldsymbol{F}]$, itself a functional of $\boldsymbol{F}$, through the fluctuation term

$$
\left\langle \boldsymbol{F} \cdot \frac{\partial}{\partial \boldsymbol{x}} \rho \right\rangle.
\tag{11}
$$

Here $\langle ., . \rangle$ means averaging over the space-time ensemble. So, to get the effective mean field evolution one needs to decouple the cross-correlation term (11). The decoupling methods strongly depend on the nature of random field $\boldsymbol{F}$.

In the Gaussian case decoupling exploits the so-called *Furutsu–Novikov* formula which we have seen at the beginning of this chapter. Namely, given a zero-mean Gaussian random vector field $\boldsymbol{F} = (F_i)$, any functional $R[\boldsymbol{F}]$ satisfies the identity,

$$\left\langle F_i(\boldsymbol{x}, t) R[\boldsymbol{F}] \right\rangle = \int d\boldsymbol{x}' \int dt' \left\langle F_i(\boldsymbol{x}, t) F_j(\boldsymbol{x}', t') \right\rangle \left\langle \frac{\delta R[\boldsymbol{F}]}{\delta F_j(\boldsymbol{x}', t')} \right\rangle. \qquad (12)$$

So, the cross-term decouples into a superposition of products of the correlation coefficients and the mean variational derivatives of R. Applying the Furutsu–Novikov formula (12) to the cross-term (11) of equation (1) we get that

$$\left(\frac{\partial}{\partial t} + \boldsymbol{v}(\boldsymbol{x}, t) \cdot \frac{\partial}{\partial \boldsymbol{x}} \right) \langle \rho(\boldsymbol{x}, t) \rangle + \int dt' \int d\boldsymbol{x}' B_{ij}(\boldsymbol{x}, t; \boldsymbol{x}', t') \frac{\partial}{\partial x_i} \left\langle \frac{\delta \rho(\boldsymbol{x}, t)}{\delta F_j(\boldsymbol{x}', t')} \right\rangle$$
$$= \kappa \frac{\partial^2}{\partial \boldsymbol{x}^2} \langle \rho(\boldsymbol{x}, t) \rangle, \qquad (13)$$

where $B_{ij}(\boldsymbol{x}, t; \boldsymbol{x}', t') = \langle F_i(\boldsymbol{x}, t) F_j(\boldsymbol{x}', t') \rangle$ is the space-time correlation of $\boldsymbol{F}$. Although equation (13) is exact for any zero-mean Gaussian field $\boldsymbol{F}$, it is not closed since the evolution of the mean field is coupled to the mean variational derivative with respect to $\boldsymbol{F}$. The variational derivative $\delta \rho / \delta \boldsymbol{F}$, itself solves a stochastic differential equation

$$\frac{\partial}{\partial t} \left(\frac{\delta \rho(\boldsymbol{x}, t)}{\delta F_j(\boldsymbol{x}', t')} \right) + [\boldsymbol{v}(\boldsymbol{x}, t) + \boldsymbol{F}(\boldsymbol{x}, t)] \cdot \frac{\partial}{\partial \boldsymbol{x}} \left(\frac{\delta \rho(\boldsymbol{x}, t)}{\delta F_j(\boldsymbol{x}', t')} \right) = \kappa \frac{\partial^2}{\partial \boldsymbol{x}^2} \left(\frac{\delta \rho(\boldsymbol{x}, t)}{\delta F_j(\boldsymbol{x}', t')} \right)$$
$$(14)$$

obtained by varying (1) in $\boldsymbol{F}$, and satisfies the initial condition

$$\frac{\delta \rho(\boldsymbol{x}, t)}{\delta F_j(\boldsymbol{x}', t')} \bigg|_{t \to t'+0} = -\delta(\boldsymbol{x} - \boldsymbol{x}') \frac{\partial}{\partial x_j} \rho(\boldsymbol{x}, t'),$$

so $\delta \rho / \delta \boldsymbol{F}$ could be viewed as a stochastic analogue of Green's function for a problem of type (1). Taking the ensemble average of (14) and then applying the Furutsu–Novikov formula would produce higher order variational derivatives $\langle \delta^2 \rho / \delta F_i \delta F_j \rangle$ coupled together. Solution of such system of moment equations would require a suitable closure hypothesis that could be rigorously implemented only in certain cases. One such case, namely the delta-correlated in time random fields $\boldsymbol{F}(\boldsymbol{x}, t)$, will be discussed below.

20.8.4. Delta-correlated random field approximation; mean tracer concentration and its correlation function.
In the delta-correlated approximation the random fluctuation field $\boldsymbol{F}(\boldsymbol{x}, t)$ is assumed to be zero-mean, Gaussian, with the covariance structure,

$$B_{ij}(\boldsymbol{x}, t; \boldsymbol{x}', t') = \langle F_i(\boldsymbol{x}, t) F_j(\boldsymbol{x}', t') \rangle = 2 B_{ij}^{\text{eff}}(\boldsymbol{x}; \boldsymbol{x}', t) \delta(t - t'), \qquad (15)$$

where
$$B_{ij}^{\text{eff}}(\boldsymbol{x};\boldsymbol{x}',t) = \frac{1}{2}\int_{-\infty}^{\infty} dt'\, B_{ij}(\boldsymbol{x},t;\boldsymbol{x}',t').$$

In this case, the integral term in (13) could be expressed through $\delta\rho(\boldsymbol{x};t)/\delta F_j(\mathbf{r}';t')$ at $t = t'$, i.e. through the initial condition of (14). As a result we get a closed-form differential equation for the mean

$$\left(\frac{\partial}{\partial t} + \boldsymbol{v}(\boldsymbol{x},t)\frac{\partial}{\partial \boldsymbol{x}}\right)\langle\rho(\boldsymbol{x},t)\rangle = \frac{\partial}{\partial x_i}\left[B_{ij}^{\text{eff}}(\boldsymbol{x};\boldsymbol{x},t)\frac{\partial}{\partial x_j} + \kappa\frac{\partial}{\partial x_i}\right]\langle\rho(\boldsymbol{x},t)\rangle.$$

So, the spatial variance $B^{\text{eff}}(\boldsymbol{x};\boldsymbol{x},t)$ of $\boldsymbol{F}$ becomes the effective diffusion coefficient for the mean concentration.

If the velocity fluctuation field $\boldsymbol{F}(\boldsymbol{x},t)$ is homogeneous and isotropic in space, and stationary in time, then the effective correlation coefficients depend just on $|\boldsymbol{x} - \boldsymbol{x}'|$, i.e.,

$$B_{ij}^{\text{eff}}(\boldsymbol{x};\boldsymbol{x}',t) = B_{ij}^{\text{eff}}(|\boldsymbol{x}-\boldsymbol{x}'|), \qquad B_{ij}^{\text{eff}}(\boldsymbol{x}-\boldsymbol{x}') = \frac{1}{2}\int_{-\infty}^{\infty} dt'\, B_{ij}(\boldsymbol{x}-\boldsymbol{x}',t-t'),$$

and their value at 0 is a scalar matrix with coefficient D_1:

$$B_{ij}^{\text{eff}}(0) = \delta_{ij}D_1, \qquad D_1 = \frac{1}{N}B_{ii}^{\text{eff}}(0). \tag{15'}$$

Here N (= 2, or 3) denotes the dimension of the space. Hence, we get

$$\left(\frac{\partial}{\partial t} + \boldsymbol{v}(\boldsymbol{x},t)\frac{\partial}{\partial \boldsymbol{x}}\right)\langle\rho(\boldsymbol{x},t)\rangle = (D_1 + \kappa)\frac{\partial^2}{\partial \boldsymbol{x}^2}\langle\rho(\boldsymbol{x},t)\rangle. \tag{16}$$

In the particular case of zero mean flow, $\boldsymbol{v}(\boldsymbol{x},t) = 0$, and the initial tracer concentration, $\rho_0(\boldsymbol{x})$, being itself a homogeneous random field, the random solution $\rho(\boldsymbol{x},t)$ will also be homogeneous and isotropic. Hence,

$$\langle\rho(\boldsymbol{x},t)\rangle = \rho_0.$$

Similarly, in this case, for correlation function

$$\Gamma(\boldsymbol{x},t) = \langle\rho(\boldsymbol{x}_1,t)\rho(\boldsymbol{x}_2,t)\rangle_{\boldsymbol{F}}, \qquad \boldsymbol{x} = \boldsymbol{x}_1 - \boldsymbol{x}_2,$$

one obtains the equation,

$$\frac{\partial}{\partial t}\Gamma(\boldsymbol{x},t) = 2\kappa\frac{\partial^2}{\partial \boldsymbol{x}^2}\Gamma(\boldsymbol{x},t) + 2\frac{\partial^2}{\partial r_i\partial r_j}D_{ij}(\boldsymbol{x})\Gamma(\boldsymbol{x},t), \tag{17}$$

where the matrix-valued structure function of the field $\boldsymbol{F}$

$$D_{ij}(\boldsymbol{x}) = B_{ij}^{\text{eff}}(0) - B_{ij}^{\text{eff}}(\boldsymbol{x}). \tag{17'}$$

Let us remark, that equations (16) and (17) have the Fokker-Planck form for the one-particle and two-particle probability densities of the Lagrangian coordinates. Furthermore, the Lagrangian relation (8) yields a Markov process. Additionally, equation (17) describes relative diffusion of two particles. For sufficiently small initial distances between two particles ($r_0 \ll l_0$, where l_0 is the spatial radius of correlation of the fluctuation field $\boldsymbol{F}$) function $D_{ij}(\boldsymbol{x})$ can be expanded in the Taylor series, and in the first approximation

$$D_{ij}(\boldsymbol{x}) = -\frac{1}{2}\frac{\partial^2 B_{ij}^{eff}(\boldsymbol{x})}{\partial x_k \partial x_l}\Big|_{\boldsymbol{x}=0} x_k x_l. \tag{18}$$

Now, let us introduce the spectral density of energy of the flow by the formula

$$B_{ij}^{eff}(\boldsymbol{x}) = \int d\boldsymbol{k}\, E(k)(\delta_{ij} - \frac{k_i k_j}{k^2})e^{i\boldsymbol{k}\cdot\boldsymbol{x}}. \tag{19}$$

Then

$$-\frac{\partial^2 B_{ij}^{eff}(\boldsymbol{x})}{\partial x_k \partial x_l}\Big|_{\boldsymbol{x}=0} = D_2\big\{(N+1)\delta_{ij}\delta_{k\ell} - \delta_{ik}\delta_{jl} - \delta_{jk}\delta_{il}\big\}, \tag{20}$$

where

$$D_2 = \frac{1}{N(N+2)}\int d\boldsymbol{k}\, k^2 E(k). \tag{21}$$

Note that the quantity D_1 introduced earlier is also determined by the spectral density $E(k)$ via the equality,

$$D_1 = \frac{N-1}{N}\int d\boldsymbol{k}\, E(k). \tag{21'}$$

In this case the diffusion tensor (18) simplifies, and can be written in the form,

$$D_{ij}(\boldsymbol{x}) = \frac{1}{2}D_2\{(N+1)\boldsymbol{x}^2\delta_{ij} - 2x_i x_j\}. \tag{22}$$

Substituting (22) into (16), multiplying both sides of the obtained equation by $\boldsymbol{x}^2$ and integrating over $\boldsymbol{x}$, we obtain the equation,

$$\frac{d}{dt}\langle \boldsymbol{x}^2(t)\rangle = 4\kappa N + 2(N+2)(N-1)D_2\langle \boldsymbol{x}^2(t)\rangle \tag{23}$$

for the variance, $\langle \boldsymbol{x}^2\rangle$, (note that the mean, $\langle \boldsymbol{x}(t)\rangle$, is conserved). Its solution has the following structure:

$$\langle \boldsymbol{x}^2\rangle = x_0^2 e^{2(N+2)(N-1)D_2 t} + \frac{2\kappa N}{(N+2)(N-1)D_2}\big\{e^{2(N+2)(N-1)D_2 t} - 1\big\}. \tag{24}$$

It is clear from (24) that, if

$$\kappa \ll D_2 x_0^2, \tag{25}$$

then the effects of molecular diffusion on the particle are not essential, and the last term in (24) can be omitted. In this case the solution becomes an exponentially growing function in time:

$$\langle \boldsymbol{x}^2 \rangle = x_0^2 e^{2(N+2)(N-1)D_2 t}. \tag{24'}$$

Expression (24′) is valid whenever expansion (18) is, that is for the time range

$$D_2 t \ll \frac{1}{(N+2)(N-1)} \ln \frac{l_0}{x_0}. \tag{26}$$

Note that the influence of the molecular diffusion for the above one-particle probability density, according to (16), can be neglected if the following condition is satisfied:

$$\kappa \ll D_1. \tag{25'}$$

Approximation (18) is, however, not valid for the turbulent fluid flow for which the structure function cannot be expanded into a Taylor series.

Remark 1. As we already mentioned in the introduction, the mean value $\langle q(r,t) \rangle$ and the correlation function $\Gamma(\boldsymbol{x},t)$ characterize the spatiotemporal scales of global passive tracer domain in the sense of statistical topography. At the same time they hide the detailed dynamics inside this domain. Clearly, the molecular diffusion coefficient has little influence on these scales and conditions (25) and (25′) are not very restrictive in the physical sense.

So far, we considered the mean concentration of the tracer and its correlation function which are described in the closed form due to the linearity of the basic equation (1). If one considers higher moments of the tracer concentration described by equation (1′) then we do not get a closed-form description. Indeed, averaging (1′) over the $\boldsymbol{F}$-ensemble we obtain the equation,

$$\left(\frac{\partial}{\partial t} + \boldsymbol{v}(\boldsymbol{x},t)\frac{\partial}{\partial \boldsymbol{x}}\right)\langle \rho^n(\boldsymbol{x},t) \rangle = (D_1 + \kappa)\frac{\partial^2}{\partial \boldsymbol{x}^2}\langle \rho^n(\boldsymbol{x},t) \rangle - \kappa n(n-1)\langle \rho^{n-2}(\boldsymbol{x},t)\boldsymbol{p}^2(\boldsymbol{x},t) \rangle, \tag{27}$$

whose righ-hand side contains an unknown covariance of the concentration field and its spatial gradient. In order to understand better the structure of the tracer gradient field one can, in the first approximation, neglect effects of the molecular diffusion, i.e., consider stochastic system (2–3).

20.8.5.   Fine structure of passive tracer fluctuations in random velocity fields. In this subsection we look at subtler characteristics of tracer fluctuations than the means and correlations considered in Subsection 20.8.4, like the joint probability density of the tracer concentration and its gradient. To this end we average Liouville equation (5) over the ensemble of realizations of fluctuation field $\boldsymbol{F}$ and use a version of the Furutsu–Novikov formula,

$$\frac{\delta}{\delta u_j(\boldsymbol{x}', t - 0)} \Phi_{t,\boldsymbol{x}}(\rho, \boldsymbol{p}) = \left\{ -\delta(\boldsymbol{x} - \boldsymbol{x}') \frac{\partial}{\partial x_j} + \frac{\partial \delta(\boldsymbol{x} - \boldsymbol{x}')}{\partial x_i} \frac{\partial}{\partial p_i} p_j \right\} \Phi_{t,\boldsymbol{x}}(\rho, \boldsymbol{p}),$$
$$(13')$$

for the variational derivative. As the result we get for the one-point joint probability density of fields $\rho(\boldsymbol{x}, t)$, and $\boldsymbol{p}(\boldsymbol{x}, t)$, the Fokker-Planck equation

$$\left\{ \frac{\partial}{\partial t} + \boldsymbol{v}(\boldsymbol{x}, t) \frac{\partial}{\partial \boldsymbol{x}} - \frac{\partial \boldsymbol{p} \boldsymbol{v}(\boldsymbol{x}, t)}{\partial \boldsymbol{x}} \frac{\partial}{\partial \boldsymbol{p}} \right\} P_{t,\boldsymbol{x}}(\rho, \boldsymbol{p}) = D_1 \frac{\partial^2}{\partial \boldsymbol{x}^2} P_{t,\boldsymbol{x}}(\rho, \boldsymbol{p}) + \quad (28)$$

$$+ D_2 \left\{ (N + 1) \frac{\partial^2}{\partial \boldsymbol{p}^2} \boldsymbol{p}^2 - 2 \frac{\partial}{\partial \boldsymbol{p}} \boldsymbol{p} - 2 \left(\frac{\partial}{\partial \boldsymbol{p}} \boldsymbol{p} \right)^2 \right\} P_{t,\boldsymbol{x}}(\rho, \boldsymbol{p})$$

with initial condition

$$P_{0,\boldsymbol{x}}(\rho, \boldsymbol{p}) = \delta(\rho_0(\boldsymbol{x}) - \rho) \delta \left(\frac{\partial}{\partial \boldsymbol{x}} \rho_0(\boldsymbol{x}) - \boldsymbol{p} \right). \quad (28')$$

Constants D_1 and D_2 introduced in (21) and (21') become, that way, the new diffusion coefficients for the Fokker-Planck equation in the $\boldsymbol{x}$- and $\boldsymbol{p}$-spaces, respectively.

Equation (28) can be written in an operator form

$$\frac{\partial}{\partial t} P_{t,\boldsymbol{x}}(\rho, \boldsymbol{p}) = \hat{L}(\boldsymbol{x}, t) P_{t,\boldsymbol{x}}(\rho, \boldsymbol{p}) + \hat{M}(\boldsymbol{x}, \boldsymbol{p}, t) P_{t,\boldsymbol{x}}(\rho, \boldsymbol{p}), \quad (29)$$

where operators $\hat{L}$ and $\hat{M}$ are defined by

$$\hat{L}(\boldsymbol{x}, t) = -\boldsymbol{v}(\boldsymbol{x}, t) \frac{\partial}{\partial \boldsymbol{x}} + D_1 \frac{\partial^2}{\partial \boldsymbol{x}^2},$$

$$\hat{M}(\boldsymbol{x}, \boldsymbol{p}, t) = \frac{\partial \boldsymbol{p} \boldsymbol{v}(\boldsymbol{x}, t)}{\partial \boldsymbol{x}} \frac{\partial}{\partial \boldsymbol{p}} + D_2 \left\{ (N + 1) \frac{\partial^2}{\partial \boldsymbol{p}^2} \boldsymbol{p}^2 - 2 \frac{\partial}{\partial \boldsymbol{p}} \boldsymbol{p} - 2 \left(\frac{\partial}{\partial \boldsymbol{p}} \boldsymbol{p} \right)^2 \right\}.$$
$$(29')$$

As was discussed earlier, operator $\hat{L}(\boldsymbol{x}, t)$ defines the spatial diffusion of the Lagrangian particle, while $\hat{M}(\boldsymbol{x}, \boldsymbol{p}, t)$ defines the diffusion of the tracer gradient and the correlation of the gradient with the position vector.

In the simplest case of zero mean flow ($\boldsymbol{v} = 0$), to be discussed below in Example 2, or in the case of shear flow with constant gradient, to be discussed in the later subsection, operators $\hat{L}$ and $\hat{M}$ commute, which reflects statistical independence of diffusions in the position $\boldsymbol{x}$-space and the gradient $\boldsymbol{p}$-space.

Also, notice that in this case for spatially homogeneous and isotropic Gaussian velocity fluctuations, the corresponding diffusion operators are also isotropic. This fact will give us additional information on the fluctuations of the tracer gradient that we shall now outline.

Example 1. Zero mean flow. Consider the case of the zero mean flow ($\boldsymbol{v}=0$). Then, the solution of equation (29) is obtained by averaging equation (8),

$$P_{t,x}(\rho, \boldsymbol{p}) = \int d\boldsymbol{\xi} \, P_t(\boldsymbol{x}|\boldsymbol{\xi}) P_t(\rho, \boldsymbol{p}|\boldsymbol{\xi}). \tag{30}$$

Here, $P_t(\boldsymbol{x}|\boldsymbol{\xi})$ denotes the probability density of the particle Lagrangian coordinate given by

$$\frac{\partial}{\partial t} P_t(\boldsymbol{x}|\boldsymbol{\xi}) = D_1 \frac{\partial^2}{\partial \boldsymbol{x}^2} P_t(\boldsymbol{x}|\boldsymbol{\xi}), \qquad P_0(\boldsymbol{x}|\boldsymbol{\xi}) = \delta(\boldsymbol{x} - \boldsymbol{\xi}), \tag{31}$$

and $P_t(\rho, \boldsymbol{p}|\boldsymbol{\xi})$ is the joint probability density of ρ and $\nabla\rho$ that satisfies

$$\frac{\partial}{\partial t} P_t(\rho, \boldsymbol{p}|\boldsymbol{\xi}) = D_2 \left\{ (N+1) \frac{\partial^2}{\partial \boldsymbol{p}^2} \boldsymbol{p}^2 - 2 \frac{\partial}{\partial \boldsymbol{p}} \boldsymbol{p} - 2 \left(\frac{\partial}{\partial \boldsymbol{p}} \boldsymbol{p} \right)^2 \right\} P_t(\rho, \boldsymbol{p}|\boldsymbol{\xi}), \tag{32}$$

$$P_0(\rho, \boldsymbol{p}|\boldsymbol{\xi}) = \delta(\rho_0(\boldsymbol{\xi}) - \rho)\delta(\boldsymbol{p}_0(\boldsymbol{\xi}) - \boldsymbol{p}).$$

Specifically, in the 2-D case the solution of equation (31) is

$$P_t(\boldsymbol{x}|\boldsymbol{\xi}) = \exp\left[D_1 t \frac{\partial^2}{\partial \boldsymbol{x}^2} \right] \delta(\boldsymbol{x} - \boldsymbol{\xi}) = (4\pi D_1 t)^{-1} \exp\left[-\frac{(\boldsymbol{x} - \boldsymbol{\xi})^2}{4 D_1 t} \right]. \tag{31'}$$

It corresponds to the Brownian particle with parameters

$$\langle \boldsymbol{x}(t|\boldsymbol{\xi}) \rangle = \boldsymbol{\xi}, \qquad \sigma_{ij}^2(t) = 2 D_1 \delta_{ij} t.$$

From (32) we derive the following moment equations for gradient $\boldsymbol{p}(t|\boldsymbol{\xi})$:

$$\frac{\partial}{\partial t} \langle |\boldsymbol{p}(t|\boldsymbol{\xi})|^n \rangle = D_2 n (N+n)(N-1) \langle |\boldsymbol{p}(t(\boldsymbol{\xi})|^n \rangle, \tag{33}$$

$$\frac{\partial}{\partial t} \langle \boldsymbol{p}|\boldsymbol{p}(t|\boldsymbol{\xi})|^n \rangle = D_2 n (N+n+2)(N-1) \langle \boldsymbol{p}|\boldsymbol{p}|^n \rangle,$$

and, in particular,

$$\frac{\partial}{\partial t} \langle \boldsymbol{p}(t|\boldsymbol{\xi}) \rangle = 0, \qquad \text{i.e.} \qquad \langle \boldsymbol{p}(t|\boldsymbol{\xi}) \rangle = \boldsymbol{p}_0(\boldsymbol{\xi}).$$

So the moment functions grow exponentially in time, with the exception of the conserved quantity $\langle \boldsymbol{p}(t|\boldsymbol{\xi}) \rangle$. Also, for an arbitrary vector $\boldsymbol{a}$, the quantity $\langle |\boldsymbol{ap}(t|\boldsymbol{\xi})| \rangle$ is conserved, i.e.,

$$\langle |\boldsymbol{ap}(t|\boldsymbol{\xi})| \rangle = |\boldsymbol{ap}_0(\boldsymbol{\xi})|. \tag{33'}$$

Notice that in the Eulerian representation, (30) implies that the exponential time growth of moments $\langle |\boldsymbol{p}(\boldsymbol{x},t)|^n \rangle$ and $\langle \boldsymbol{p}(\boldsymbol{x},t)|\boldsymbol{p}(\boldsymbol{x},t)|^n \rangle$ is accompanied by their spatial dissipation with the diffusion rate D_1.

Equation (33) also implies that the normalized quantity $|\boldsymbol{p}(t|\boldsymbol{\xi})|/|\boldsymbol{p}_0(\boldsymbol{\xi})|$ has a lognormal probability distribution, i.e.,

$$\chi(t) = \ln \frac{|\boldsymbol{p}(t|\boldsymbol{\xi})|}{|\boldsymbol{p}_0(\boldsymbol{\xi})|}$$

is Gaussian with parameters

$$\langle \chi(t) \rangle = D_2 N(N-1)t, \qquad \sigma_\chi^2(t) = 2D_2(N-1)t. \tag{34}$$

So, as we have shown in previous sections, the typical realization of process $|\boldsymbol{p}(t)|$ has an exponential growth

$$|\boldsymbol{p}(t|\boldsymbol{\xi})| \sim |\boldsymbol{p}_0(\boldsymbol{\xi})| \exp[D_2 N(N-1)t],$$

accompanied by large excursions relative to the above exponential curve. In addition, there exist several lower probabilistic estimates for the quantity $\chi(t)$. Note that this situation is fundamentally different from the one-dimensional problem (where the fluid flow is always compressible). There, the gradient conserves its sign and a typical realization of the gradient process is an exponentially decaying curve.

We also get from (32) the evolution equation

$$\frac{\partial}{\partial t}\langle p_i(t|\boldsymbol{\xi})p_j(t|\boldsymbol{\xi}) \rangle = -4D_2\langle p_i(t|\boldsymbol{\xi})p_j(t|\boldsymbol{\xi}) \rangle + 2D_2(N+1)\delta_{ij}\langle \boldsymbol{p}^2(t)|\boldsymbol{\xi} \rangle, \tag{35}$$

for the covariance $\langle p_i(t|\boldsymbol{\xi})p_j(t|\boldsymbol{\xi}) \rangle$. Clearly, cross-terms $i \neq j$ of the correlation of different components of gradient $\boldsymbol{p}(t|\boldsymbol{\xi})$ converge rapidly (exponentially) to zero. So, for large time values $D_2 t \gg 1/4$ the vector $\boldsymbol{p}(t|\boldsymbol{\xi})$ undergoes full statistical isotropization independent of the initial conditions.

20.8.6. Geometric interpretation of the fine structure: statistical topography aspects. In the previous subsections we have obtained a series of general equations which, in principle, permit us to obtain information about the time evolution of one-point, two-point, etc., probability densities. The complete set of these equations, obviously, will give also the exhaustive description of the behavior of separate realizations of solutions of the initial stochastic equations. However, in practice, even the one-point probability densities for solutions of the stochastic equations can be calculated only in a few special cases. So in our case of mean flow it becomes impossible. Even in cases when this could be done, and various statistical characteristics of solutions could be computed they would behave differently than individual realizations of the original stochastic system.

In this context an important question has to be addressed: how can we obtain information on geometric properties of individual realizations of random fields from partial information of their statistical characteristics? This question is especially pertinent for real physical systems like oceans and the atmosphere, where one, generally speaking, deals with concrete realizations rather than ensembles. The study of these problems is the subject matter of *statistical topography* of random fields.

The structure of the spatial random field $\rho(\boldsymbol{x}, t)$ of the passive tracer is highly chaotic, its individual realizations constantly change their shape, and are characterized by "sharp peaks," saddles, ridges, etc. Averaging clearly smoothes out all special features of individual realizations. The level curves of such a "rough system" driven by stochastic flows also obey a stochastic time evolution determined by the equation $\rho(\boldsymbol{x}, t) = \rho$. The mean values of distributional indicator functions $\Phi_{t,\boldsymbol{x}}(\rho) = \delta(\rho(\boldsymbol{x}, t) - \rho)$ of these level curves define the corresponding probability density. The function $\Phi_{t,\boldsymbol{x}}(\rho)$ determines a surface S of constant values of, for example, concentration, temperature, etc., in the 3-D space, and an analogous contour ℓ in the 2-D space.

In this subsection, we will restrict our attention to the case of two-dimensional fluid flows.

Example 2. Level Curve Length Statistics.   Consider the following auxiliary integral related to function $\Phi_{t,\boldsymbol{x}}(\rho)$:

$$A_n(t) = \int d\boldsymbol{x} |\boldsymbol{p}(\boldsymbol{x}, t)|^{n+1} \delta[\rho(\boldsymbol{x}, t) - \rho] = \oint |p(\boldsymbol{x}, t)|^n d\ell. \qquad (36)$$

where $\boldsymbol{p}(\boldsymbol{x}, t) = \partial\rho(\boldsymbol{x}, t)/\partial\boldsymbol{x}$ is the spatial gradient of the random field $\rho(\boldsymbol{x}, t)$. These integrals are moments of density gradients integrated over contours

In particular, for $n = 0$, formula (36) gives the length of the contour

$$A_0(t) = \ell(t) = \int d\boldsymbol{x} |\boldsymbol{p}(\boldsymbol{x}, t)| \delta[\rho(\boldsymbol{x}, t) - \rho] = \oint d\ell. \qquad (36')$$

The mean tracer concentration gradient over the planar level sets is given by the contour integral

$$\boldsymbol{A}(t) = \int_S \nabla\rho(\boldsymbol{x}, t) dS = \rho \int d\boldsymbol{x}\, \boldsymbol{p}(\boldsymbol{x}, t) \delta[\rho(\boldsymbol{x}, t) - \rho] = \rho \oint \frac{\boldsymbol{p}(\boldsymbol{x}, t)}{|\boldsymbol{p}(\boldsymbol{x}, t)|} d\ell. \quad (36'')$$

Equation (36–36″) can be rewritten in terms of the probability distribution function $\Phi_{t,\boldsymbol{x}}(\rho, \boldsymbol{p})$ as follows:

$$A_n(t) = \int d\boldsymbol{x} \int d\boldsymbol{p}\, p^{n+1} \Phi_{t,\boldsymbol{x}}(\rho, \boldsymbol{p}), \qquad \boldsymbol{A}(t) = \rho \int d\boldsymbol{x} \int d\boldsymbol{p}\, \boldsymbol{p}\, \Phi_{t,\boldsymbol{x}}(\rho, \boldsymbol{p}).$$
$$(37)$$

Consequently, their averages are determined by one-point probability density $P_{t,x}(\rho, \boldsymbol{p})$ via formulas

$$\langle A_n(t) \rangle = \int d\boldsymbol{x} \int d\boldsymbol{p}\, p^{n+1} P_{t,x}(\rho, \boldsymbol{p}), \qquad \langle \boldsymbol{A}(t) \rangle = \rho \int d\boldsymbol{x} \int d\boldsymbol{p}\, \boldsymbol{p} P_{t,x}(\rho, \boldsymbol{p}).$$

$$(37')$$

Substituting (30) for $P_{t,x}(\rho, \boldsymbol{p})$ and taking into account that $\int d\boldsymbol{x}\, P_t(\boldsymbol{x}|\boldsymbol{\xi}) = 1$, we express the quantities of interest

$$\langle A_n(t) \rangle = \int d\boldsymbol{\xi} \int d\boldsymbol{p}\, p^{n+1} P_t(\rho, \boldsymbol{p}|\boldsymbol{\xi}), \qquad \langle \boldsymbol{A}(t) \rangle = \rho \int d\boldsymbol{\xi} \int d\boldsymbol{p}\, \boldsymbol{p} P_t(\rho, \boldsymbol{p}|\boldsymbol{\xi}),$$

$$(37'')$$

in terms of Lagrangian probability density.

In the 2-D case, those satisfy Equation (32)

$$\frac{\partial}{\partial t} P_t(\rho, \boldsymbol{p}|\boldsymbol{\xi}) = D_2 \left\{ 3\frac{\partial^2}{\partial \boldsymbol{p}^2} p^2 - 2\frac{\partial}{\partial \boldsymbol{p}} \boldsymbol{p} - 2(\frac{\partial}{\partial \boldsymbol{p}} \boldsymbol{p})^2 \right\} P_t(\rho, \boldsymbol{p}|\boldsymbol{\xi}),$$

$$P_0(\rho, \boldsymbol{p}|\boldsymbol{\xi}) = \delta(\rho_0(\boldsymbol{\xi}) - \rho)\delta(\boldsymbol{p}_0(\boldsymbol{\xi}) - \boldsymbol{p}), \qquad \boldsymbol{p}_0(\xi) = \frac{\partial \rho_0(\boldsymbol{\xi})}{\partial \boldsymbol{\xi}}.$$

Differentiating expression $(37'')$ with respect to time and applying equality (32) with $N = 2$, we obtain differential equations for the mean values

$$\frac{d}{dt}\langle A_n(t) \rangle = (n+1)(n+3)D_2\langle A_n(t) \rangle, \quad \langle A_n(0) \rangle = A_n(0),$$

$$\frac{d}{dt}\langle \boldsymbol{A}(t) \rangle = 0, \qquad \langle \boldsymbol{A}(0) \rangle = \boldsymbol{A}(0).$$

Their solutions give exponentially growing in time functions

$$\langle \ell(t) \rangle = \ell_0 e^{3D_2 t}, \quad \langle A_n(t) \rangle = A_n(0)e^{(1+n)(n+3)D_2 t}, \tag{38}$$

whereas the mean concentration gradient, averaged over the area, is conserved, i.e., $\langle \boldsymbol{A}(t) \rangle = \boldsymbol{A}(0)$.

The exponential growth of (38) indicates strong roughening of the tracer level curves with time, that leads to "fractal-like" structures. The situation is similar both for the tracer density and its gradient. Examples of numerical modeling of this phenomenon are shown in Fig. 20.7.1.

Example 3. Area Statistics for Level Curves. Let us note the following expression for the area bounded by the level curve of the field $\rho(\boldsymbol{x}, t)$:

$$S(t, \rho) = \pm \frac{1}{2} \int d\boldsymbol{x}\, (\boldsymbol{x}\boldsymbol{p})\Phi_{t,x}(\rho), \tag{39}$$

where the choice of the $\pm$ sign is determined by the value of $S(t, q)$ at $t = 0$, or by the type of monotonicity of ρ (in the case $\rho(\boldsymbol{x}, t)$ varies monotonically). Thus, if the field $\rho(\boldsymbol{x}, t)$ is radial, i.e., $\rho(\boldsymbol{x}, t) = \rho(r, t)$, then

$$\boldsymbol{p}(\boldsymbol{x}, t) = \frac{\boldsymbol{x}}{r} \frac{\partial}{\partial r} \rho(r, t),$$

and the sign depends on the sign of the derivative $\partial \rho(r, t)/\partial r$. One can integrate (39) by parts to get the relation

$$\frac{\partial}{\partial \rho} S(t, \rho) = \pm \int d\boldsymbol{x} \, \Phi_{t,\boldsymbol{x}}(\rho) = \pm A_{-1}(t, \rho). \tag{40}$$

It is obvious that in the more general case

$$F_S(t; \rho) = \int_S F(t, \boldsymbol{x}; \rho(\boldsymbol{x}, t)) \, d\boldsymbol{x}$$

integrated over the level set S bounded by the curve $\rho(\boldsymbol{x}, t) = q$ we have

$$\frac{\partial}{\partial \rho} F_S(t, \rho) = \pm \int d\boldsymbol{x} \, F(t, \boldsymbol{x}; \rho) \Phi_{t,\boldsymbol{x}}(\rho). \tag{40'}$$

In particular

$$\frac{\partial}{\partial \rho} \int_{S(t,\rho)} d\boldsymbol{x} \, F(\rho(\boldsymbol{x}, t)) = \pm F(\rho) \frac{\partial}{\partial \rho} S(t, \rho) \tag{40''}$$

and, consequently, the integral $\int_{S(t,\rho)} d\boldsymbol{x} \, F(\rho(\boldsymbol{x}, t))$ is independent of time, because $S(t, \rho) = S(0, \rho)$ in view of the incompressibility of the flow.

Example 4. Mean Number of Level Contours. Here we consider the time evolution of the level contours

$$\rho(\boldsymbol{x}, t) = \rho = const. \tag{41}$$

The dynamics of ρ increases the complexity of level curves and leads to their fragmentation into disconnected contours. This process is partly described by statistics of the contour number $\mathcal{N}(t, \rho)$ which admits the following geometric estimate expressed in terms of ρ and $\nabla \rho$:

$$\mathcal{N}(t, \rho) \leq \int_0^\infty dr \left| \frac{\partial \rho(\boldsymbol{x}, t)}{\partial r} \right| \delta(\rho(\boldsymbol{x}, t) - \rho). \tag{43}$$

The estimate is written in polar coordinates (r, ϕ). We look at each direction ϕ and count the number of intersections along the ray (r, ϕ) with the level set (41). The right-hand side of (43) assigns to each level curve the maximal number of its radial branches, hence provides an obvious estimate of $\mathcal{N}$ (Fig. 20.8.1).

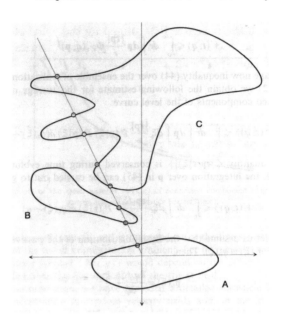

FIGURE 20.8.1.
Calculation of the estimate for the number of disconnected contours
of a level curve. The number of intersections for contours $A, B,$ and
C is, respectively, 1,2, and 6.

One could write an exact expression for $\mathcal{N}$ in terms of curvature κ, namely,

$$\mathcal{N}(t,\rho) = (1/2\pi) \int d\boldsymbol{x}\, \kappa(\boldsymbol{x},t)\, |\nabla\rho(\boldsymbol{x},t)|\, \delta(\rho(\boldsymbol{x},t) - \rho).$$

However, κ involves a complicated expression in terms of first and second derivatives of ρ, that are too complicated to analyze statistically.

Taking into account the fact that

$$\frac{\partial}{\partial r}\rho(\boldsymbol{x},t) = \frac{\boldsymbol{x}}{r}\nabla\rho(\boldsymbol{x},t),$$

one can rewrite formula (43) in the form

$$\mathcal{N}(t,\rho) \leq \int_0^\infty dr \int d\boldsymbol{p}\, \frac{|\boldsymbol{x}\boldsymbol{p}|}{r}\Phi_{t,\boldsymbol{x}}(\rho,\boldsymbol{p}). \tag{44}$$

If we average now inequality (44) over the ensemble of realizations, in view of (30)–(32) we obtain the following estimate for the average number of disconnected components of the level curve:

$$\langle \mathcal{N}(t,\rho) \rangle \leq \int_0^\infty dr \int d\boldsymbol{p} \int d\boldsymbol{\xi}\, \frac{|\boldsymbol{x}\boldsymbol{p}|}{r} P_t(\boldsymbol{x}|\boldsymbol{\xi})\, P_t(\boldsymbol{p}|\boldsymbol{\xi})\, \delta(\rho_0(\boldsymbol{\xi}) - \rho). \tag{45}$$

Since the quantity $\langle |\boldsymbol{x}\boldsymbol{p}(t|\boldsymbol{\xi})| \rangle$ is conserved during time evolution (see, e.g., (33')), the integration over $\boldsymbol{p}$ in (45) can be carried out to give

$$\langle \mathcal{N}(t,\rho) \rangle \le \int_0^\infty dr \int d\boldsymbol{\xi} \, \frac{|\boldsymbol{x}\boldsymbol{p}_0(\boldsymbol{\xi})|}{r} P_t(\boldsymbol{x}|\boldsymbol{\xi}) \, \delta(\rho_0(\boldsymbol{\xi}) - \rho). \tag{46}$$

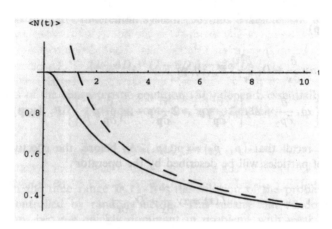

FIGURE 20.8.2.
Estimate of the mean number $\mathcal{N}(t,q)$ of connected component contours of level curves and its asymptotics (dashed line).

Now, let us assume that the initial distribution of the passive tracer is radial, i.e., $\rho_0(\boldsymbol{\xi}) \equiv \rho_0(\xi)$. Then,

$$\boldsymbol{p}_0(\boldsymbol{\xi}) = \frac{\partial \rho_0(\xi)}{\partial \xi} \cdot \frac{\boldsymbol{\xi}}{\xi},$$

and the inequality (46), taking into account (31'), can be rewritten in the form

$$\langle \mathcal{N}(\tau) \rangle \le \frac{4}{\pi \tau} \int_0^{\pi/2} d\phi \int_0^\infty dr \, \cos\phi \, e^{-(1+r^2)/\tau} \cosh[2r \cos\phi / \tau], \tag{47}$$

where the dimensionless time $\tau = 4D_1 t / r_0^2(\rho)$ has been introduced and where $r_0(\rho)$ is the radius of the initial level concentration curve. It follows from (47) that, for $\tau \gg 1$ one has the following asymptotics

$$\langle \mathcal{N}(\tau) \rangle = \frac{2}{\sqrt{\pi \tau}}, \tag{48}$$

i.e., the mean number of connected components of the concentration level curve decreases in time according to the power law. Fig. 20.8.2 shows the dependence of the estimate of $\langle \mathcal{N}(\tau) \rangle$ on the dimensionless time τ as well as the asymptotic expression (48). This dependence is totally determined by the diffusion coefficient D_1 and is independent of the fine structure of fluctuations of the tracer concentration gradient. Let us remark that the exact mean contour number $\langle N(t, \rho) \rangle$ would depend on the diffusion coefficient D_2 that describes the fine structure of the tracer field.

20.8.7. Linear shear flow. In the preceding subsection, we have provided a detailed statistical analysis of tracer concentration in random velocity fields and, in absence of the mean flow, the analysis of tracer gradient. The presence of a mean flow (even a deterministic one) leads to steepening of the tracer gradient and the deformation of its level sets. The presence of even small fluctuations of the velocity field quickly accelerates these processes. In the present subsection we will illustrate this phenomenon by the example of the simplest two-dimensional linear shear mean flow,

$$v_x = \alpha y, \qquad v_y = 0,$$

and analyze the effect of the shear on the statistics of tracer concentration.

The probability distribution $P_{t,x}(\rho, \boldsymbol{p})$ is described by equation (28) for $N = 2$, which now takes the form

$$\frac{\partial}{\partial t} P_{t,x}(\rho, \boldsymbol{p}) = \left\{ -\alpha y \frac{\partial}{\partial x} + D_1 \frac{\partial^2}{\partial x^2} \right\} P_{t,x}(\rho, \boldsymbol{p}) \tag{49}$$

$$+ \left\{ \alpha p_1 \frac{\partial}{\partial p_2} + D_2 \left[3 \frac{\partial^2}{\partial \boldsymbol{p}^2} \boldsymbol{p}^2 - 2 \frac{\partial}{\partial \boldsymbol{p}} \boldsymbol{p} - 2 \left(\frac{\partial}{\partial \boldsymbol{p}} \boldsymbol{p} \right)^2 \right] \right\} P_{t,x}(\rho, \boldsymbol{p}),$$

where, we recall that $(p_1, p_2) = (p_x, p_y)$. As before, the effective spatial diffusion of particles will be described by the operator

$$\hat{L}(\boldsymbol{x}) = -\alpha y \frac{\partial}{\partial x} + D_1 \frac{\partial^2}{\partial x^2}.$$

The corresponding probability distribution is Gaussian with parameters

$$\langle x(t) \rangle = x_0 + \alpha y_0 t, \quad \langle y(t) \rangle = y_0,$$

$$\sigma_{xx}^2(t) = 2D_1 t \left(1 + \frac{1}{3} \alpha^2 t^2 \right), \quad \sigma_{yy}^2 = 2D_1 t, \quad \sigma_{xy}^2 = \alpha D_1 t^2.$$

The operator

$$\hat{M}(\boldsymbol{p}) = \alpha p_1 \frac{\partial}{\partial p_2} + D_2 \left[3 \frac{\partial^2}{\partial \boldsymbol{p}^2} \boldsymbol{p}^2 - 2 \frac{\partial}{\partial \boldsymbol{p}} \boldsymbol{p} - 2 \left(\frac{\partial}{\partial \boldsymbol{p}} \boldsymbol{p} \right)^2 \right]$$

describes the diffusion of the tracer concentration gradient and is now anisotropic. In this case the mean value of vector $\boldsymbol{p}$ is not conserved and, in the case of zero velocity fluctuations, it is a linear function of t,

$$\langle p_1(t) \rangle = p_1(0), \quad \langle p_2(t) \rangle = p_2(0) - \alpha p_1(0) t.$$

We consider second moments of vector $\boldsymbol{p}$ and write the Lagrangian equations for them:

$$\frac{d}{dt}\langle\boldsymbol{p}^2\rangle = 8D_2\langle\boldsymbol{p}^2\rangle - 2\alpha\langle p_1 p_2\rangle,$$

$$\frac{d}{dt}\langle p_1 p_2\rangle = -4D_2\langle p_1 p_2\rangle - \alpha\langle p_1^2\rangle, \tag{50}$$

$$\frac{d}{dt}\langle p_1^2\rangle = -4D_2\langle p_1^2\rangle + 6D_2\langle\boldsymbol{p}^2\rangle.$$

The linear ordinary differential system (50) has characteristic exponents λ that obey

$$(\lambda + 4D_2)^2(\lambda - 8D_2) = 12\alpha^2 D_2. \tag{51}$$

The roots of the characteristic equation (51) depend, essentially, on α/D_2. For small $\alpha/D_2 \ll 1$, these roots are, approximately,

$$\lambda_1 = 8D_2 + \frac{1}{12}\frac{\alpha^2}{D_2}, \quad \lambda_2 = -4D_2 + i|\alpha|, \quad \lambda_3 = -4D_2 - i|\alpha|. \tag{52}$$

Hence, in the time range $D_2 t \gg 1/4$, solution of the problem is completely controlled by random factors. That means that random velocity fluctuations become quickly dominant in problems with weak mean field gradients.

In the case of large $\alpha/D_2 \gg 1$, the characteristic equation (51) has approximate roots

$$\lambda_1 = (12\alpha^2 D_2)^{1/3}, \quad \lambda_2 = (12\alpha^2 D_2)^{1/3}e^{i(2/3)\pi}, \quad \lambda_3 = (12\alpha^2 D_2)^{1/3}e^{-i(2/3)\pi}. \tag{53}$$

Since the real parts of λ_2 and λ_3 are negative, for $(12\alpha^2 D_2)^{1/3}t \gg 1$, solutions are asymptotic to

$$\langle\boldsymbol{p}^2(t)\rangle \sim \exp\{(12\alpha^2 D_2)^{1/3}t\}, \tag{54}$$

so even small velocity fluctuations have significant effect on the second moment in sufficiently strong mean gradient flows.

20.9 Exercises

1. Show that in the case of the "Gaussian" shape (20.1.1c) of functions $F_\perp$, and $G_\|$, coefficients D, and B, appearing in the equation can be expressed through the variance σ_v^2 of the velocity field, and its characteristic spatial and temporal scales l_v, and τ_v, as follows:

$$D = \sqrt{2\pi}\sigma_v^2\tau_v/3, \quad \text{and} \quad B = 5\kappa\sqrt{2\pi}\sigma_v^2\tau_v/l_v^2.$$

2. Extend the formulas (2), (15), (18), and (20), of Section 20.4, which connect Lagrangian and Eulerian probability distributions, to the case of multipoint probability distributions describing joint statistical properties of Lagrangian and Eulerian fields at different points of space. As an example we shall quote two formulas. In the first,

$$f_E(j_i, j_2; \boldsymbol{x}_1, \boldsymbol{x}_2, t_1, t_2) = j_1 j_2 \int \int f_L(\boldsymbol{x}_1, \boldsymbol{x}_2, j_1, j_2; \boldsymbol{y}_1, \boldsymbol{y}_2, t_1, t_2) d\boldsymbol{y}_1 d\boldsymbol{y}_2,$$

(1)

the joint, two-point and two-time, probability distribution

$$f_E(j_i, j_2; \boldsymbol{x}_1, \boldsymbol{x}_2, t_1, t_2) = \Big\langle \delta(j_1 - j(\boldsymbol{x}_1, t_1))\delta(j_2 - j(\boldsymbol{x}_2, t_2)) \Big\rangle,$$

of the field $j(\boldsymbol{x}, t)$ appears on the left-hand side, and the Lagrangian probability distribution

$$f_L(\boldsymbol{x}_1, \boldsymbol{x}_2, j_1, j_2; \boldsymbol{y}_1, \boldsymbol{y}_2, t_1, t_2) =$$

$$\Big\langle \delta(\boldsymbol{x}_1 - \boldsymbol{X}(\boldsymbol{y}_1, t_1))\delta(\boldsymbol{x}_2 - \boldsymbol{X}(\boldsymbol{y}_2, t_2))\delta(j_1 - J(\boldsymbol{y}_1, t_1))\delta(j_2 - J(\boldsymbol{y}_2, t_2)) \Big\rangle$$

appears on the righ-hand side.

The second assumes that the velocity field of the medium is statistically homogeneous in space and that $t_1 = t_2 = t$. Then, formula (1) takes a simpler form,

$$f_E(j_1, j_2; \boldsymbol{s}, t) = j_1 j_2 \int \int f_L(\boldsymbol{s}, j_1, j_2; \boldsymbol{s}_0, t) d\boldsymbol{s}_0,$$

where $\boldsymbol{s} = \boldsymbol{x}_1 - \boldsymbol{x}_2$,

$$f_L(\boldsymbol{s}, j_1, j_2; \boldsymbol{s}_0, t) = \Big\langle \delta(\Delta\boldsymbol{X}(\boldsymbol{y}, \boldsymbol{s}_0, t) - \boldsymbol{s})\delta(j_1 - J(\boldsymbol{y} + \boldsymbol{s}_0, t))\delta(j_2 - J(\boldsymbol{y}, t)) \Big\rangle,$$

and

$$\Delta\boldsymbol{X} = \boldsymbol{X}(\boldsymbol{y} + \boldsymbol{s}_0, t) - \boldsymbol{X}(\boldsymbol{y}, t).$$

3. Complete the calculations showing that the one-point probability distribution

$$w(z; t) = \langle \delta(Z(t) - z) \rangle$$

of the random process $Z(t)$ introduced in Section 20.7 satisfies, in the diffusion approximation, the equation,

$$\frac{\partial w}{\partial \tau} + \frac{\partial}{\partial z}(\coth zw) = \frac{\partial^2 w}{\partial z^2}, \qquad w(z; 0) = \delta(z),$$

where $\tau = 4Bt$ denotes the dimensionless time.

Part VI

Anomalous Fractional Dynamics

Chapter 21

Lévy Processes and Their Generalized Derivatives

This chapter discusses the Lévy-type classical and generalized processes and related problems of infinitely divisible distributions. In a sense it is a continuation of the material discussed in Chapters 16 and 17. The ideas regarding generalized Lévy processes are mostly due to K. Urbanik (see references [8–10] in Chapter 17). A comprehensive exposition of the classical theory of infinitely divisible distributions and Lévy processes can be found in the monographs [1–5]. In the following chapters we will employ these tools to a study of anomalous diffusions, both linear and nonlinear.

21.1 Infinitely divisible and stable probability distributions

A (generalized) p.d.f. $f(x)$ is said to be *infinitely divisible* if, for any $\theta > 0$, the function $\tilde{f}(u; \theta) \stackrel{\text{def}}{=} \tilde{f}^{\theta}(u)$ is also the characteristic function of a random variable which we will denote by $X^{(\theta)}$.

Infinitely divisible p.d.f. $f(x)$ of a random variable X can always be represented as the distribution of the sum of an arbitrary number of independent, identically distributed random variables. Indeed, for any $n = 1, 2, \ldots,$

$$X \stackrel{\text{d}}{=} X_1^{(1/n)} + X_2^{(1/n)} + \cdots + X_n^{(1/n)},$$

where the random variables on the right-hand side are independent, and each has the p.d.f. $f(x; 1/n)$; hence the term *infinitely divisible p.d.f.*. The notation $\stackrel{\text{d}}{=}$ signifies here the equality of the probability distributions of random variables.

© Springer International Publishing AG, part of Springer Nature 2018
A. I. Saichev and W. A. Woyczynski, *Distributions in the Physical and Engineering Sciences, Volume 3*, Applied and Numerical Harmonic Analysis, https://doi.org/10.1007/978-3-319-92586-8_21

244   Chapter 21. *Lévy Processes and Their Generalized Derivatives*

The simplest nontrivial infinitely divisible distribution is the Poisson distribution with the p.d.f.

$$f(x) = \sum_{n=0}^{\infty} e^{-\lambda} \frac{\lambda^n}{n!} \delta(x-n),$$

and the characteristic function

$$\tilde{f}(u) = \exp\left[\lambda(e^{iu}-1)\right]. \tag{1}$$

Parameter λ is assumed to be positive; it is the mean value of the Poissonian random variable. The Poissonian random variable takes only nonnegative integer values with the probability of value n being $e^{-\lambda}\lambda^n/n!$.

If $T \geq 0$ is a nonnegative random variable, it is often more convenient to use the real-valued Laplace transform

$$\hat{f}(s) = \langle e^{-sT}\rangle = \int_0^\infty e^{-st} f(t)\, dt.$$

of its p.d.f. $f(x)$ instead of the characteristic function which, in general, is complex-valued. The Poisson random variable is nonnegative, and substituting $u = is$ in (1), we obtain the Laplace transform of the Poissonian random variable

$$\hat{f}(s) = \exp\left[\lambda(e^{-s}-1)\right].$$

Using Poissonian random variables as building blocks one can construct other infinitely divisible distributions which play a key role in the theory of anomalous diffusion. Indeed, let us consider a random variable

$$X = \sum_m a_m K_m,$$

where $\{K_m\}$ are independent random integers with the Poisson distributions with parameters $\{\lambda_m\}$, respectively, while $\{a_m\}$'s are deterministic real-valued coefficients. Obviously, the characteristic function of X is equal to

$$\tilde{f}(u) = \prod_m \exp\left[\lambda_m(e^{iua_m}-1)\right] = \exp\left[\sum_m \lambda_m\left(e^{iua_m}-1\right)\right].$$

Note that the random variable X, a mixture of the rescaled Poisson random variables, can take as its values arbitrary integer multiplicities of the constants a_m and their sums.

Taking a continuum limit of the above mixtures of Poissonian distributions by putting

$$a_m = m\,\Delta, \qquad \lambda_m = \psi(m\Delta)\,\Delta,$$

for a given nonnegative function $\psi(u)$, usually called the *intensity*, or *rate* function, and letting $\Delta \to 0$, we arrive at an infinitely divisible random variable with the characteristic function of the form

$$\tilde{f}(u) = e^{\Psi(u)}, \qquad \text{where} \qquad \Psi(u) = \int \psi(z)(e^{iuz} - 1)\, dz, \qquad (2)$$

as long as the above integral is well defined for a given intensity function $\psi(u)$. This $\tilde{f}(u)$ is the characteristic function of the random variable

$$X =_{\mathrm{d}} \lim_{\Delta \to 0} \Delta \sum_m m\, K_m, \qquad (3)$$

which we will call the generalized (or compound) Poissonian random variable with the intensity function $\psi(u)$.

If, for $u < 0$, the intensity function $\psi(u) = 0$, then the corresponding generalized Poissonian random variable T is nonnegative and, in analogy with (2), its Laplace transform is

$$\hat{f}(s) = e^{\Phi(s)}, \qquad \text{where} \qquad \Phi(s) = \Psi(is) = \int_0^\infty \psi(z)(e^{-s} - 1)\, dz. \qquad (4)$$

In the particular case of the intensity function

$$\psi(z) = \frac{\beta}{\Gamma(1 - \beta)}\, z^{-\beta - 1}\, e^{-\delta z} \chi(z),$$

where $\chi(x)$ is the unit step function, and $0 < \beta < 1$, the integral $\Phi(s)$ in (4) exists and can be explicitly evaluated with the help of the formula

$$\int_0^\infty \frac{e^{-az} - e^{-bz}}{z^{\gamma + 1}}\, dz = \Gamma(-\gamma)\,(a^\gamma - b^\gamma) \qquad (\gamma < 1) \qquad (5)$$

to give

$$\hat{f}_\beta(s; \delta) = \exp\left(\delta^\beta - (s + \delta)^\beta\right) \qquad (0 < \beta < 1). \qquad (6)$$

an example of the Laplace transform of an infinitely divisible p.d.f. $f_\beta(s; \delta)$. The p.d.f. described by the Laplace transform (6) is called the *tempered β-stable p.d.f.*[1]

Recall that a p.d.f. $f(x)$ is called *strictly stable* if the p.d.f. $f_n(x)$ of the sum of n independent random variables, each with p.d.f. $f(x)$, is of the form

$$f_n(x) = \frac{1}{c_n}\, f\left(\frac{x}{c_n}\right),$$

for a certain sequence of constants c_n. In the language of the characteristic functions and the Laplace transforms the above condition can be written as follows:

$$\tilde{f}^n(u) = \tilde{f}(c_n u), \qquad \hat{f}^n(s) = \hat{f}(c_n s).$$

---

[1] For more information about the tempered stable distributions, see, e.g., Rosinski [6].

It is clear that, for $\delta = 0$, formula (A.6) produces the Laplace transform

$$\hat{f}_\beta(s) = e^{-s^\beta}, \qquad (0 < \beta < 1) \tag{7}$$

of a strictly stable distribution $f_\beta(t)$. A random variable T with the Laplace transform (7) has an infinite mean. Relying on the general Tauberian theorems describing the relationship between asymptotics of p.d.f.s and their Laplace transforms one can show that

$$\hat{f}_\beta(s) \sim 1 - s^\beta \quad (s \to 0) \quad \Leftrightarrow \quad f_\beta(t) \sim \frac{\beta}{\Gamma(1-\beta)} t^{-\beta-1} \quad (t \to \infty) \tag{8}$$
$$(0 < \beta < 1).$$

The strictly stable p.d.f. $f_\beta(u)$ corresponding to the Laplace transform (7) is of the form

$$f_\beta(t) = \frac{1}{\pi} \int_0^\infty \exp\left[-\cos\left(\frac{\pi\beta}{2}\right) u^\beta\right] \cos\left[ut - u^\beta \sin\left(\frac{\pi\beta}{2}\right)\right] du. \tag{9}$$

Since the absolute value of the integrand in (9) converges to zero exponentially as $|u| \to \infty$, the p.d.f. $f_\beta(t)$ is infinitely differentiable. Graphs of the stable p.d.f. $f_\beta(t)$, for several values of parameter β, are given in Fig. 21.1.1.

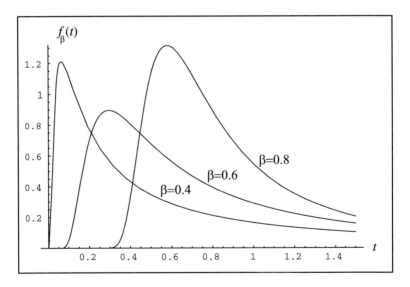

FIGURE 21.1.1.
Plots of the stable p.d.f.s $f_\beta(t)$, for different values of β

In general, the integral in (9) cannot be evaluated in a closed form. However, for $\beta = 1/2$, and $\beta = 1/3$, this can be done:

$$f_{1/2}(t) = \frac{1}{2t\sqrt{\pi t}} \exp\left(-\frac{1}{4t}\right), \quad \text{and} \quad f_{1/3}(t) = \frac{1}{t^3\sqrt{3t}} \operatorname{Ai}\left(\frac{1}{\sqrt[3]{3t}}\right), \tag{10}$$

where Ai(x) is the so-called Airy function.

The characteristic function of a symmetric α-stable random variable is of the form

$$\Theta_\alpha(u) = e^{-|u|^\alpha}, \qquad 0 < \alpha < 2, \tag{11}$$

and its p.d.f. is given by the integral formula

$$f_\alpha(x) = \frac{1}{\pi} \int_0^\infty e^{-u^\alpha} \cos(ux)\, dx.$$

As in the case of (8), the power asymptotics of the characteristic function (11),

$$\Theta_\alpha(u) \sim 1 - |u|^\alpha, \qquad u \to 0,$$

implies the power law of decay of the p.d.f. itself:

$$f_\alpha(x) \sim \frac{1}{\pi}\Gamma(\alpha+1)\sin\left(\frac{\pi\alpha}{2}\right)|x|^{-\alpha-1}, \qquad (|x| \to \infty). \tag{12}$$

Hence, the variances of all α-stable random variables, $0 < \alpha < 2$, are infinite.

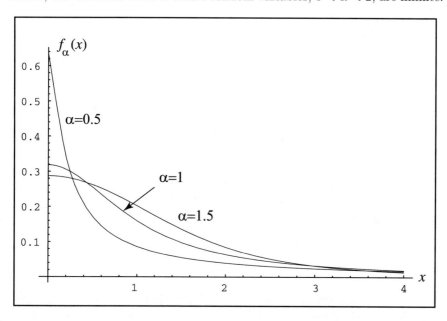

FIGURE 21.1.2.
Plots of the symmetric α-stable p.d.f.s for different values of α. They are presented here for positive values of x only

On the other hand, the exponential decay of the characteristic function (11), for $u \to \infty$, makes it possible to expand the p.d.f. in the Taylor series

$$f_\alpha(x) = \frac{1}{\pi\alpha} \sum_{n=0}^\infty (-1)^{2n} \frac{\Gamma\left(\frac{2n+1}{\alpha}\right)}{(2n)!} x^{2n}.$$

Plots of the symmetric stable p.d.f.s $f_\alpha(x)$, for different values of α, are shown in Fig. 21.1.2.

Although the second moment of a symmetric α-stable distribution is infinite, moments of order κ less than α are finite. We shall calculate them relying on the Parseval equality

$$\langle g(X) \rangle_\alpha = \int g(x) f_\alpha(x)\, dx = \frac{1}{2\pi} \int \tilde{g}(u)\, e^{-|u|^\alpha}\, du \,, \tag{13}$$

with $g(x) = |x|^\kappa$ and $\tilde{g}(u) = \int g(x)\, e^{iux}\, dx$ being the generalized Fourier transform. For $u \neq 0$, the values of the latter can be found by the summation of the divergent Fourier integral (see, for instance, Chapter 8 in Volume 1 of this monograph series):

$$\tilde{g}(u) = -\frac{2\Gamma(1+\kappa)}{|u|^\kappa} \sin\left(\frac{\pi\kappa}{2}\right), \qquad (u \neq 0)\,.$$

Substituting this expression in (13), we obtain the divergent integral

$$\int |x|^\kappa f(x)\, dx = -\frac{2\Gamma(1+\kappa)}{\pi} \sin\left(\frac{\pi\kappa}{2}\right) \int_0^\infty \frac{e^{-u^\alpha}}{u^{\kappa+1}}\, du \,. \tag{14}$$

However, since

$$\int |x|^\kappa f(x)\, dx = \int |x|^\kappa [f(x) - \delta(x)]\, dx, \qquad \kappa > 0\,,$$

the right-hand side of the equality (A.14) can be replaced by the regularized integral

$$\int |x|^\kappa f(x)\, dx = -\frac{2\Gamma(1+\kappa)}{\pi} \sin\left(\frac{\pi\kappa}{2}\right) \int_0^\infty \frac{e^{-u^\alpha} - 1}{u^{\kappa+1}}\, du \,. \tag{15}$$

Finally, using the formula (5), we obtain

$$\langle |X|^\kappa \rangle_\alpha = \frac{2}{\pi} \Gamma(\kappa)\, \Gamma\left(1 - \frac{\kappa}{\alpha}\right) \sin\left(\frac{\pi\kappa}{2}\right) \qquad (\kappa < \alpha)\,. \tag{16}$$

We will conclude this subsection by constructing a symmetric tempered $1/2$-stable random variable, that is a generalized Poissonian random variable which for "small values" behaves like the symmetric α-stable random variable with $\alpha = 1/2$, but which has a finite variance. For this purpose, substitute into (2) the intensity function

$$\psi(z) = \frac{e^{-\delta|z|}}{2\sqrt{2\pi}|z|\sqrt{|z|}} \,,$$

to obtain the logarithm of the characteristic function

$$\Psi(u) = \sqrt{2\delta} - \sqrt{\delta + \sqrt{\delta^2 + u^2}} \,. \tag{17}$$

For $u \to 0$, its main asymptotics is as follows:

$$\Psi(u) \sim -\frac{1}{4\sqrt{2}\,\delta^{3/2}}\,u^2, \qquad (u \to 0).$$

Thus, the variance of the corresponding tempered 1/2-stable random variable X can be explicitly calculated:

$$\sigma_X^2 = \frac{1}{\sqrt{8}}\,\delta^{-3/2} \simeq 0.354\,\delta^{-3/2} < \infty. \tag{18}$$

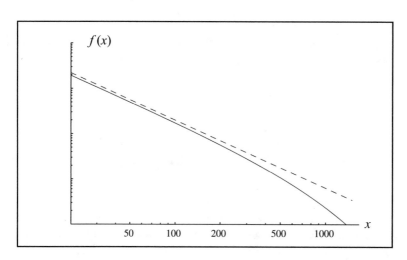

FIGURE 21.1.3.
A log-log plot of the p.d.f. of the tempered 1/2-stable random variable (19) (continuous line) compared to the asymptotics of the 1/2-stable p.d.f. $f_{1/2}(x)$ (dashed line)

On the other hand, for $u \to \infty$, the asymptotics of $\Psi(u)$ in (17) is as follows:

$$\Psi(u) \sim -\sqrt{|u|}, \qquad (u \to \infty),$$

which corresponds to the asymptotics of the α-stable p.d.f. with index $\alpha = 1/2$.
A log-log plot of the p.d.f. $f(x)$ with the characteristic function

$$\tilde{f}(u) = \exp\left(\sqrt{2\delta} - \sqrt{\delta + \sqrt{\delta^2 + u^2}}\right) \tag{19}$$

and $\delta = 10^{-3}$ is shown in Fig. 21.1.3. The upper, dashed straight line corresponds to the asymptotic formula (12):

$$f_{1/2}(x) \sim \frac{1}{2\sqrt{2\pi}}\,|x|^{-3/2}, \qquad (|x| \to \infty). \tag{20}$$

One can see that almost up to the value $x \approx 1/\delta$ the p.d.f. $f(x)$ remains close to the asymptotics of the stable p.d.f. Afterward it quickly goes to zero to assure the finiteness of the variance (18).

In the general case, the characteristic function of an arbitrary infinitely divisible distribution can be written in the form of the so-called *Lévy-Khinchine formula*,[2]

$$\tilde{f}(u) = \exp\left(\int_{-\infty}^{\infty} \left(e^{iua} - 1 - iuaI_{[-1,1]}(a) \right) L(da) \right)$$

where the Lévy measure L has to satisfy the following weaker integrability condition:

$$\int_{-\infty}^{\infty} \min(1, a^2) L(da) < \infty.$$

In the case of a symmetric infinitely divisible distribution this general characteristic function is real-valued and of a simpler form,

$$\tilde{f}(u) = \exp\left(\int_{-\infty}^{\infty} \left(\cos ua - 1 \right) L(da) \right).$$

21.2   Physical phenomena described by infinitely divisible probability distributions

There are instances when the appearance of α-stable laws, with $0 < \alpha < 2$, can be rigorously deduced from what we, for a lack of better word, call "first principles." What we mean by that is a description of random phenomena where one assumes only the basic physical mechanism which inexorably leads to a description in terms of an α-stable law with a particular α, although no such parameter seems to appear explicitly in the description of the original phenomenon. Below, we selected four physical situations, starting with some textbook examples, which illustrate what we have in mind (see, also, Woyczyński [10]).

21.2.1.   The first hitting time for the Brownian particle has a 1/2-stable probability distribution.   Consider a Brownian particle moving in $\mathbf{R}^1$ whose trajectory $X(t), t \geq 0$, starts at the origin, that is $X(0) = 0$. The first time, $T_b > 0$, it hits the barrier located at $x = b > 0$ is a random quantity that can be defined by the formula

$$T_b = \inf\{t \geq 0 : X(t) = b\}. \tag{1}$$

Our goal is to find the probability distribution of T_b. Note that because

$$\mathbf{P}[T_b > t] = \mathbf{P}\left[\sup_{0 \leq \tau \leq t} X(\tau) < b \right], \tag{2}$$

the problem is equivalent to the problem of finding the probability distribution of the maximum position of the Brownian particle in a given time interval.

---

[2]See, e.g., Bertoin [2] for details.

The answer may be obtained via the following intuitive argument which can be made rigorous using the strong Markov property of the Brownian motion. For any $t > 0$,

$$\mathbf{P}[T_b < t] = \mathbf{P}[T_b < t, X(t) > b] + \mathbf{P}[T_b < t, X(t) < b], \tag{3}$$

and obviously $\mathbf{P}[T_b < y, X(t) > b] = \mathbf{P}[X(t) > b]$. But, after hitting the barrier b, the Brownian motion can be thought of as "starting afresh," so its symmetry property gives

$$\mathbf{P}[T_b < t, X(t) > b] = \mathbf{P}[T_b < t, X(t) < b] = \mathbf{P}[X(t) > b], \tag{4}$$

so that

$$\mathbf{P}[T_b < t] = 2\mathbf{P}[X(t) > b] = \sqrt{\frac{2}{\pi}} \int_{bt^{-1/2}}^{\infty} e^{-x^2/2} \, dx \tag{5}$$

which, finally, gives the desired probability density function

$$f_{T_b}(t) = \frac{b}{\sqrt{2\pi}} t^{-3/2} e^{-b^2/2t}, \qquad t > 0. \tag{6}$$

This is an infinitely divisible probability distribution, and its characteristic function is

$$\tilde{f}_{T_b}(\lambda; b) = \exp\left\{ b\left(-(2\lambda)^{1/2} + i2\lambda|2\lambda|^{-1/2} \right) \right\}. \tag{7}$$

So, the first hitting time T_b has a totally asymmetric $1/2$-stable probability distribution.

21.2.2. Particles emitted from a point source have a 1-stable distribution in the detection plane. Consider a source located at the point $(0, \eta)$ in the (x, y)-plane emitting particles into the right half-space with random directions (angles) Θ uniformly distributed over the interval $[-\pi/2, \pi/2]$. The particles are being detected by a flat panel device represented by the vertical line $x = \tau$ at the distance τ from the source. What is the probability distribution function of the random variable representing the position Y of particles on the detecting device? Clearly,

$$F_Y(y; \eta, \tau) = \mathbf{P}[Y \le y] = \mathbf{P}[\tan \Theta \le (y - \eta)/\tau] = \tag{8}$$

$$\mathbf{P}[\Theta \le \arctan((y - \eta)/\tau)] = \frac{1}{2} + \frac{1}{\pi} \arctan((y - \eta)/\tau).$$

The corresponding 1-stable probability density function (also called *Cauchy density* function (see Chapter 16) , and often labeled as the *Lorentz density* in the physical sciences) with the *location parameter* η and the *scale parameter* τ is given by the formula

$$f_Y(x; \eta, \tau) = \frac{1}{\pi\tau\left(1 + ((y - \eta)/\tau)^2\right)}, \qquad y \in \mathbf{R}, \tag{9}$$

with the characteristic function

$$\tilde{f}_Y(\lambda; \eta, \tau) = \exp\{i\eta\lambda - |\tau\lambda|\}. \tag{10}$$

21.2.3. Stars, uniformly distributed in space, generate a gravitational field with 3/2-stable (Holtsmark) probability distribution.

Consider a model of the universe in which the stars with masses $M_i \geq 0$, $i = 1, 2, \ldots$, located at positions $X_i \in \mathbf{R}^3, i = 1, 2, \ldots$, interact via the Newtonian gravitational potential, exerting force

$$G_i = gM_i \frac{X_i}{|X_i|^3} \in \mathbf{R}^3, \qquad i = 1, 2, \ldots, \tag{11}$$

on a unit mass located at the origin $(0, 0, 0)$. Here g is the universal gravitational constant. The assumptions are:

• The locations $X_i, i = 1, 2, \ldots$, form a Poisson point process in $\mathbf{R}^3$ with density ρ. In particular, if D is a bounded domain in $\mathbf{R}^3$ then the random number N_D of stars located in D has expectation $\mathbf{E}(N_D) = \rho|D|$, where $|D|$ is the volume of the domain D.

• The masses $M_i, i = 1, 2, \ldots$, of stars form a sequence of independent, identically distributed random variables (with density $f_M(m)$) which is independent of the star location Poisson point process $X_i, i = 1, 2, \ldots$

The total gravitational force G_R on a unit mass located at the origin, exerted by stars located inside the ball B_R, centered at $(0,0,0)$ and of radius R, is

$$G_R = \sum_{\{i: |X_i| \leq R\}} G_i, \tag{12}$$

and this random quantity has the characteristic function

$$
\begin{aligned}
\tilde{f}_{G_R}(\lambda) &= \mathbf{E} \exp\left\{i\left(\lambda, \sum_{\{j: |X_j| \leq R\}} G_j\right)\right\} \\
&= \sum_{k=0}^{\infty} \mathbf{P}\left[N_{B_R} = k\right] \cdot \mathbf{E} \exp\left\{i\left(\lambda, \sum_{j=1}^{k} G_j\right)\right\} \\
&= \sum_{k=0}^{\infty} e^{-\rho|B_R|} \frac{(\rho|B_R|)^k}{k!} \left(\mathbf{E} e^{i\langle \lambda, G_1 \rangle}\right)^k \\
&= \exp\left\{\rho|B_R|\left(\mathbf{E} e^{i\langle \lambda, G_1 \rangle} - 1\right)\right\} \\
&= \exp\left\{\rho \int_{|x| \leq R} \int_0^{\infty} \left(\exp\left\{i\left(\lambda, gm\frac{x}{|x|^3}\right)\right\} - 1\right) f_M(m)\, dm\, dx\right\}
\end{aligned}
$$

in view of the conditional properties of the Poisson point process. In the case of a degenerate density $f_M(m)$ concentrated at a single point, that is with all the star masses identical, say $M_i = 1$, taking the limit $R \to \infty$ leads to

$$\tilde{f}_{G_\infty}(\lambda) = \exp\left\{ \rho \int_{\mathbf{R}^3} \left(\exp\{i(\lambda, g\frac{x}{|x|^3})\} - 1 \right) dx \right\}$$

which, after changing to spherical coordinates, gives

$$\tilde{f}_{G_\infty}(\lambda) = \exp\left\{ \frac{\rho}{2} \int_0^\infty \int_{S_r} (\exp\{i(\lambda, gsr)\} - 1) \, ds \, \frac{dr}{r^{3/2}} \right\} = \exp\{c\rho|g\lambda|^{3/2}\} \tag{13}$$

with $c = -8\sqrt{2}/15$. For an arbitrary distribution $f_M(m)$ of star masses, the constant c has to be multiplied by $\int_0^\infty |m|^{3/2} f_M(m) \, dm$ which, of course, needs to be finite for the formula to make sense.

Thus the random force G_∞ has a three-dimensional, spherically symmetric 3/2-stable distribution which is traditionally called the *Holtsmark distribution*. It was one of the first nontrivial infinitely divisible distributions encountered by the physical scientists (see Holtsmark [11]).

21.2.4. The size of large polymerized molecules also has the Holtsmark 3/2-stable distribution. The work summarized here is in the tradition of random graph models of polymerization *á la* Flory–Stockmayer–Whittle–Spouge. Despite certain similarities, the behavior of this model is rather different from the well-known Erdös–Renyi random graph models. The details can be found in a series of papers by Pittel, Woyczynski, and Mann [12–15].

The main features of the model of polymerization discussed here can be intuitively described as follows:

- The matter is in a graph-like state: monomers are vertices of a graph, bonds are edges, and polymerized molecules form connected components.

- The assumption of equireactivity is made: monomers are allowed to form bonds without any regard for mutual spatial positions.

- A Markov process approach with forbidden rings is used: the state space is the collection of forests of trees on n vertices, and the stationary distribution on this set is studied in the thermodynamic limit as $n \to \infty$,

- Variable association and dissociation rates are permitted: bond formation and bond breaking rates depend on bonds already formed.

The objective is to study the limit distribution of "shapes" of connected components, and of various numerical parameters of the resulting random tree (such as sizes of components, degrees of vertices) Certain physical quantities (e.g., viscosity) of polymers modeled by such random trees can be directly (and empirically) related to such distributions. However, other physical properties depend also on the geometric configurations of molecules, so, in those cases, an additional effort has to be made to develop an "hybrid" approach by injecting some "manageable" Euclidean geometry into a purely structural graph-theoretical model.

The formalism is as follows: Let V_n be a set of n labeled vertices, the state space $\mathcal{M}$ be a set of all multigraphs on V_n and let $M(t), t \geq 0$, be a Markov, continuous time stochastic process (called polymerization process) with association (bond making) and dissociation (bond breaking) rates as follows:

If $a, b \in V_n$, the degree of a is j and the degree of b is k then the intercomponent bond formation rate is $\lambda A_{j+1}A_{k+1}/(A_j A_k)$, the intracomponent bond formation rate is $\lambda' A_{j+1}A_{k+1}/(A_j A_k)$, and the bond breaking rate is $\mu D_{j-1}D_{k-1}/(D_j D_k)$, where $\lambda, \lambda', \mu \geq 0$, and $A_1, A_2, \ldots$ and $D_1, D_2, \ldots$ are sequences of nonnegative numbers known in advance from physical considerations.

In particular, the case $\lambda = \lambda'$ corresponds to the situation when interpolymer and intrapolymer bond formation rates are identical. The case $\lambda' = 0$ allows formation of trees with only single bonding. In the case $A_j = 0$ for $j > m$, each unit may form at most m bonds (valency of units is at most m). The often considered special case is when $A_j = 0$ for $j > m$ and $A_j = m!/(m-j)!$ for $j \leq m$. Then $A_{j+1}A_{k+1}/(A_j A_k) = (m-j)(m-k), j, k \leq m$, and the rate of bond formation between units a and b is proportional to the number of still "available" bonds of both a and b.

Stationary Distributions. It turns out that, under some extra technical conditions the stationary distribution $\mathbf{P}[M]$ of the process $M(t)$ for the "rings forbidden" model is $q(M)/Q, M \in \mathcal{M}$, where Q is a normalizing factor (statistical mechanical partition function), and, for $\lambda > 0$,

$$q(M) = \left(\frac{\mu}{\lambda}\right)^{C(M)} \prod_{a \in V_n} H_{\deg(a)}, \tag{14}$$

with $H_j := A_j D_j, j \geq 0$, and $C(M)$ being the total number of trees in forest M.

Keeping in mind that $M = M_n, \mathbf{P} = \mathbf{P}_m, Q = Q_n$, we seek information on the asymptotic behavior of the stationary distribution $\mathbf{P}[M]$ in the thermodynamic limit as $n \to \infty$, i.e., under the assumption that $\{H_j : j \geq 0\}$ is fixed, but μ and λ change with n in such a way that

$$\mu/\lambda = n/\sigma, \tag{15}$$

where σ is a "temperature" parameter. Intuitively, this means that if n units interact in a certain volume Ω and the dissociation rate parameter μ is independent of the volume then the association rate parameter λ is of the order Ω^{-1}.

Critical transition. The gelation phenomenon in polymers (the sol phase to gel phase transition) occurs when the connected components of polymerized monomers form giant molecules of the order of the whole solution. The existence of such a critical point can be studied within our model.

Consider the moment generating function $H(y) = \sum_j H_j y^j / j!$ of the structural sequence $H_j = A_j D_j, j = 1, 2, \ldots$. Denote by $\bar{y}$ the positive root of the equation $H''(y)y - H'(y) = 0$. Equation $y = xH'(y)$ determines implicitly a function $y = R(x)$ which is analytic for $|x| < \bar{x}$, where $\bar{y} = R(\bar{x})$.

It turns out that the thermodynamic limit behavior of our polymerization model varies dramatically depending on the value of the "temperature" parameter σ. To be more precise, there exists a critical temperature

$$\bar{\sigma} = \bar{x}H(R(\bar{x})) \tag{16}$$

such that:

- In the subcritical case $\sigma < \bar{\sigma}$, the size $L_n^{(1)}$ of the largest (random) component of M_n is of the order $\log n - (5/2) \log \log n$, where n is the size of the whole system.

- In the nearcritical case $\sigma \sim \bar{\sigma}$, or more precisely if $\sigma - \bar{\sigma} = O(n^{-1/3})$, the size $L_n^{(1)}$ of the largest (random) component of M_n is of the order $n^{2/3}$.

- In the supercritical case $\sigma > \bar{\sigma}$, the size $L_n^{(1)}$ of the unique largest (random) component of M_n is of the order n.

Fluctuations in molecular weight distributions. In the subcritical case the sizes of connected components have a Gaussian joint distribution. More precisely, if c_{nj} denotes the number of tree components of size j in the random forest M_n then $\{n^{-1/2}(c_{nj} - nm_j) : j = 1, 2, \ldots\}$ converges, as $n \to \infty$, to a zero-mean Gaussian vector g on the Banach sequence space ℓ_1. The centering constants m_j, and the covariance operator for g can be explicitly computed.

In the nearcritical case, say with $\bar{\sigma}/\sigma = 1 - an^{-1/3}$, the size $L_n^{(k)}$ of the kth largest components of M_n has, asymptotically, the distribution

$$\lim_{n \to \infty} \mathbf{P}\left[L_n^{(k)} < xn^{2/3}\right] = \frac{1}{2\pi p(a)} \int_{-\infty}^{\infty} e^{-iau} \left(e^{\Lambda(x,u)} \sum_{j=0}^{k-1} \frac{\Lambda^j(x,u)}{j!}\right) du, \tag{17}$$

where $p(.)$ is the 3/2-stable (Holtsmark) probability density function with the characteristic function $\exp(i\Xi(u))$ with

$$\Xi(u) = 4(3\bar{\sigma})^{-1}\pi^{1/2}\beta e^{-i3\pi/4}u^{3/2}, \quad \text{for} \quad u \geq 0$$

and

$$= 4(3\bar{\sigma})^{-1}\pi^{1/2}\beta e^{i3\pi/4}|u|^{3/2}, \quad \text{for} \quad u < 0,$$

and

$$\Lambda(x, u) = \frac{\beta}{\sigma} \int_x^\infty e^{iuy} y^{-5/2}\, dy, \qquad \beta = \sqrt{\frac{\bar{x}\bar{y}}{2\pi H^{(3)}(\bar{y})}} H'(\bar{y}).$$

Curiously, the limit is a mixture of Poisson-type "probabilities" with the complex-valued parameter Λ, taken with the complex-valued weights which, however, add up to 1, since $p(a) = (2\pi)^{-1} \int \exp\{i(\Xi(u) - au)\}\, du$.

In the supercritical case $\sigma > \bar{\sigma}$ one can prove that, with probability approaching 1 as $n \to \infty$, the random forest M_n has a unique component of order of magnitude $\sim n$, and $L_n^{(1)}/n \to 1 - \bar{\sigma}/\sigma$. The most surprising result is that, in this case, the size of the largest connected component has, asymptotically, a Holtsmark probability distribution. More precisely, the distribution of

$$\frac{n(1 - \bar{\sigma}/\sigma) - L_n^{(1)}}{((\bar{\sigma}/\sigma)n)^{2/3}} \tag{18}$$

converges, as $n \to \infty$, to the 3/2-stable (Holtsmark) distribution with the density $p(.)$ described above.

21.2.5.   Lévy laws from experimental data: Flight time of particles trapped in vortices.

Data with heavy tails are encountered in many experimental situations. This clearly eliminates the Gaussian distribution as a model for the phenomena under investigation and claims are sometimes made that those data come from an α-stable distribution. Such claims are often tenuous, especially in cases when one cannot offer a supporting evidence of the self-similarity being forced by the physical constraints present. Moreover, with all the measured physical events being finite in scale, the assertion of a particular tail behavior at infinity, or self-similarity at all scales, has to be taken with a grain of salt. Every once in a while the obvious case of wishful thinking can be diagnosed, following an old adage "When all you have is a hammer, everything looks like a nail". Below, we discuss such an example.

Weeks et al. [16] conducted an experiment in a rotating annular tank filled with a mixture of water and glycerol. The setup is described in Fig. 21.2.1. In some flow regimes chaotic advection is present and the typical pattern is that of a chain of several vortices enclosed in between azimuthal jets. The passive tracer particles are then caught in vortices for an irregular period of time (trapping event) before being ejected into the jets where they travel at high velocity (a flight event) only to be trapped into another vortex after a random time period. The typical trajectory of a tracer particle is shown in Fig. 21.2.2.

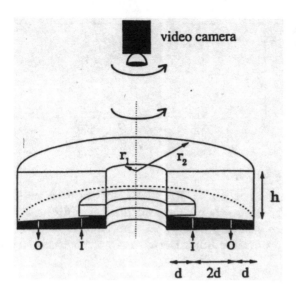

FIGURE 21.2.1.
The annulus rotates at **1.5 Hz.** Flow is produced by pumping fluid in through the inner ring of holes (marked **I**) and out through the outer ring (marked **O**). The rotation guarantees that the flow is essentially two dimensional. A plexiglass ring sits above the inner ring of holes. The flow is observed through a video camera that rotates overhead. The conical bottom models the *beta-plane effect* (from Weeks et al. [16])

The angular position of the particle as a function of time is shown in Fig. 21.2.3. The flat oscillating portions correspond to trapping events, and the steep parts reflect the flight events. Weeks and his collaborators measured then the probability distribution of the duration of flight events and the trapping events, and in two flow regimes found that the former follow a power law which indicated the α-stable behavior with $\alpha = 1.3$ and, respectively, 1.6 (see. Fig. 21.2.4).

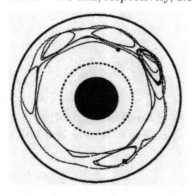

FIGURE 21.2.2.
A typical trajectory of the passive tracer particle in a chaotic regime. Trapping in one of the vortices is followed by a fast flight in the azimuthal jet (from Weeks et al. [16])

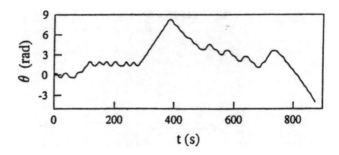

FIGURE 21.2.3.
The azimuthal coordinate $\theta(t)$ is shown for the trajectory shown in Fig. 21.2.2. The steep diagonal lines are flight events, and the oscillations are sticking events. In the time interval between 500 s and 700 s the particle hops between four vortices (from Weeks et al. [16]).

The picture for the probability density function of the trapping events, which should be exponential, was however less conclusive, partly due to imperfections of the experimental setup; the flow was not perfectly planar and it was hard to keep track of the tracer particles for long periods of time.

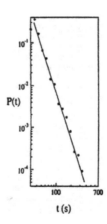

FIGURE 21.2.4.
The measurements of the probability distribution of the duration of flight events clearly indicate power law behavior (from Weeks et al. [16])

Whatever the shortcomings of the above experiments, the results provide the first direct repeatable experimental indication of the existence of Lévy flights in the fluid mechanics context.

21.3 Lévy infinitely divisible processes

Let $\tilde{f}(u)$ be the characteristic function of an infinitely divisible p.d.f. In the context of infinitely divisible processes we shall call it the *mother characteristic function*. Consider the stochastic process

$$X(t) = \lim_{\Delta t \to 0} \sum_{k=1}^{\lfloor t/\Delta t \rfloor} X_k^{(\gamma \, \Delta t)}, \tag{1}$$

where the summands $X_k^{(\gamma \Delta t)}$ are statistically independent and have the same characteristic function $\tilde{f}^{\gamma \Delta t}(u)$. The characteristic function of the process $X(t)$ is

$$\tilde{f}(u;t) = \tilde{f}^{\gamma t}(u) = \exp[\gamma \, t \, \Psi(u)], \tag{2}$$

where $\Psi(u)$ is the logarithm of the mother characteristic function. Applying the Laplace transformation to (2) we obtain the following expression for the Laplace–Fourier transform of the marginal p.d.f. $f(x;t)$ of the process $X(t)$: :

$$\hat{\tilde{f}}(u;s) = \int_0^\infty dt \, e^{-st} \int dx \, e^{iut} \, f(x;t) = \frac{e^{iux_0}}{s - \gamma \Psi(u)}. \tag{3}$$

The equality (3) permits a direct derivation of an equation for the p.d.f. $f(x;t)$. Indeed, rewrite (3) in the form

$$s\hat{\tilde{f}} - \gamma \Psi(u)\hat{\tilde{f}} = 1 \tag{4}$$

and assume that

$$f(x;t=0) \equiv 0.$$

Then the inverse Laplace–Fourier transform of the first summand on the left-hand side of the equality (4) is the derivative:

$$s\,\hat{\tilde{f}}(u;s) \qquad \longmapsto \qquad \frac{\partial f(x;t)}{\partial t}.$$

In what follows we will encounter expression of the form $s^\beta \, \hat{\tilde{f}}$ to which we will formally attach the fractional derivatives[3]

$$s^\beta \qquad \longmapsto \qquad \frac{\partial^\beta}{\partial t^\beta}.$$

In the spatial domain one can consider so-called Weyl multiplier operators corresponding to multiplication of the Fourier transform by $\Psi(u)$. Let $\tilde{f}(u)$ be the characteristic function of a symmetric α-stable random variable,

$$\tilde{f}(u) = \exp\left(\Psi(u)\right), \qquad \text{where} \qquad \Psi(u) = -\sigma^\alpha \, |u|^\alpha. \tag{5}$$

---

[3]The concept of fractional derivatives will be studied in detail in Chapter 22.

Definition: The operator in the x-space corresponding to multiplication by a fractional power of $-iu$ in the Fourier space will be called the *right fractional derivative (in the sense of Weyl)*. In other words,

$$(-iu)^\alpha \qquad \longmapsto \qquad \frac{\partial^\alpha}{\partial x^\alpha}\,,$$

or, more precisely,

$$\left[\frac{\partial^\alpha}{\partial x^\alpha} g(x)\right]^{\widetilde{}}(u) := (-iu)^\alpha \tilde{g}(u).$$

The analogous operator corresponding to multiplication by a fractional power of iu will be called the *left fractional derivative*:

$$(iu)^\alpha \qquad \longmapsto \qquad \frac{\partial^\alpha}{\partial(-x)^\alpha}\,.$$

The logarithm of the mother characteristic function in (5) is a power of the absolute value of u, and the operator in the x-space corresponding to this multiplier will be called the *symmetric fractional derivative*:

$$|u|^\alpha \qquad \longmapsto \qquad -\frac{\partial^\alpha}{\partial|x|^\alpha}\,.$$

Relations between these fractional differentiation operators can be easily established relying on algebraic identities

$$(\pm iu)^\alpha = |u|^\alpha \cos\left(\frac{\pi\alpha}{2}\right) \pm i\,s\,|u|^\alpha \sin\left(\frac{\pi\alpha}{2}\right), \qquad s = \frac{u}{|u|}\,,$$

and

$$|u|^\alpha = \frac{(-iu)^\alpha + (iu)^\alpha}{2\cos(\pi\alpha/2)}\,.$$

Consequently,

$$\frac{\partial^\alpha}{\partial|x|^\alpha} = \frac{-1}{2\cos(\pi\alpha/2)}\left[\frac{\partial^\alpha}{\partial x^\alpha} + \frac{\partial^\alpha}{\partial(-x)^\alpha}\right].$$

If the mother characteristic function of the summands of the infinitely divisible process (1) is equal to (5), then the Laplace–Fourier transform $\hat{\tilde{f}}(u;s)$ of the p.d.f. $f(x;t)$ of the infinitely divisible process $X(t)$ satisfies equation

$$s\,\hat{\Theta} + \gamma\,\sigma^\alpha\,|u|^\alpha\,\hat{\Theta} = 1\,.$$

Applying the inverse Laplace and Fourier transformations to both sides of the above equation we get the following equation in fractional partial derivatives for $f(x;t)$:

$$\frac{\partial f}{\partial t} = \gamma\,\sigma^\alpha\,\frac{\partial^\alpha f}{\partial|x|^\alpha} + \delta(x)\,\delta(t)\,. \tag{6}$$

This equation can be solved by finding the inverse Fourier transform of the corresponding characteristic function

$$\tilde{f}(u;t) = \exp\left(-\gamma\,\sigma^\alpha\,|u|^\alpha\,t\right).$$

As a result we obtain the p.d.f. of the so-called *Lévy flights* or *α-stable Lévy processes*:

$$f_\alpha(x;\tau) = \frac{1}{\sigma\tau^{1/\alpha}} f_\alpha\left(\frac{x}{\sigma\tau^{1/\alpha}}\right), \qquad \tau = \gamma\,t. \tag{7}$$

As we already observed, the mean square of the Lévy flights is infinite but moments of order $\kappa < \alpha$ are finite:

$$\langle |X(\tau)|^\kappa\rangle = \langle |X|^\kappa\rangle\,\tau^{\kappa/\alpha}, \qquad -1 < \kappa < \alpha, \tag{8}$$

where $\langle |X|^\kappa\rangle$ is given by the expression (21.1.16).

Observe that Lévy flights can describe the intermediate asymptotics of stochastic processes with finite variance. To explain what we mean by this statement let us consider the following example of a tempered Lévy flight: Take

$$\psi(z) = -\frac{|z|^{-\alpha-1}}{2\Gamma(-\alpha)\cos\left(\frac{\pi\alpha}{2}\right)}\,e^{-\delta|z|}$$

as the intensity function in the generalized Poissonian random variable (21.1.3). Substituting this intensity function into (21.1.2) we obtain the logarithm of the characteristic function corresponding to the tempered α-stable p.d.f.:

$$\Psi(u,\alpha,\delta) = \frac{1}{\cos\left(\frac{\pi\alpha}{2}\right)}\left[\delta^\alpha - (\delta^2 + u^2)^{\alpha/2}\cos\left(\alpha\arctan\left(\frac{|u|}{\delta}\right)\right)\right]. \tag{9}$$

In particular, for $\alpha = 1$, we obtain

$$\Psi(u,\alpha=1,\delta) = -\frac{2}{\pi}\arctan\left(\frac{|u|}{\delta}\right)|u| + \frac{\delta}{\pi}\ln\left(1 + \frac{u^2}{\delta^2}\right),$$

while, for $\alpha = 1/2$, we get the expression (21.1.17).

For $u \to 0$, the main asymptotics of $\Psi(u,\alpha,\delta)$ in (9) is

$$\Psi(u,\alpha,\delta) \sim -\frac{\alpha(1-\alpha)}{2\cos(\pi\alpha/2)\,\delta^{2-\alpha}}\,u^2, \qquad (u \to 0).$$

Consequently, in this case, the Lévy infinitely divisible process with the characteristic function

$$\tilde{f}(u;\tau;,\alpha;\delta) = \exp\left(\tau\Psi(u,\alpha,\delta)\right)$$

has a finite mean square which grows linearly; indeed, for any $\tau > 0$, and small $\delta > 0$,

$$\langle X(\tau)\rangle \approx \frac{\alpha(1-\alpha)}{\cos(\pi\alpha/2)\,\delta^{2-\alpha}}\,\tau \tag{10}$$

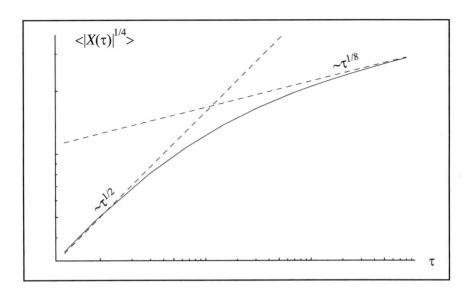

FIGURE 21.3.1.
Dependence of the moment $\langle |X(\tau)|^{1/4} \rangle$ on τ for the tempered Lévy flight corresponding to $\alpha = 1/2$. Dashed lines correspond to the law $\tau^{1/2}$, characteristic of the Lévy flight, and $\tau^{1/8}$, characteristic of the Wiener process. One can observe that, initially, the moment of the order $1/4$ grows like that of the Lévy flight

On the other hand, the tails of the p.d.f. of $X(t)$ have the power asymptotics (21.1.12), so that the moments $\langle |X(\tau)|^\kappa \rangle$ of order $\kappa < \alpha$ depend on the time in the way characteristic for Lévy flights. This behavior is clear from Fig. 21.3.1 which shows dependence of $\langle |X(\tau)|^{1/4} \rangle$ on τ obtained, as in (21.1.15), by numerical integration of the formula

$$\langle |X(\tau)|^\kappa \rangle = -\frac{2\kappa\Gamma(\kappa)}{\pi} \sin\left(\frac{\pi\kappa}{2}\right) \int_0^\infty \frac{\tilde{f}(u;\tau;\alpha;\delta) - 1}{u^{\kappa+1}} \, du \,.$$

21.4   Fractional exponential distribution

In addition to the β-stable p.d.f. $f_\beta(\tau)$ discussed in Section 21.1, the one-sided p.d.f. $\varphi_\beta(\tau)$, with an infinite mean, and the Laplace transform

$$\hat{\varphi}(s) = \frac{1}{1 + s^\beta} \qquad (0 < \beta < 1), \tag{1}$$

plays an essential role in the study of anomalous diffusion. We will call it *fractional exponential distribution*. Let us explore its properties.

First, notice that the equality (1) is equivalent to the following fractional differential equation:

$$\frac{d^\beta \varphi}{dt^\beta} + \varphi = \delta(t).$$

The solution of this equation is

$$\varphi_\beta(\tau) = -\frac{1}{\tau} D_\beta(-\tau^\beta)\,\chi(\tau), \quad \text{where} \quad D_\beta(z) = \beta\,z\,\frac{dE_\beta(z)}{dz},$$

and $E_\beta(z)$ is the Mittag-Leffler function. It is easy to see that the function $D_\beta(z)$ is expressed by a contour integral

$$D_\beta(z) = \frac{z}{2\pi i} \int_{\mathcal{H}} \frac{e^y\,dy}{y^\beta - z},$$

where the integration is carried out along the Hankel loop depicted in Fig. 21.4.1.

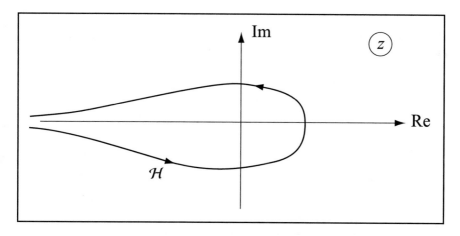

FIGURE 21.4.1.Plot of the Hankel loop

For the negative real values of its argument, it is easy to obtain the following integral representation for the fractional exponential p.d.f.:

$$\varphi_\beta(\tau) = \frac{\sin(\pi\beta)}{\pi}\,t^{\beta-1} \int_0^\infty \frac{x^\beta\,e^{-x}\,dx}{x^{2\beta} + \tau^{2\beta} + 2x^\beta\tau^\beta \cos(\pi\beta)}, \qquad 0 < \beta < 1. \quad (2)$$

The above formula yields the following asymptotics

$$\varphi_\beta(\tau) \sim \frac{\tau^{\beta-1}}{\Gamma(\beta)} \quad (\tau \to 0); \qquad \varphi_\beta(\tau) \sim \frac{\tau^{-\beta-1}}{\Gamma(1-\beta)} \quad (\tau \to \infty). \quad (3)$$

Fractional exponential p.d.f.s are sometimes written as a mixture of standard exponential distributions. To find this mixture let us return to the integral (2) and introduce a new variable of integration $\mu = \tau/x$. As a result, (2) assumes the form

$$\varphi_\beta(\tau) = \int_0^\infty \frac{1}{\mu} \exp\left(-\frac{\tau}{\mu}\right) \xi_\beta(\mu)\, d\mu, \tag{4}$$

where

$$\xi_\beta(\mu) = \frac{1}{\pi\mu} \frac{\sin(\pi\beta)}{\mu^\beta + \mu^{-\beta} + 2\cos(\pi\beta)}. \tag{5}$$

The latter function can be thought of as the spectrum of the mean values of the standard exponential p.d.f.

$$\varphi(\tau) = \frac{1}{\mu} \exp\left(-\frac{\tau}{\mu}\right).$$

For $\beta \to 1$, the spectrum (5) of the random times weakly converges to the Dirac delta, while $\varphi_\beta(t)$ approaches the standard exponential p.d.f.

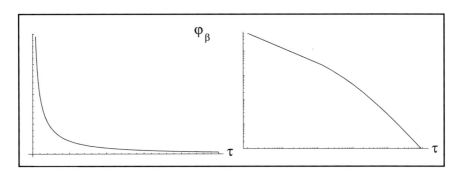

FIGURE 21.4.2.
Plots of the p.d.f. $\varphi_\beta(\tau)$, for $\beta = 1/2$, in the linear (left) and the log-log (right) scales. The latter clearly shows the power asymptotics (3) of the fractional exponential p.d.f

Finally, observe that, for $\beta = 1/2$, the fractional exponential p.d.f. can be explicitly written in terms of the complementary error function:

$$\varphi_{1/2}(\tau) = \sqrt{\frac{\pi}{\tau}} - \pi\, e^\tau \operatorname{erfc}\left(\sqrt{\tau}\right).$$

A plot of the above p.d.f. is given in Fig. 21.4.2.

21.5 Local characteristics of Lévy processes and their generalized derivatives

In this section we will consider local characteristics of Lévy processes $\xi(t)$, that is processes with homogeneous and independent increments and their generalized derivatives. We will assume additionally, without loss of generality that $\xi(0) = 0$, and that the realization of $\xi(t)$ is locally integrable; the class of such processes will be denoted by $\mathcal{K}$, the classes of their derivatives of order k, all of them generalized processes with independent values at each point, will be denoted by $\mathcal{K}_s$.

Since, all the moments of $\xi(t) \in \mathcal{K} \cap \mathcal{M}$ are finite, its characteristic function can be written in the Lévy-Khinchine form[4]

$$R_t(z) = \exp\left(ic_\xi tz + t \int_{-\infty}^{\infty} (e^{iuz} - 1 - iuz)\frac{1}{u^2}dG_\xi(u) \right), \tag{1}$$

where c_ξ is a constant, and $G_\xi(u)$ is a nondecreasing cad-lag (continuous on the right, and with limits on the left) function, normalized by the condition $G_\xi(-\infty) = 0$, and satisfying the condition

$$\int_{-\infty}^{\infty} |x|^n dG_\xi(u) < \infty, \qquad n = 1, 2, \ldots. \tag{2}$$

Indeed, condition (2) is necessary and sufficient for the existence of all moments for the process $\xi(t)$.

The following definitions are justified by the examples considered in Section 17.6.

Definition 1. (i) A local characteristic $[a_n, L_n]$ is said to be *singular* if, for a certain constant c

$$L_n(\varphi) = \frac{\varphi(c)}{1 + |c|^n}, \qquad n = 0, 1, 2, \ldots,$$

(ii) A local characteristic (a_n, L_n) is said to be *Poissonian*, if for certain constants c_0, c_1, a_n, b_n,

$$L_0(\varphi) = \frac{\varphi(c_0)}{2}, \quad L_1(\varphi) = c_1\varphi(c_0) + a_1 G_1^+(\varphi) + b_1 G_1^-(\varphi),$$

and

$$L_n(\varphi) = a_n G_n^+(\varphi) + b_n G_n^-(\varphi), \qquad n = 2, 3, \ldots.$$

---

[4]For the validity of this version of the Lévy-Khinchine formula it is sufficient that the second moment of $\xi(t)$ is finite.

(iii) A local characteristic $[a_n, L_n]$ is said to be *quasi-Poissonian*, if for certain constants c, a_n, b_n,

$$L_0(\varphi) = \frac{\varphi(c)}{2}, \quad \text{and} \quad L_n(\varphi) = a_n G_n^+(\varphi) + b_n G_n^-(\varphi), \qquad n = 1, 2, \ldots.$$

(iv) A local characteristic $[a_n, L_n]$ is said to be *Cauchy*, if for certain constants a, b, a_n, b_n, and $n = 1, 2, \ldots$,

$$L_0(\varphi) = \frac{a}{2\pi} \int_{-\infty}^{\infty} \frac{\varphi(x)}{a^2 + (x - b)^2} dx, \quad \text{and} \quad L_n(\varphi) = a_n G_n^+(\varphi) + b_n G_n^-(\varphi).$$

and, finally,

(v) A local characteristic $[a_n, L_n]$ is said to be *uniform*, if the functionals $L_n, n = 0, 1, 2, \ldots$, are invariant under translations, that is, for each a, we have $L_n(\varphi) = L_n(\varphi_a)$, where $\varphi_a(x) = \varphi(x + a)$, or , equivalently, that

$$L_n(\varphi) = a_n G_n^+(\varphi) + b_n G_n^-(\varphi), \qquad n = 1, 2, \ldots$$

Although the above list may seem arbitrary at the first sight, it turns out that is exhausts all the possibilities of the forms of local characteristics for generalized processes in $\mathcal{K}_1 \cup \mathcal{K}_2 \cup \ldots$. Indeed we have the following

THEOREM 1. *(a) The local characteristics of generalized processes in class $\mathcal{K}_1$, that is, the first-order generalized derivatives of Lévy processes in class $\mathcal{K}$, are always of one of the following types:*
        singular, Poissonian, quasi-Poissonian, Cauchy, or uniform.

(b) The local characteristics of generalized processes in class $\mathcal{K}_2 \cup \mathcal{K}_3 \cup \ldots$, that is, the second or higher-order generalized derivatives of Lévy processes in class $\mathcal{K}$, are always of one of the following types:
        singular, quasi-Poissonian, or uniform.

The proof of the above theorem is beyond the scope of this book[5], and we have already noted examples of particular processes with the local characteristics from the above list. One of them was, of course, the Brownian motion and its derivatives which have uniform local characteristics. But here is a general criterion for the derivatives of a Lévy process to have quasi-Poissonian local characteristics;

Criterion for quasi-Poissonian local characteristics for derivatives of a Lévy process. Consider a Lévy process $\xi(t) \in \mathcal{K} \cap \mathcal{M}$ with the Lévy function $G_\xi(u)$.

(i) If , for some x_0, we have $\int_0^{x_0} u^{-1} dG_\xi(u) = \infty$, and the limits

$$c_n = \lim_{h \downarrow 0} \frac{\int_0^\infty x^n dP(D_h \xi(t) < x)}{\int_{-\infty}^\infty |x|^n dP(D_h \xi(t) < x)}, \quad \Lambda(\varphi) = \lim_{h \downarrow 0} \int_{-\infty}^\infty \varphi(x) dP(D_h(f) < x),$$

---

[5]Urbanik [9] , pp 231–241, see references to Chapter 17.

exist for $n = 1, 2, \ldots$, then the derivative $\xi'(t)$ has the local characteristics $[a_n, L_n]$, with

$$L_0(\varphi) = \frac{1}{2}\Lambda(\varphi), \quad and \quad L_n(\varphi) = c_n G_n^+(\varphi) + (1 - c_n)G_n^-(\varphi), \ n = 1, 2, \ldots \quad (3)$$

(ii) If , for some x_0, we have $\int_0^{x_0} u^{-1} d(G_\xi(u) - G_\xi(-u)) = \infty$, and the limits

$$c_n = \lim_{h_1, \ldots h_k \downarrow 0} \frac{\int_0^\infty x^n dP(D_{h_1, \ldots h_k}\xi(t) < x)}{\int_{-\infty}^\infty |x|^n dP(D_{h_1, \ldots h_k}\xi(t) < x)},$$

and

$$\Lambda(\varphi) = \lim_{h_1, \ldots h_k \downarrow 0} \int_{-\infty}^\infty \varphi(x) dP(D_{h_1, \ldots h_k}(f) < x),$$

exist for $n = 1, 2, \ldots$, then the kth derivative $\xi^{(k)}(t), k \geq 2$, has the local characteristics $[a_n, L_n]$ given in (3).

(iii) In the case of a process $\xi(t) \in \mathcal{K} \cap \mathcal{M}$ with symmetrically distributed increments, if, for some x_0, the integral $\int_0^{x_0} u^{-1} dG_\xi(u) = \infty$, and the limit

$$\Lambda(\varphi) = \lim_{h \downarrow 0} \int_{-\infty}^\infty \varphi(x) dP(D_h(f) < x),$$

exists, the local characteristic $[a_n, L_n]$ of $\xi'(t)$ is of the form

$$L_0(\varphi) = \frac{1}{2}\Lambda(\varphi), \quad and \quad L_n(\varphi) = \frac{1}{2}G_n^+(\varphi) + \frac{1}{2}G_n^-(\varphi), \ n = 1, 2, \ldots, \quad (4)$$

and, if $\int_0^{x_0} u^{-1} d(G_\xi(u) - G_\xi(-u)) = \infty$, and

$$\Lambda(\varphi) = \lim_{h_1, \ldots h_k \downarrow 0} \int_{-\infty}^\infty \varphi(x) dP(D_{h_1, \ldots h_k}(f) < x)$$

exist, then the kth derivative $\xi^{(k)}(t), k \geq 2$, has the local characteristics $[a_n, L_n]$ given in (4).

21.6 Examples of processes whose derivatives have local characteristics in different classes

In this section we will provide concrete examples of generalized processes whose local characteristics belong to one of the four categories described in the previous section.

21.6.1.   A process whose all derivatives have quasi-Poissonian local characteristics.   Consider a Lévy process $\xi(t)$ with the Lévy function of the form

$$G_\xi(x) = \int_{-\infty}^x \frac{e^{-|u|}}{1 + |\log|u||}du.$$

Because all the moment

$$\int_{-\infty}^x |u|^n \frac{e^{-|u|}}{1 + |\log|u||}du, \qquad n = 1, 2, \ldots$$

are finite, $\xi(t) \in \mathcal{K} \cap \mathcal{M}$, and the characteristic function of the differential ratios $D_{h_1} \ldots D_{h_k}\xi(t)$ is given by the formula

$$Q_{h_1,\ldots,h_k}(z) = \exp\left[2^k \int_0^\infty \left(\cos\frac{zu}{h_1 \cdots h_k} - 1\right)\frac{e^{-h_k u}}{u^2(1 + |\log h_k u|)}du\right].$$

Hence, uniformly on every finite interval, $Q_{h_1,\ldots,h_k}(z) \to 1$, as $h_1, \ldots, h_k \downarrow 0$. As a result, for any $\varphi \in \mathcal{X}_0$,

$$\lim_{h_1,\ldots,h_k \downarrow 0} \int_{-\infty}^\infty \varphi(x)dP(D_{h_1,\ldots,h_k}\xi(t) < x) = \varphi(0).$$

Since

$$\frac{1}{2}\int_0^1 \frac{1}{u}d(G_\xi(u) - G_\xi(-u)) = \int_0^1 \frac{1}{u}dG_\xi(u) = \int_0^1 \frac{e^{-|u|}}{u(1 + |\log|u||)}du = \infty$$

in view of the results in the previous section, the derivatives $f^{(k)}(t), k = 1, 2, \ldots$, have quasi-Poissonian local characteristics $[a_n^{(k)}, L_n^{(k)}]$, that is,

$$L_0^{(k)}(\varphi) = \varphi(0)/2, \qquad L_n^{(k)}(\varphi) = (G_n^+(\varphi) + G_n^-(\varphi))/2, \qquad n = 1, 2, \ldots$$

21.6.2.   A process whose first derivative has a Cauchy local characteristics.   In this case take

$$G_\xi(x) = \frac{1}{2}\int_{-\infty}^x e^{-|u|}du.$$

Again $\xi(t) \in \mathcal{K} \cap \mathcal{M}$, because $\int_{-\infty}^x |u|^n e^{-|u|}du < \infty$, $n = 1, 2, \ldots$. The characteristic function of $D_h(f)$,

$$Q_h(z) = \exp\left[h\int_0^\infty \left(\cos\frac{zu}{h} - 1\right)\frac{e^{-u}}{u^2}du\right] = \exp\left[\int_0^z \exp\left(-\frac{h^2}{2u^2}\right)du\right] \to e^{-|z|}.$$

As $h \downarrow 0$, for each $\varphi \in \mathcal{X}_0$,

$$\lim_{h \downarrow 0}\int_{-\infty}^\infty \varphi(x)dP(D_h\xi(t) < x) = \frac{1}{\pi}\int_{-\infty}^\infty \frac{\varphi(x)}{1 + x^2}dx.$$

Since $Q_h(z)$ is real, and

$$\int_0^1 u^{-1} dG_\xi(u) = \int_0^1 u^1 e^{-u} du = \infty,$$

in view of the previous section we see that $\xi'(t)$ has the Cauchy local characteristic $[a_n, L_n]$ with

$$L_0(\varphi) = \frac{1}{2\pi} \int_{-\infty}^{\infty} \frac{\varphi(x)}{1+x^2} dx, \qquad L_n(\varphi) = (G_n^+(\varphi) + G_n^-(\varphi))/2, \qquad n = 1, 2, \ldots$$

21.6.3.   Processes whose all derivatives have uniform local characteristics.
We have observed before that all the derivatives of the Brownian motion process have uniform local characteristics. One can demonstrate easily that if the Lévy process, $\xi(t)$, has a nontrivial Gaussian component, that is $G_\xi(+0) - G_\xi(-0) > 0$, and the derivative $\xi^{(k)}(t)$ possesses a local characteristic, this characteristic is necessarily uniform.

Here we provide an example of a process $\xi(t) \in \mathcal{K} \cap \mathcal{M}$ without a Gaussian component whose first derivative $\xi'(t)$ has a uniform local characteristic. Let us begin with

$$G_\xi(x) = \frac{1}{2} \int_{-\infty}^{x} \frac{e^{-|u|}}{|u|^{1/2}} du.$$

Clearly, $G_\xi(+0) - G_\xi(-0) = 0$ so there is no Gaussian component here, and since $\int_{-\infty}^{\infty} |x|^n dG_\xi(x) < \infty$, $n = 1, 2, \ldots$ we have $\xi(t) \in \mathcal{K} \cap \mathcal{M}$, indeed. The characteristic function of the rescaled increment $\Delta_h \xi(t)/h^{2/3}$ is

$$R_h(z) = \exp\left(h \int_{-\infty}^{\infty} (e^{izuh^{-2/3}} - 1 - izuh^{-23/}) \frac{e^{-|u|}}{|u|^{5/2}} du \right)$$

$$= \exp\left(-2|z|^{5/2} \int_0^{\infty} \frac{1 - \cos u}{u^{5/2}} \exp\left(-\frac{uh^{2/3}}{|z|} \right) du \right)$$

so that the (uniform on every finite interval) limit of the characteristic function

$$\lim_{h \downarrow 0} R_h(z) = \exp\left(-2|z|^{5/2} \int_0^{\infty} \frac{1 - \cos u}{u^{5/2}} du \right)$$

is integrable over the whole real line and, consequently, there exists a symmetric density function $r(x)$ such that

$$\lim_{h \downarrow 0} P(\Delta_h \xi(t)/h^{2/3} < x) = \int_{-\infty}^{x} r(y) dy. \tag{1}$$

In view of (1) and the symmetry of $r(x)$, for each $\varphi \in \mathcal{X}_0$, and $\epsilon > 0$, we have the inequality,

$$\limsup_{h \downarrow 0} \left| \int_{-\infty}^{\infty} \varphi(x) dP(D_h \xi(t) < x) - \frac{1}{2}(G_0^+(\varphi) + G_0^-(\varphi)) \right|$$

$$\leq |G_0^+(\varphi)| \int_0^{\epsilon} r(u) du + |G_0^-(\varphi)| \int_{-\epsilon}^0 r(u) du + \|\varphi\| \int_{-\epsilon}^{\epsilon} r(u) du,$$

so that, as $\epsilon \to 0$, and $h \downarrow 0$,

$$\int_{-\infty}^{\infty} \varphi(x) dP(D_h \xi(t) < x) \to \frac{1}{2}(G_0^+(\varphi) + G_0^-(\varphi)).$$

Since $D_h \xi(t)$ have symmetric distributions, and

$$\int_0^1 u^{-1} dG_\xi(u) = \int_0^1 u^{-3/2} e^u du = \infty,$$

we infer that $\xi'(t)$ has a uniform local characteristic.

21.6.4. More exotic examples. In an analogous fashion one can demonstrate that for the Lévy process $\xi(t)$ with the Lévy function

$$G_\xi(x) = \int_{-\infty}^x (2 - \sin(\log |u|)) e^{-|u|} du,$$

no derivative has a local characteristic, and for $\xi(t)$ with

$$G_\xi(x) = \int_{-\infty}^x (2 - \operatorname{sgn} u \cdot \sin(\log |u|)) e^{-|u|} du,$$

the first derivative does not have a local characteristic but all the higher derivatives do (see, Exercises). Both processes are in $\mathcal{K} \cap \mathcal{M}$.

21.6.5. Compound Poisson processes. A measurable homogeneous stochastic process $\xi(t)$ with independent increments is called a *compound Poisson process* if the characteristic function of its increments is of the form,

$$R_h(z) = E \exp(iz(\xi(t-h) - \xi(t)) = \exp\left(ih\gamma_\xi^* z + h \int_{u \neq 0} (e^{iuz} - 1) d\mu_\xi(u) \right),$$

for some real constant γ_ξ^*, and a σ-finite measure μ_ξ such that $0 < \int_{-\infty}^{\infty} |z|(1 + |z|)^{-1} d\mu_\xi(z) < \infty$. The condition $\int_{-\infty}^{\infty} |z|^n d\mu_\xi(z) < \infty$ is necessary and sufficient for $\xi(t) \in \mathcal{M}$.

In terms of the previous description of the Lévy process we have equalities

$$\gamma_\xi = \gamma_\xi^* + \int_{-\infty}^{\infty} z d\mu_\xi(z), \quad and \quad G_\xi(x) = \int_{-\infty}^x z^2 d\mu_\xi(z),$$

and we can conclude that a process $\xi(t) \in \mathcal{K} \cap \mathcal{M}$ is a compound Poisson process if, and only if,

$$0 < \int_{-\infty}^{\infty} |u|^{-1} dG_\xi(u) < \infty.$$

The following fundamental result on generalized stochastic processes with Poissonian local characteristics is due to K. Urbanik (for the proof see the reference [9] to Chapter 17) . Denote the nth absolute moment of μ_ξ by κ_n:

$$\kappa_n := \int_{-\infty}^{\infty} |u|^m d\mu_\xi(u).$$

THEOREM 1: *A generalized stochastic process* $\Xi(t) \in \mathcal{K}_1 \cup \mathcal{K}_2 \cup \ldots$ *has a Poissonian local characteristic if and only if it is the first derivative of a compound Poisson process in* $\mathcal{M}$, *i.e.,* $\Xi(t) = \xi'(t)$. *In such a case the local characteristic* (a_n, L_n) *of* $\Xi(t)$ *is given by the following formulas:*

$$a_0 = \frac{1}{2}, \quad a_1(\lambda) = \frac{1}{1 + \kappa_1}, \quad a_n(\lambda) = \frac{\lambda^{n-1}}{\kappa_n}, n \geq 2, \quad L_0(\varphi) = \frac{\varphi(\gamma_\xi^*)}{2},$$

$$L_1(\varphi) = \frac{1}{1 + \kappa_1} \left(\varphi(\gamma_\xi^*) + \int_0^\infty u d\mu_\xi(u) \cdot G_1^+(\varphi) + \int_{-\infty}^0 |u| d\mu_\xi(u) \cdot G_1^-(\varphi) \right),$$

$$L_n(\varphi) = \frac{1}{\kappa_n} \left(\int_0^\infty u^n d\mu_\xi(u) \cdot G_1^+(\varphi) + \int_{-\infty}^0 |u|^n d\mu_\xi(u) \cdot G_1^-(\varphi) \right), \quad n \geq 2.$$

As a corollary to the above result, and a Ryll-Nardzewski Theorem [7], one can prove that the derivative of a homogeneous process $\xi(t)$ with independent increments whose sample paths are of bounded variation has a Poissonian local characteristic if, and only if, $\xi(t) \in \mathcal{M}$.

21.7   Relationship between moments of process' increments and local characteristics of its generalized derivatives

In this section we will explain in what sense the knowledge of the local characteristics of a generalized process provides information about its "probability distribution" or, more exactly, about the moment sequence for approximations of that generalized process. Here, as in the case of the classical probability theory, there are some constraints; in some cases moment sequences do not completely determine the probability distribution of the regular random variable. When that does happens the issue is handled by the solution of the so-called Stieltjes moment problem. We shall begin by a seemingly simple statement.

PROPOSITION 1. *If* $\xi \in \mathcal{K} \cap \mathcal{M}$ *and* $\int_{-\infty}^{\infty} |x| dG_f(x) > 0$, *and* $\xi'(t)$ *has the local characteristic* (a_n, L_n), *then for* $n \geq 2$,

$$L_n(x^n) = 0, \qquad \text{if and only if} \qquad m_n := \int_{-\infty}^{\infty} u^{n-2} dG_\xi(u) = 0.$$

Proof. It suffices to consider only odd n's. For even n's, by assumption, $m_n > 0$, and thus $L_n(x^n) = L_n(1 + |x|^n) = 1$. Now, since

$$\lim_{h \downarrow 0} a_n(h) \int_{-\infty}^{\infty} |x|^n d\mathbf{P}(D_h \xi(\omega, t) < x) = L_n(|x|^n),$$

in view of Theorem 21.4.1,

$$\lim_{h \downarrow 0} a_n(h) h^{-n} \mathbf{E} |\Delta_h \xi(\omega, t)|^n = L_n(1 + |x|^n) = 1, \qquad (1)$$

for $n \geq 2$.

Taking into account the inequality

$$\mathbf{E} |\Delta_h \xi(\omega, t)|^{2s+1} \leq \mathbf{E} |\Delta_h \xi(\omega, t)|^{2s} + \mathbf{E} |\Delta_h \xi(\omega, t)|^{2s+2},$$

we obtain, for each $s = 1, 2, \ldots$, that

$$\mathbf{E} |\Delta_h \xi(\omega, t)|^{2s+1} \leq h \left(\int_{-\infty}^{\infty} u^{2s-2} dG_\xi(u) + \int_{-\infty}^{\infty} u^{2s} dG_\xi(u) + o(1) \right).$$

Hence, in view of (1),

$$\liminf_{h \downarrow 0} a_{2s+1}(h) h^{-2s} > 0, \quad \text{for} \quad s = 1, 2, \ldots . \qquad (2)$$

Similarly, one obtains that

$$\lim_{h \downarrow 0} a_{2s+1}(h) h^{-2s} (m_{2s+1} + o(1)) = L_{2s+1}(x^{2s+1}), \quad \text{for} \quad s = 1, 2, \ldots .$$

Consequently, in view of (2), $L_{2s+1}(x^{2s+1}) = 0$ implies $m_{2s+1} = 0, s = 1, 2, \ldots$.
Next, since

$$\lim_{h \downarrow 0} a_{2s+1}(h) \mathbf{E}(D_h \xi(\omega, t))^{2s+1} = L_{2s+1}(x^{2s+1}), \qquad (3)$$

in view of Theorem 1,

$$\lim_{h \downarrow 0} a_{2s+1}(h) \mathbf{E}(D_h \xi(\omega, t))^{2s} = L_{2s+1}(x^{2s}) = 0. \qquad (4)$$

Now, let us suppose, to the contrary, that $m_{2s+1} = 0$, but

$$L_{2s+1}(x^{2s+1}) \neq 0, \qquad \text{for some} \quad s \geq 1. \qquad (5)$$

Then

$$\lim_{h \downarrow 0} h^{2s-1} \mathbf{E}(D_h \xi(\omega, t))^{2s} = \int_{-\infty}^{\infty} u^{2s-2} dG_\xi(u), \tag{6}$$

and the limit

$$\lim_{h \downarrow 0} h^{2s-1} \mathbf{E}(D_h \xi(\omega, t))^{2s+1} \tag{7}$$

exists. From (3), (5), and (7), it follows that

$$\liminf_{h \downarrow 0} h^{1-2s} a_{2s+1} > 0.$$

Hence, in view of (4),

$$\lim_{h \downarrow 0} \frac{a_{2s+1}(h) \mathbf{E}(D_h \xi(\omega, t))^{2s}}{a_{2s-1}(h) h^{1-2s}} = \lim_{h \downarrow 0} h^{2s-1} \mathbf{E}(D_h \xi(\omega, t))^{2s} = 0,$$

so that, according to (6),

$$\int_{-\infty}^{\infty} u^{2s-2} dG_\xi(u) = 0,$$

which contradicts the assumption $\int_{-\infty}^{\infty} |x| dG_f(x) > 0$. Consequently, the equality $m_{2s+1} = 0$ implies that $L_{2s+1}(x^{2s+1}) = 0$, for $s = 1, 2, \ldots$, which completes the proof of the Proposition.

In the next theorem we will provide sufficient conditions for when the local characteristics determine the moments of the increments of the process. For a generalized process with the local characteristic (a_n, L_n) define the nonnegative integer

$$k_0 = \begin{cases} s_0, & \text{if } \liminf_{\lambda \downarrow 0} \frac{\lambda^2 a_2(\lambda)}{a_4(\lambda)} = 0; \\ \lfloor (s_0 + 1)/2 \rfloor, & \text{if } \liminf_{\lambda \downarrow 0} \frac{\lambda^2 a_2(\lambda)}{a_4(\lambda)} > 0, \end{cases}$$

where

$$s_0 = \lim_{\lambda \downarrow 0} \frac{\log a_2(\lambda)}{\log \lambda}$$

if the limit exists.

THEOREM 1: *If a generalized stochastic process $\Xi(t) \in \mathcal{K}_1 \cup \mathcal{K}_2 \cup \ldots$ has the local characteristic (a_n, L_n) then the limit s_0 exists and*
   (a) If $k_0 = 0$ then $\Xi(t)$ is a constant process;
   (b) If $k_0 \geq 1$ then

$$\Xi(t) = \xi^{(k_0)}(t)$$

for some process $\xi \in \mathcal{K} \cap \mathcal{M}$, and the local characteristic (a_n, L_n) determines all the moments of the increments $\Delta_{h_1} \Delta_{h_2} \xi(t), h_1, h_2 > 0$,
   (c) If $k_0 = 1$ then the local characteristic (a_n, L_n) determines all the moments of the centered increments $\Delta_h \xi(t) - E\Delta_h \xi(t)$.

Proof. In view of the basic properties of generalized processes there exists an integer k, and a process $\xi(t) \in \mathcal{K} \cap \mathcal{M}$, such that $\xi^{(k)}(t) = \Xi(t)$.

Suppose that $\xi(t)$ is a deterministic process, that is $\xi(t) = ct$, where c is a constant. Then $\Xi(t)$ is a constant process, and $\lim_{\lambda \downarrow 0} a_n(\lambda) > 0$, for $n = 0, 1, 2, \ldots$, so that

$$s_0 = 0, \quad \lim_{\lambda \downarrow 0} \frac{\lambda^2 a_2(\lambda)}{a_4(\lambda)} = 0, \quad \text{and} \quad k_0 = 0.$$

Now, we shall prove that in the other cases

$$k_0 = k, \tag{8}$$

which implies that $k_0 \geq 1$.

First, let us assume that $\xi(t)$ is a Brownian motion process in which case

$$\lim_{\lambda \downarrow 0} \frac{\sqrt{\pi} \lambda^{k-1+n/2}}{a_n(\lambda) 2^{kn/2} \sigma^n \Gamma((n+1)/2)} = 1, \qquad n = 1, 2, \ldots. \tag{9}$$

As a result,

$$s_0 = k, \quad \lim_{\lambda \downarrow 0} \frac{\lambda^2 a_2(\lambda)}{a_4(\lambda)} = 0,$$

and, consequently, $k_0 = k$. Now, we may suppose that

$$\int_{-\infty}^{\infty} |x| \, dG_\xi(x) > 0. \tag{10}$$

Since

$$L_{2s}(x^{2s}) = L_{2s}(1 + x^{2s}) = 1, \quad \text{for} \quad s = 1, 2, \ldots,$$

we have, for $s = 1, 2, \ldots$,

$$\lim_{\lambda \downarrow 0} \lambda^{1-2sk} a_{2s}(\lambda) = \frac{1}{2^{k-1} \int_{-\infty}^{\infty} x^{2s-1} \, dG_\xi(x)}, \tag{11}$$

which implies that

$$s_0 = 2k - 1, \quad \liminf_{\lambda \downarrow 0} \frac{\lambda^2 a_2(\lambda)}{a_4(\lambda)} > 0,$$

so that, consequently, $k_0 = k$.

Next, since the characteristic function of the increments $\Delta_{h_1} \Delta_{h_2} \xi(t), h_1, h_2 > 0$, is given by the formula

$$R_{h_1 h_2}(z) = \exp\left(2 \min(h_1, h_2) \int_{-\infty}^{\infty} (\cos zu - 1) \, dG_\xi(u)\right)$$

we see that the odd moments $\mathbf{E}(\Delta_{h_1}\Delta_{h_2}\xi(t))^n = 0, n = 1, 3, \ldots$, and the even moments $\mathbf{E}(\Delta_{h_1}\Delta_{h_2}\xi(t))^n, n = 0, 2, \ldots$, are determined by the moments

$$\int_{-\infty}^{\infty} x^{2s}\, dG_\xi(x), \qquad s = 0, 1, 2, \ldots. \tag{12}$$

Thus to prove that the local characteristic $[a_n, L_n]$ determines the moments of $\Delta_{h_1}\Delta_{h_2}\xi(t), h_1, h_2 > 0$, it suffices to show that the moments (12) are determined by the local characteristic. Indeed, from (9) and (11), we obtain for $s = 0, 1, 2, \ldots$ that the moments (12) are equal to

$$\lim_{\lambda\downarrow 0} \frac{\lambda^2(s+1)k_0 - 1}{2^{k_0-1}a_{2(s+1)}(\lambda)} \qquad \text{if} \qquad \liminf_{\lambda\downarrow 0} \frac{\lambda^2 a_2(\lambda)}{a_4(\lambda)} > 0,$$

and

$$\lim_{\lambda\downarrow 0} \frac{\lambda}{a_2(\lambda)}, \qquad \text{if} \qquad s = 0, \quad \text{and} \quad \liminf_{\lambda\downarrow 0} \frac{\lambda^2 a_2(\lambda)}{a_4(\lambda)} = 0.$$

They are equal to 0 in other cases.

Thus the local characteristic $[a_n, L_n]$ determines the moments (12), and, consequently, determines the moments of $\Delta_{h_1}\Delta_{h_2}\xi(t), h_1, h_2 > 0$.

Analogously, to prove that (a_n, L_n) determines the moments of centered differences $\Delta_h\xi(t) - \mathbf{E}(\Delta_h\xi(t))$ in the case $k_0 = 1$, it suffices to show that the moments

$$\int_{-\infty}^{\infty} u^n\, dG_\xi(u), \qquad n = 0, 1, 2, \ldots,$$

are determined by (a_n, L_n).

Hitherto we have proved that the condition

$$\liminf_{\lambda\downarrow 0} \frac{\lambda^2 a_2(\lambda)}{a_4(\lambda)} > 0,$$

is equivalent to the condition (10). Consequently, in this case we obtain that the moments

$$\int_{-\infty}^{\infty} u^n\, dG_\xi(u) = L_{n+2}(x^{n+2}) \lim_{\lambda\downarrow 0} \frac{\lambda^{n+1}}{a_{n+2}(\lambda)},$$

if $L_{n+2}(x^{n+2}) > 0$, and they are equal to 0, if $L_{n+2}(x^{n+2}) = 0$. But, in view of (9),

$$\int_{-\infty}^{\infty} u^n\, dG_\xi(u) = \lim_{\lambda\downarrow 0} \frac{\lambda a_2(\lambda)}{3 a_4(\lambda)}$$

for $n = 0$, and is equal to 0, for $n = 1, 2, \ldots$, and the theorem is thus proved.

Finally, we provide an example of two processes whose generalized derivatives have the same local characteristic but for which the probability distribution functions of the first-order centered increments are different. So, the tool of local characteristics while extremely useful cannot serve as a universal discriminant for "probability distributions" of generalized processes.

Example 1. Indeterminate moment problem. Consider two functions

$$s_1(x) = x^{-\log x}, \ x \geq 0, \qquad \text{and} \qquad s_2(x) = x^{\log x}(1 + \sin(2\pi \log x)).$$

The moment problem is indeterminate for these two functions, that is,

$$\kappa_n = \int_0^\infty x^n s_1(x)dx = \int_0^\infty x^n s_2(x)dx, \quad n = 1, 2, \ldots, \tag{13}$$

and yet the functions are different.[6]

Now, consider the compound Poisson processes $\xi_1(t), \xi_2(t) \in \mathcal{M}$, with $\gamma_{\xi_1}^* = \gamma_{\xi_2}^* = 0$, with their Lévy measures vanishing on the negative half-line, and such that, for $u \geq 0$,

$$\mu_{\xi_1}(u) := \int_0^u s_1(u)du, \qquad \text{and} \qquad \mu_{\xi_2}(u) := \int_0^u s_2(u)du.$$

Obviously, the probability distribution of $\Delta_h \xi_1(t) - E\Delta_h \xi_1(t)$ is different from that of $\Delta_h \xi_2(t) - E\Delta_h \xi_2(t)$. But, the local characteristics $(a_{n,1}, L_{n,1})$, $(a_{n,2}, L_{n,2})$ of the first derivatives $\xi_1'(t), \xi_2'(t)$ are the same. Indeed, in view of the last Theorem of the previous section, and the equality of moments (13), a direct calculation shows that

$$a_{0,1} = a_{0,1} = \frac{1}{2}, \qquad L_{0,1}(\varphi) = L_{0,2}(\varphi) = \frac{1}{2}\varphi(0),$$

$$a_{1,1} = a_{1,2} = \frac{1}{1+\kappa_1}, \qquad L_{1,1}(\varphi) = L_{1,2}(\varphi) = \frac{1}{1+\kappa_1}\left(\varphi(0) + \kappa_1 G_1^+(\varphi)\right),$$

and

$$a_{n,1} = a_{n,2} = \frac{\lambda^{n-1}}{\kappa_n}, \qquad L_{n,1}(\varphi) = L_{n,2}(\varphi) = G_n^+(\varphi), \qquad n \geq 2,$$

where

$$\kappa_n = \int_0^\infty u^n s_1(u)\, du = \int_0^\infty u^n s_2(u)\, du.$$

---

[6]This classical example is due to Stieltjes [8].

21.8   Exercises

1. Prove that α-stable distributions have infinite absolute moments of order greater than or equal to α and that all the absolute moments of order less than α are finite by demonstrating the tail behavior of the *alpha* stable random variable X is as follows

$$\lim_{x\to\infty} x^\alpha \mathbf{P}(|X| > x) = C < \infty.$$

2. Show that the Brownian motion process is 2-stable and that it can be viewed as a limit of the Lévy processes with the Lévy measure

$$\lambda_\epsilon(a)\, da = \frac{\delta(a = \epsilon)}{a^2},$$

for $\epsilon \to 0$.

3. *Chambers–Mallows–Stuck simulation algorithm.* Since , with few exceptions, it is impossible to explicitly calculate the CDF of the α-stable distribution, and its inverse, the usual standard method of generating random variables is not available. In this context there were many efforts to produce such an algorithm and one of them, due to Chambers, Mallows, and Stuck [9], depends on the observation that, if $0 < \alpha \le 2$, U is a random variable uniformly distributed on $[-\pi/2, \pi/2]$, and V is an independent exponential random variable with mean 1, then

$$S_\alpha = \frac{\sin(\alpha U)}{[\cos(U)]^{1/\alpha}} \cdot \left[\frac{\cos(U - \alpha U)}{V} \right]^{\frac{1-\alpha}{\alpha}},$$

is a standard α-stable random variable.

4. Demonstrate that for the Lévy process $\xi(t)$ with the Lévy function

$$G_\xi(x) = \int_{-\infty}^x (2 - \sin(\log |u|)) e^{-|u|} du,$$

no derivative has a local characteristic, and for $\xi(t)$ with

$$G_\xi(x) = \int_{-\infty}^x (2 - \operatorname{sgn} u \cdot \sin(\log |u|)) e^{-|u|} du,$$

the first derivative does not have a local characteristic but all the higher derivatives do.

Chapter 22

Linear Anomalous Fractional Dynamics in Continuous Media

The chapter discusses a model for anomalous diffusion processes. Their one-point probability density functions (p.d.f.) are exact solutions of fractional diffusion equations. The model reflects the asymptotic behavior of a jump (anomalous random walk) process with random jump sizes and random inter-jump time intervals with infinite means (and variances) which do not satisfy the law of large numbers. In the case when these intervals have a fractional exponential p.d.f., the fractional Kolmogorov–Feller equation for the corresponding anomalous diffusion is provided and methods of finding its solutions are discussed. Finally, some statistical properties of solutions of the related Langevin equation are studied. The subdiffusive case is explored in detail. The emphasis is on a rigorous presentation which, however, would be accessible to the physical sciences audience.

22.1 Basic concepts

The simplest way to introduce fractional scaling in stochastic processes is by applying fractional integral transformation to the standard Brownian motion process $B(t), t \geq 0$, to define

$$B_H(t) := \int_{-\infty}^{\infty} \left[(t - s)_+^{H-1/2} - (-s)_+^{H-1/2} \right] dB(s),$$

where the above integral is understood as the classical Wiener random integral with respect to the Brownian motion. The parameter H introduced above is traditionally called the *Hurst parameter* and it is assumed to take values in

© Springer International Publishing AG, part of Springer Nature 2018
A. I. Saichev and W. A. Woyczynski, *Distributions in the Physical and Engineering Sciences, Volume 3*, Applied and Numerical Harmonic Analysis, https://doi.org/10.1007/978-3-319-92586-8_22

the interval $(0, 1)$. The resulting Gaussian process B_H, satisfies the condition $B_H(0) = 0$, is called the *fractional Brownian motion* with Hurst parameter H, and in the case $H = 1/2$ it is obviously the standard Brownian motion, $B(t)$.

By definition, the mean value

$$\langle B_H(t) \rangle = 0,$$

the variance of its increments, $t_1 < t_2$,

$$\langle [B_H(t_2) - B_H(t_1)]^2 \rangle = c_H^{-1} \int_{-\infty}^{\infty} \left[(t_2 - s)_+^{H-1/2} - (t_1 - s)_+^{H-1/2} \right] ds = (t_2 - t_1)^{2H},$$

where $t_+ = t$ for positive t, and 0 for negative t. For $H = 1/2$ we obtain the classical linear growth of variances of the ordinary Brownian motion.

The above formula permits an easy calculation of the covariance structure of the fractional Brownian motion, which, in view of its Gaussian probability distribution, completely determines all of its finite-dimensional probability distributions. Indeed,

$$\mathrm{Cov}(B_H(t_2), B_H(t_1)) = \langle B_H(t_2), B_H(t_1) \rangle$$

$$= \frac{1}{2} \left(\langle B_H^2(t_2) \rangle + \langle B_H^2(t_1) \rangle - \langle [B_H(t_2) - B_H(t_1)]^2 \rangle \right) = \frac{1}{2} \left(|t_1|^{2H} + |t_2|^{2H} - |t_2 - t_1|^{2H} \right).$$

One can also verify that the probability density function $f_H(x, t)$ of the process $B_H(t)$ satisfies the linear (but not homogeneous) partial differential equation

$$\frac{\partial f_H(x, t)}{\partial t} = 2H t^{2H-1} \frac{\partial^2 f_H(x, t)}{\partial x^2}.$$

The Hurst parameter H is here also the *self-similarity exponent*, since the finite-dimensional probability distributions of the random vectors

$$\left(B_H(t_1), \ldots, B_H(t_n) \right), \quad \text{and} \quad \left(c^H B_H(t_1), \ldots, c^H B_H(t_n) \right),$$

are equal, for any $c > 0, t_1, \ldots, t_n \in \mathbf{R}$, and $n = 1, 2, \ldots$.

Over the last several decades, beginning with the work of Montroll and his collaborators, see, Montroll and Weiss [1], the physical community showed a steady interest in the anomalous diffusions, a somewhat vague term that describes diffusive behavior in the absence of the second moments, with scaling different than that of the classical Gaussian diffusion; see, e.g., Shlesinger [2], Saichev and Zaslavsky [3], and Metzler and Klafter [4]. More recently, this model found applications in several other areas, including econophysics and mathematical finance; see e.g., Scalas et al. [5] and Mainardi et al. [6]. The standard Lévy processes and their calculus have an enormous mathematical literature including comprehensive monographs by Kwapień and Woyczyński [7], Bertoin [8], and Sato [9]. The material in this chapter is also based on the articles by Piryatinska, Saichev, and Woyczyński [10], and Terdik, Woyczyński, and Piryatinska [11].

Despite all these efforts and very active current research in the area of dynamical features of nonlinear anomalous diffusions there are still many open problems to be solved. Moreover, although it is well known that one-point probability density functions (p.d.f.) of such processes can be described by the diffusion equations involving derivatives of fractional orders, their physically and computationally useful theory is still developing and the basic facts will be presented in Chapter 23.

The goal of the present chapter is to provide a lucid and rigorous analysis of linear anomalous diffusions which provides some new insights into the dynamics of their statistical properties and serves as a basis for further developments both in the mathematical and the physical literatures.

To motivate our work on anomalous diffusion it is worth recalling a few elementary facts regarding the classical Brownian motion diffusion (the Wiener process) $B(t), t \geq 0$, which can be defined as a Gaussian process with zero mean and the covariance function $\langle B(t)B(s) \rangle = \min(t, s)$. The physicist usually thinks about it as a limit (in distribution) of a process of independent Gaussian (say) jumps taken at smaller and smaller time intervals of (nonrandom) length $1/n$. In other words,

$$B(t) =_d \lim_{n \to \infty} \sum_{k=1}^{\lfloor nt \rfloor} B_k \left(\frac{1}{n} \right) \tag{1}$$

where $\lfloor a \rfloor$ denotes the integer part of number a, and $B_1(1/n), B_2(1/n), \ldots$, are independent copies of the random variable $B(1/n)$.

The key observation is that this mental picture can be extended to the situation when the fixed inter-jump times $\Delta t = 1/n$ are replaced by random inter-jump times of an accelerated Poissonian process $N(nt), t \geq 0$, independent of $B(t)$. Here, $N(t)$ is the standard Poisson process with mean $\langle N(t) \rangle = t$, and variance $\langle (N(t) - t)^2 \rangle = t$. Since the mean $\langle N(nt) \rangle = nt$ is finite, in view of the law of large numbers,

$$\frac{N(nt)}{nt} \to 1, \quad \text{as} \quad t \to \infty, \tag{2}$$

so that the Brownian diffusion $B(t)$ can also be viewed as a limit of the process of random Gaussian jumps at random Poissonian jump times:

$$\lim_{n \to \infty} B\left(\frac{N(nt)}{n} \right) =_d \lim_{n \to \infty} \sum_{k=1}^{N(nt)} B_k \left(\frac{1}{n} \right) =_d \lim_{n \to \infty} \sum_{k=1}^{\lfloor tn \rfloor} B_k \left(\frac{1}{n} \right) =_d B(t). \tag{3}$$

It is a worthwhile exercise to see directly that the process

$$Z(t) = \lim_{n \to \infty} B\left(\frac{N(nt)}{n} \right) \tag{4}$$

represents a Brownian motion diffusion. Indeed, $Z(t)$ is clearly a process with independent and time-homogeneous (stationary) increments and its one-point cumulative distribution function (c.d.f.)

$$\mathbf{P}(Z(t) \le z) = \lim_{n \to \infty} \sum_{k=0}^{\infty} \mathbf{P}(B(k/n) \le z) \cdot \mathbf{P}(N(nt) = k)$$

$$= \lim_{n \to \infty} \int_{-\infty}^{z} \sum_{k=0}^{\infty} \frac{e^{-x^2/(2k/n)}}{\sqrt{2\pi k/n}} e^{-nt} \frac{(nt)^k}{k!} dx = \int_{-\infty}^{z} \frac{e^{-x^2/(2t)}}{\sqrt{2\pi t}} dx,$$

because the Poissonian distribution in question has mean nt and standard deviation $\sqrt{nt}$. Asymptotically, for large n, it is concentrated around the value $k = nt$ and acts, in the limit, as the Dirac delta centered at $k/n = t$.

The jump times process $T(n), n = 1, 2, \ldots$, the random time of the nth jump, and the Poissonian process $N(t)$ are mutually inverse random function as

$$T(n) = \min\{t : N(t) \ge n\}, \tag{5}$$

and the inter-jump random time intervals $T(n) - T(n-1), n = 1, 2, \ldots$, are independent and have identical exponential p.d.f. $e^{-t}, t \ge 0$.

The main thrust of this chapter is to replace the Brownian motion $B(t)$ by a general Lévy diffusion $X(t)$, the random times $T(n)$ by an increasing infinite mean continuous parameter process $T(\theta)$ (sometimes called the *subordinator*) with the inverse random time $\Theta(t)$ replacing the Poisson process $N(t)$, and consider the composite process $X(\Theta(t))$ which we call here *anomalous diffusion*. It is the infinite mean of the random time $T(\theta)$ which makes the diffusion $X(\Theta(t))$ anomalous and non-Gaussian since the above law of large numbers argument no longer applies.

The chapter is organized as follows: Sections 22.2 and 22.3 introduce the concepts, and study statistical properties of fractional random time and inverse random time. These concepts are based on the principles of infinite divisibility and self-similarity under rescaling. The role of the Mittag-Leffler functions as generalizations of the exponential distribution is here explored. Section 22.4 investigates anomalous diffusions evolving in such inverse fractional random time, and it is followed by a study of the special case of fractional drift processes in Section 22.5. Section 22.6 discusses the fractional subdiffusion in the case when the usual Brownian diffusion is run in the inverse fractional random time; the multidimensional case is discussed in Section 22.7. Then, Section 22.8 introduces the concept of tempered anomalous subdiffusion which has small jumps similar to those of the fractional anomalous diffusion while preserving the important finiteness of all moments that is characteristic of the Brownian motion.

Finally, Sections 22.9 and 22.10 show that anomalous diffusions studied in sections 22.2–22.8 can be interpreted as limit cases of pure jump processes which we call *anomalous random walks*. We first look at their statistical properties, related Kolmogorov–Feller equations, and then at the Langevin-type equations driven by them. We conclude, in the spirit of Section 21.7, with analysis of fractional- and integer-order moments, and multiscaling for smoothly truncated Lévy diffusions.

Essential concepts of the theory of infinitely divisible Lévy distributions have been explained in Chapter 21.

22.2 Integrals and derivatives of fractional order

In this section we review the basic concepts of fractional calculus. A more complete exposition of this material can be found in Chapter 6 of Volume 1.[1]

22.2.1. Principal-value distribution. To begin let us take a look at the Cauchy integral integral,

$$\int \frac{\varphi(s)}{s-x}ds, \tag{1}$$

via the notion of its *principal value* which, for $x = 0$, is defined as a symmetric limit

$$\mathcal{PV}\int \frac{\varphi(s)}{s}ds = \lim_{\varepsilon \to 0}\left[\int_{-\infty}^{-\varepsilon} + \int_{+\varepsilon}^{\infty}\right]\frac{\varphi(s)}{s}ds, \tag{2}$$

with an analogous definition for other values of x. The letters $\mathcal{PV}$ in front of the integral indicate that the integral is taken in the sense of its principal value. In terms of the distribution theory, equality (2) defines distribution $T_{1/s}$, which acts on a test function φ using the expression on the right-hand side of (2).

Another approach to evaluation of integral (1) depends on its interpretation as the limit

$$\int \frac{\varphi(s)}{s-x}ds = \lim_{y \to 0}\int \frac{\varphi(s)}{s-z}ds, \tag{3}$$

where

$$z = x + iy$$

is a complex parameter. Moving into the complex plane vicinity of the real axis removes the singularity.

Let us find the above limit by separating explicitly the real and the imaginary parts of the expression

$$\frac{1}{s-z} = \frac{s-x}{(s-x)^2+y^2} + \frac{iy}{(s-x)^2+y^2}.$$

Substituting this sum into (3) and observing that the integral of the real part converges to the principal value as $y \to 0$, we obtain that

$$\lim_{y \to 0}\int \frac{\varphi(s)}{s-z}ds = \mathcal{PV}\int \frac{\varphi(s)}{s-z}ds + i\lim_{y \to 0}\int \frac{y}{(s-x)^2+y^2}\varphi(s)ds. \tag{4}$$

---

[1] For a modern and elegant exposition of work on fractional calculus in the context of stochastic processes we refer to Meerschaert and Sikorskii [21], Meerschaert, Nane and Vellaisamy [22], Leonenko, Meerschaert and Sikorskii [23], Leonenko, Meerschaert, Schilling and Sikorskii [24], and Leonenko and Merzbach [25].

Notice that the factor in front of function $\varphi(s)$ in the last integral coincides, up to π, with the familiar Lorentz curve

$$\frac{1}{\pi}\frac{y}{(x-s)^2+y^2},$$

which weakly converges to $\delta(x-s)$ as $y \to 0+$. Hence, the evaluation of the integral (1) using the limit procedure (3) leads to an identity

$$\int \frac{\varphi(s)}{s-x}ds = \mathcal{PV}\int \frac{\varphi(s)}{s-x}ds \pm i\pi\varphi(x).$$

The plus sign corresponds to the limit $y \to 0+$ with y's restricted to the upper half-plane and the minus—to the limit $y \to 0-$ with y's restricted to the lower half-plane. Thus equality (4) determines two distributions:

$$\frac{1}{s-x-i0} = \mathcal{PV}\frac{1}{s-x} + i\pi\delta(x-s) \tag{5}$$

and

$$\frac{1}{s-x+i0} = \mathcal{PV}\frac{1}{s-x} - i\pi\delta(x-s). \tag{6}$$

Although assigning complex values to real-valued integrals may seem strange at the first glance, these formulas often give the correct physical answer. The point is that their imaginary parts reflect the *causality principle* which was not spelled out explicitly when the original physical problem was posed but which, as we will see later on, plays an important role. Obvious physical arguments then permit us to indicate which of the formulas (5–6) exactly correspond to the physical problem under consideration.

22.2.2.   The Cauchy formula.   The results of Section 22.2 are closely related to the *Cauchy formula* from the theory of functions of a complex variable. The formula asserts that for any function $f(z)$, analytic in a simply connected domain D in the complex plane $\mathbf{C}$ and continuous on its closure $\bar{D}$ including the boundary contour C,

$$f(z) = \frac{1}{2\pi i}\int_C \frac{f(\zeta)d\zeta}{\zeta - z},$$

where z is a point in the interior of D and contour C is oriented counterclockwise. If z is an arbitrary point of the complex plane then the above integral defines a new function

$$F(z) = \frac{1}{2\pi i}\int_C \frac{f(\zeta)d\zeta}{\zeta - z}. \tag{7}$$

For z inside the contour of integration

$$F(z) = f(z), \qquad z \in D.$$

If $z \notin \bar{D}$ then the integrand is analytic everywhere in D and, by the *Cauchy's Theorem*, $F(z) \equiv 0$. For a boundary point $z = \zeta_0 \in C$ the integral (7) is singular and $F(z)$ will be understood in the principal-value sense. In the present context, this will mean that

$$F(\zeta_0) = \mathcal{PV}[F(\zeta_o)] = \lim_{r \to 0} \frac{1}{2\pi i} \int_{C_r} \frac{f(\zeta) d\zeta}{\zeta - \zeta_0}, \qquad (8)$$

where C_r is a curve obtained from contour C by removing its part contained in a disc of radius $r \to 0$ and centered at ζ_0.

Observe that, for any function f continuous on contour, function $F(z)$ is well defined everywhere.

Formula (8) generalizes the concept of the principal value of a singular integral on the real axis and we will study it for a function $f(z)$ analytic inside contour C and continuous on it. To that end substitute an identity

$$f(\zeta) = [f(\zeta) - f(\zeta_0)] + f(\zeta_0),$$

into (8) and split the integral into two parts:

$$\mathcal{PV}[F(\zeta_0)] = \frac{1}{2\pi i} \int_C \frac{[f(\zeta) - f(\zeta_0)] d\zeta}{\zeta - \zeta_0} + \lim_{r \to 0} \frac{f(\zeta_0)}{2\pi i} \int_{C_r} \frac{d\zeta}{\zeta - \zeta_0}.$$

We deliberately replaced C_r by C in the first integral since for $f(z)$ satisfying the Lipschitz condition the integral is no longer singular and one can integrate over the whole closed contour C. Since its integrand is analytic inside the contour and continous on the contour, the first integral vanishes by Cauchy's theorem. Thus

$$\mathcal{PV}[F(\zeta_0)] = \lim_{r \to 0} \frac{f(\zeta_0)}{2\pi i} \int_{C_r} \frac{d\zeta}{\zeta - \zeta_0}. \qquad (9)$$

The last integral can be evaluated assuming that contour C is smooth in the vicinity of point ζ_0, which is called the *regular point of the contour*. Let us add and subtract from (3) the integral over portion c_r of the located in D circle of radius r with center at ζ_0. Since the integral over the closed contour $C_r + c_r$ is equal to zero, equality (9) can be rewritten in the form

$$\mathcal{PV}[F(\zeta_0)] = - \lim_{r \to 0} \frac{f(\zeta_0)}{2\pi i} \int_{c_r} \frac{d\zeta}{\zeta - \zeta_0}.$$

The latter integral is easy to evaluate:

$$\int_{c_r} \frac{d\zeta}{\zeta - \zeta_0} = \int_\pi^0 \frac{i r e^{i\varphi} d\varphi}{r e^{i\varphi}} = -i\pi.$$

Hence, we obtain that

$$\mathcal{PV}[F(\zeta_0)] = f(\zeta_0)/2,$$

and the principal value of the analytic function is equal to

$$\mathcal{PV}\int_C \frac{f(\zeta)d\zeta}{\zeta - \zeta_0} = i\pi f(\zeta_0), \tag{10}$$

that is, the value of function f at the singular point of the integrand multiplied by $i\pi$. Thus, the behavior of the Cauchy integral in the neighborhood of a regular point ζ_0 of contour C can be summarized as follows:

$$\int_C \frac{f(\zeta)d\zeta}{\zeta - z} = 2\pi i f(\zeta_0), \tag{11}$$

as $z \to \zeta_0+$, equals $\frac{1}{2}f(\zeta_0)$, for $z = \zeta_0$, and equals 0, for $z \to \zeta_0-$, where the plus sign corresponds to the limit value of the integral while ζ_0 is approached from the inside of contour C, and the minus sign corresponds to the approach from outside.

Example 1. Let us consider a contour C consisting of the interval $[-R, R]$ on the real axis and the semicircle C_R of radius R with the center at point $z = 0$ located in the upper half-plane. If $f(z)$ is analytic in the upper half-plane and such that the integral over C_R uniformly converges to zero as $R \to \infty$, then the Cauchy formula is transformed into

$$\int \frac{f(s)}{s - z}ds = 2\pi i f(z), \qquad y > 0, \tag{12}$$

and equality (4) assumes the form

$$\mathcal{PV}\int \frac{f(s)}{s - x}ds = i\pi f(x). \tag{13}$$

Recall that, by the well known *Jordan Lemma* in the complex functions theory, function

$$f(z) = \exp(i\lambda z), \qquad \lambda > 0, \tag{14}$$

satisfies the conditions mentioned earlier so that the above formula implies that

$$\mathcal{PV}\int \frac{e^{i\lambda s}}{s - x}ds = \pi i e^{i\lambda x}. \tag{15}$$

In particular, it follows that for $x = 0$

$$-i\mathcal{PV}\int \frac{e^{i\lambda s}}{s}ds = \int \frac{\sin(\lambda s)}{s}ds = \pi. \tag{16}$$

Notice that the Cauchy formula (12) interpreted in the spirit of the distribution theory defines a new distribution

$$\hat{T}(s - z) = \frac{1}{2\pi i} \frac{1}{s - z}, \tag{17}$$

which is called the *analytic representation of the Dirac delta*. Its functional action assigns to a function $f(s)$, analytic in the upper half-plane and rapidly decaying at infinity, its value at the point z. Applied to any usual "well-behaved" function of the real variable s, it defines a new function of complex variable z which is analytic everywhere with the possible exception of the real axis. Crossing the real axis at the point $z = x$ the functional $\hat{T}(s - z)[f]$ has a jump of size $f(x)$; this follows from formulas of Subsection 22.2.1. We will extract this jump by introducing a new distribution

$$\hat{\delta}(s - z) = \hat{T}(s - z) + \hat{T}^*(s - z) = \frac{1}{\pi} \frac{y}{(s - x)^2 + y^2},$$

which is harmonic for $y \neq 0$ and which converges to the usual Dirac delta $\delta(s - x)$ as $y \to 0+$. As we have seen in Volume 2, the corresponding functional $\hat{\delta}(s - z)[f(s)]$ solves the Dirichlet problem for the 2-D Laplace's equation in the upper half-plane $y > 0$. ∎

22.2.3. Fractional integration. In the next three subsections we will develop another class of important singular integrals which arise when one tries to extend the notion of n-tuple integrals and of nth order derivatives of classical calculus to noninteger (or fractional) n. We begin with the concept of *fractional integration*. It is natural to introduce it as a generalization of the *Cauchy formula*

$$(I^n g)(t) = \int_{-\infty}^{t} dt_1 \int_{-\infty}^{t_1} dt_2 \ldots \int_{-\infty}^{t_{n-1}} dt_n \, g(t_n)$$

$$= \frac{1}{(n-1)!} \int_{-\infty}^{t} (t - s)^{n-1} g(s) \, ds, \tag{18}$$

which expresses the result of n-tuple integration of function $g(t)$ of a single variable via the single integration operator.

Before we move on to fractional integrals, let us take a closer look at the Cauchy formula (18). It is valid for absolutely integrable functions $g(t)$ which decay for $t \to -\infty$ sufficiently rapidly to guarantee the existence of the integral on the right-hand side of (18). Assuming that the integrand $g(t)$ vanishes for $t < 0$, the Cauchy formula can be rewritten in the form

$$(I^n g)(t) = \int_{0}^{t} dt_1 \int_{0}^{t_1} dt_2 \ldots \int_{0}^{t_{n-1}} dt_n \, g(t_n)$$

$$= \frac{1}{(n-1)!} \int_0^t (t-s)^{n-1} g(s) \, ds. \tag{19}$$

Remark 1. The Cauchy formula can be viewed as an illustration of the general *Riesz Theorem* about representation of any linear continuous (in a certain precise sense) operator L transforming function g of one real variable into another function $L[g]$ as an integral operator

$$L[g](t) = \int h(t,s) g(s) \, ds \tag{20}$$

with an appropriate kernel $h(t,s)$. In our case, the n-tuple integration linear operator in (18) has a representation via the single integral operator with kernel $h(t,s) = (t-s)^{n-1}/(n-1)!$.

Let us check the validity of the Cauchy formula (18) by observing that the n-tuple integral in (18) is a solution of the differential equation

$$\frac{d^n}{dt^n} x(t) = g(t) \tag{21}$$

satisfying the causality principle. Such a solution can be written as convolution

$$x(t) = \int k_n(t-s) g(s) \, ds, \tag{22}$$

where

$$k_n(t) = \chi(t) y(t),$$

and $y(t)$ is the solution of the corresponding homogeneous equation

$$\frac{d^n}{dt^n} x(t) = 0$$

with the initial conditions

$$y(0) = y'(0) = \cdots = y^{(n-2)}(0) = 0, \quad y^{(n-1)} = 1.$$

Solving the above initial-value problem we get

$$k_n(t) = \frac{1}{(n-1)!} t^{n-1} \chi(t),$$

the kernel that appears in the Cauchy formula (18).

This is a good point to introduce fractional integrals. Replacing integer n in kernel k_n by an arbitrary positive real number α and the factorial $(n-1)!$ by the gamma function $\Gamma(\alpha)$ we arrive at the generalized kernel

$$k_\alpha(t) = \frac{1}{\Gamma(\alpha)} t^{\alpha-1} \chi(t). \tag{23}$$

So it is natural to call the convolution operator

$$(I^\alpha g)(t) = k_\alpha(t) * g(t),\tag{24}$$

the *fractional integration operator of order* α.

In the case when function $g(t) \equiv 0$ for $t < 0$ the convolution (24) reduces to the integral

$$(I^\alpha g)(t) = \frac{1}{\Gamma(\alpha)} \int_0^t (t - s)^{\alpha-1} g(s)\, ds\tag{25}$$

where the upper integration limit reflects the causality property of the operator of fractional integration.

Let us establish some of the important properties of the fractional integration operator assuming, for simplicity, that function $g(t)$ in (25) is bounded and continuous.

Existence of fractional integrals. For $\alpha \geq 1$ the integrand in (25) is bounded and continuous, and the integral exists in the Riemann sense. For $0 < \alpha < 1$, the kernel $(t - s)^{\alpha-1}$ is singular but the singularity is integrable and the integral is an absolutely convergent improper integral.

Zero-order integration. As $\alpha \to 0+$ the operators I^α tend to the identity operator. Indeed, in view of the recurrent formula $\Gamma(\alpha + 1) = \alpha\Gamma(\alpha)$ for the gamma function and the fact that $\Gamma(1) = 1$ we have the asymptotics $\Gamma(\alpha) \sim 1/\alpha$, $(\alpha \to 0+)$. Hence,

$$\lim_{\alpha \to 0+} (I^\alpha g)(t) = \lim_{\alpha \to 0+} \int g(s)\chi(t - s)\alpha(t - s)^{\alpha-1}\, ds.$$

Function $\chi(t-s)\alpha(t-s)^{\alpha-1}$ weakly converges to the Dirac delta $\delta(t-s-0)$ as $\alpha \to 0+$.[2] Consequently,

$$(I^0 g)(t) = \lim_{\alpha \to 0+} (I^\alpha g)(t) = g(t),$$

or, equivalently, in the distributional language,

$$k_0(t) = \delta(t).\tag{26}$$

Iteration of fractional integrals. As in the case of usual n-tuple integrals, repeated application of fractional integrals is subject to the rule

$$I^\alpha I^\beta = I^\alpha+\beta.\tag{27}$$

To see this it suffices to check that, for any $\alpha, \beta > 0$,

---

[2]This notation emphasizes that the support of this Dirac delta lies inside the interval $(0, t)$ so that $\int_0^t \delta(t - s - 0)\, ds = 1$, and not $1/2$ as in the standard definition.

$$k_\alpha * k_\beta = k_{\alpha+\beta}. \tag{28}$$

Indeed, the left-hand side of (28), in view of (23), equals

$$k_\alpha * k_\beta = \frac{\chi(t)}{\Gamma(\alpha)\Gamma(\beta)} \int_0^t s^{\alpha-1}(t-s)^{\beta-1}ds,$$

so that, passing to the new dimensionless variable of integration, $\tau = s/t$,

$$k_\alpha * k_\beta = \frac{\chi(t)t^{\alpha+\beta-1}}{\Gamma(\alpha)\Gamma(\beta)} B(\alpha,\beta), \tag{29}$$

where

$$B(\alpha,\beta) = \int_0^1 \tau^{\alpha-1}(1-\tau)^{\beta-1}d\tau$$

is the *beta function*. It can be expressed in terms of the gamma function via the formula,

$$B(\alpha,\beta) = \frac{\Gamma(\alpha)\Gamma(\beta)}{\Gamma(\alpha+\beta)}.$$

Substituting it into (29) we obtain equality (28).

Fractional integrals as continuous operators. The following two inequalities show that integration of fractional order has some continuity properties as a linear operation. We will assume that $0 < \alpha < 1$. First, observe that, for $\alpha > \frac{1}{2}, I^\alpha$ is a continuous operator from $\mathbf{L}^2[0,1]$ into $\mathbf{L}^p[0,1]$ for each $p < \infty$.[3] Indeed, by the *Schwarz Inequality*,

$$\|I^\alpha f\|_p^p = \int_0^1 |(I^\alpha f)(t)|^p dt$$

$$\leq \int_0^1 \left(\int_0^t f^2(s)ds \right)^{p/2} \cdot \left(\int_0^t k_\alpha^2(s)ds \right)^{p/2} dt \tag{30}$$

$$\leq c(\alpha,p)\left(\int_0^1 f^2(s)ds \right)^{p/2} = c(\alpha,p)\|f\|_2^p,$$

where $c(\alpha,p)$ is a constant depending only on α and p.

Additionally, by a similar argument, but using the *Hölder Inequality* with $1/p + 1/q = 1$,

$$\|I^\beta f\|_\infty = \sup_{0\leq t\leq 1}\left| \int_0^t k_\beta(t-s)f(s)ds \right|$$

$$\leq \left(\int_0^1 |k_\beta(s)|^q ds \right)^{1/q}\left(\int_0^1 |f(s)|^p ds \right)^{1/p} = c(\beta,p)\|f\|_p, \tag{31}$$

---

[3]Recall the $\mathbf{L}^p[0,1]$ denotes the *Lebesgue space* of functions f on the interval $[0,1]$ which have pth powers integrable, i.e., for which the norm $\|f\|_p := (\int_0^1 |f(s)|^p ds)^{1/p} < \infty$.

so that for any $\beta > 1/p$, the operator I^β is *continuous* from $\mathbf{L}^p[0,1]$ into the space of continuous functions $\mathbf{C}[0,1]$.

22.2.4. Fractional differentiation. The operator D^α of *fractional differentiation* is defined as the inverse of the operator I^α of the fractional integration, that is, via the operator equation

$$D^\alpha I^\alpha = \mathrm{Id}, \tag{32}$$

where Id denotes the identity operator. Similarly to the fractional integration operator, the fractional differentiation operator has an integral representation

$$(D^\alpha g)(t) = r_\alpha(t) * g(t), \tag{33}$$

where $r_\alpha(t)$ is the convolution kernel which will be identified next. To do that notice that the operator equation (32) is equivalent to the convolution algebra equation

$$r_\alpha(t) * k_\alpha(t) = \delta(t), \tag{34}$$

where k_α is the fractional integration kernel (23). Denote by γ the solution of the equation $\alpha + \gamma = n$, where $n = \lceil \alpha \rceil$ is the smallest integer greater than or equal to α. In other words,

$$n - 1 < \alpha \le n, \qquad \gamma = n - \alpha, \qquad 0 \le \gamma < 1. \tag{35}$$

Applying the fractional integration operator I^γ to both sides of (34) we get

$$r_\alpha(t) * k_n(t) = k_\gamma(t).$$

The expression on the left-hand side represents the usual n-tuple integral of the fractal differentiation kernel r_α. Thus, if we differentiate it n times we arrive at the explicit formula

$$r_\alpha(t) = k_\gamma^{(n)}(t), \qquad \gamma = n - \alpha > 0 \tag{36}$$

for the fractional differentiation kernel. The corresponding fractional differentiation operator is then given by the convolution

$$(D^\alpha g)(t) = k_\gamma^{(n)}(t) * g(t). \tag{37}$$

For $t > 0$, the nth derivative of the kernel $k_\gamma(t)$ appearing in (36) exists in the classical sense and

$$r_\alpha(t) = \frac{1}{\Gamma(-\alpha)} t^{-\alpha-1} = k_{-\alpha}(t), \qquad t > 0. \tag{38}$$

So, the fractional differentiation kernel is equal to the fractional integration kernel with the opposite index $-\alpha$. Here, the gamma function $\Gamma(-\alpha)$ of negative noninteger variable is defined via the above-mentioned recurrence property as follows:

$$\Gamma(-\alpha) = \Gamma(n - \alpha)/(n - \alpha - 1) \cdots (-\alpha).$$

Therefore, the fractional differentiation operator can be treated as the fractional integration operator of the negative order:

$$D^\alpha = I^{-\alpha}, \tag{39}$$

which adds attractive symmetry to the fractional calculus.

Consequently, for a function $g(t)$ which vanishes on the negative half-axis, equation (37) can be rewritten in the following symbolic integral form:

$$(D^\alpha)(t) = (I^{-\alpha}g)(t) = \frac{1}{\Gamma(-\alpha)} \int_0^t (t - s)^{-\alpha-1} g(s)\, ds. \tag{40}$$

For $\alpha \geq 0$, the above improper integral diverges in view of the nonintegrable singularity of the integrand in the vicinity of the upper limit of integration. Therefore, its values have to be taken as the values of the corresponding regularized integral which can be found treating equality (37) as the convolution of distributions. In view of properties of the distributional convolution, the operation of n-tuple integration can be shifted from the first convolution factor to the second so that

$$(D^\alpha g)(t) = k_\gamma(t) * g^{(n)}(t), \tag{41}$$

with converging integral on the right-hand side. In particular, for integer $\alpha = n$, taking (40) into account, we obtain (as expected) that

$$(D^n g)(t) = k_0(t) * g^{(n)}(t) = g^{(n)}(t).$$

Example 1. Let us now consider the special case of a function $g(t)$ which vanishes identically for $t < 0$ and is of the form

$$g(t) = \chi(t)\phi(t), \tag{42}$$

where $\phi(t)$ is an arbitrary infinitely differentiable function. Differentiating (42) n times and taking into account the multiplier probing property of the Dirac delta, we obtain

$$g^{(n)}(t) = \chi(t)\phi^{(n)}(t) + \sum_{m=0}^{n-1} \delta^{(m)}(t)\phi^{(n-m-1)}(0).$$

Substituting the above formula in (41), and remembering that $\gamma = n - \alpha$ and

$$k_\gamma^{(m)}(t) = k_{\gamma-m}(t), \qquad t > 0,$$

we finally get that

$$(D^\alpha g)(t) = \frac{1}{\Gamma(n-\alpha)} \int_0^t (t-s)^{n-\alpha-1} \phi^{(n)}(s)\, ds$$

$$+ \sum_{m=0}^{n-1} \phi^{(n-m-1)}(0) k_{n-m-\alpha}(t), \qquad t > 0. \qquad (43)$$

In particular,

$$D^\alpha \chi = k_{-\alpha-1}, \qquad D^\alpha k_\beta = k_{\beta-\alpha}.$$

Also

$$D^\alpha(\chi(s)s^\beta \log|s|) = \frac{\Gamma(\beta+1)}{\Gamma(\beta-\alpha+1)} \chi(s)s^{\beta-\alpha} \Big[\log|s| + C\Big],$$

where the constant (see the literature in the bibliography for its derivation and other formulas of the fractal calculus)

$$C = (\log\Gamma)'(\beta+1) - (\log\Gamma)'(\beta-\alpha+1).$$

Furthermore, for $0 < \alpha < 1$, we have

$$(D^\alpha g)(t) = \frac{1}{\Gamma(1-\alpha)} \int_0^t \frac{\phi'(s)}{(t-s)^\alpha}\, ds + \phi(0)\frac{1}{\Gamma(1-\alpha)} t^{-\alpha}, \qquad t > 0. \quad (44)$$

Observe that, in contrast to (40), the singular integral on the right-hand sides of (43) and (44) converges absolutely, and that the regularizations (41), (43) define a new distribution—the principal value of function $r_\alpha(t)\chi(t)$ (38):

$$\mathcal{PV}\chi(t)\frac{1}{\Gamma(-\alpha)} t^{-\alpha-1}, \qquad \alpha \geq 0. \qquad (45)$$

Its convolution

$$(D^\alpha g)(t) = \frac{1}{\Gamma(-\alpha)} \mathcal{PV} \int_0^t (t-s)^{-\alpha-1} g(s)\, ds$$

with any function of the form (11) has a distributional interpretation via the right-hand side of formula (43). ∎

The following properties of the operation of differentiation of fractional order highlight its peculiarities.

Nonlocal character. Values $D^n g(t)$ of the usual derivatives of integer orders depend only on values of function $g(t)$ in the immediate and arbitrarily small (infinitesimal) vicinity of the point t. By contrast, the fractional (noninteger) derivatives are *nonlocal operators* since the value $D^\alpha g(t)$ depends on the values of $g(\tau)$ for all $\tau < t$. In particular, this fact explains why a function's discontinuity at a certain point ($t = 0$ for function (42)) generates slowly decaying "tails" in its fractional derivatives (the last sum on the right-hand side of formula (43)).

Causality. Fractal derivatives enjoy the causality property: If function $g(t)$ is identically equal to zero for $t < t_0$ then so does its fractional derivative.

Scale invariance. Like usual derivatives, fractional derivatives are scale invariant. This means that differentiation of the compressed ($\kappa > 1$) function $g_\kappa(t) = g(\kappa t)$ just requires multiplication of the compressed derivative by the compression factor:

$$(D^\alpha g_\kappa)(t) = \kappa^\alpha (D^\alpha g)(\kappa t). \tag{46}$$

Fourier transform. Under Fourier transformation, fractional derivatives behave just like the ordinary derivatives. In the distributional sense

$$(D^\alpha g)(t) \longmapsto (i\omega)^\alpha \tilde{g}(\omega). \tag{47}$$

This formula follows directly from (41) and from the results of the previous subsections.

Remark 1. Fractional Laplacians. The above definitions of fractional differential operators for functions of one variable can be extended to fractional partial differential operators for functions of several variables (see the literature at the end for further details). For example the *fractional Laplacian* can be defined through the Fourier transform approach as follows: For any $\phi \in \mathcal{D}(\mathbf{R}^d)$,

$$(\Delta_\alpha \phi)(\boldsymbol{x}) \longmapsto -\|\boldsymbol{\omega}\|^\alpha \tilde{\phi}(\boldsymbol{\omega})$$

The following integration by parts formula is then obtained via the Parseval equality

$$\int |\Delta_{\alpha/2}\phi(\boldsymbol{x})|^2 d\boldsymbol{x} = -\int \phi(\boldsymbol{x})\Delta_\alpha \phi(\boldsymbol{x})d\boldsymbol{x}$$

It is also clear that the fundamental solution (Green's function) G_α of the equation $-\Delta_\alpha u = \delta$ has the Fourier transform

$$\tilde{G}_\alpha(\boldsymbol{\omega}) = \|\boldsymbol{\omega}\|^{-\alpha}.$$

The explicit inversion depends on the dimension d of the space. If $\alpha=2$, $d \geq 3$, or $0 < \alpha < 2$, $d \geq 2$, or $0 < \alpha < 1$, $d = 1$, then

$$G_\alpha(\boldsymbol{x}) = \frac{\Gamma((d-\alpha)/2)}{\pi^d \Gamma(\alpha/2)} \|\boldsymbol{x}\|^{\alpha-d}.$$

22.3 Random time and its inverse

The basic element in the construction of our model of anomalous diffusion is the concept of random time $T(\theta)$, which is a stochastic process depending on a nonnegative parameter θ. We will impose physically justifiable requirements that the random time be nonnegative, nondecreasing as a function of parameter θ, and that its increments be independent and time-homogeneous, with infinitely divisible and self-similar probability distributions.[4]

More precisely, to satisfy the time-homogeneity and infinite divisibility assumptions, we will suppose that $T(\theta)$ is a Lévy infinitely divisible process (see Chapter 21) with $T(\theta = 0) = 0$, nondecreasing trajectories, and such that the p.d.f. of $T(\theta)$ is equal to $f(t; \theta)$ (in brief, $T(\theta) \sim_d f(t, \theta)$). In this context, denoting $f(t; 1) = f(t)$, we have that, for every $n = 2, 3, \ldots$,

$$T(1) =_d T_1(1/n) + \cdots + T_n(1/n),$$

where the random quantities on the right-hand side are independent copies of $T(1/n)$ and $=_d$ denotes the equality of probability distributions. In terms of the "mother" Laplace transform

$$\hat{f}(s) = \int_0^\infty e^{-ts} f(t)\, dt$$

the above infinite divisibility condition can be written in the form

$$\hat{f}(s; 1/n) = \int_0^\infty e^{-ts} f(t, 1/n)\, dt = \hat{f}^{1/n}(s).$$

More generally, we thus obtain that, for any $\theta > 0$,

$$T(\theta) =_d \lim_{n \to \infty} \sum_{k=1}^{\lfloor \theta n \rfloor} T_k(1/n), \tag{1}$$

where $\lfloor x \rfloor$ denotes the integer part of number x.

The monotonicity of the trajectories of random time $T(\theta)$ permits us to introduce the *inverse random time*

$$\Theta(t) = \min\{\theta : T(\theta) \geq t\}.$$

The p.d.f. $g(\theta; t)$ of the inverse random time $\Theta(t)$, which is a probability density with respect to the variable t whereas θ plays the role of a parameter,

---

[4]In numerous works on the subject, the random time is called the subordinator process, and its inverse —the inverse subordinator process. In this book we use the more transparent physically term "random time." In the case of stable subordinator, both the subordinator and the inverse subordinator are self-similar.

can be calculated from the p.d.f. $f(t; \theta)$ of the random time $T(\theta)$, which is a probability density with respect to the variable θ with t playing the role of a parameter. Indeed, observing the equivalence

$$T(\theta) < t \qquad \Longleftrightarrow \qquad \Theta(t) \geq \theta ,$$

we get the relationship

$$F(t; \theta) = \mathbf{P}(T(\theta) < t) = \mathbf{P}(\Theta(t) \geq \theta) = \int_{\theta}^{\infty} g(\theta'; t)\, d\theta' .$$

which gives the following formula for the p.d.f. $g(\theta; t)$ of $\Theta(t)$ in terms of the p.d.f. $f(t; \theta)$:

$$g(\theta; t) = -\frac{\partial F(t; \theta)}{\partial \theta} = -\frac{\partial}{\partial \theta} \int_{-\infty}^{t} f(t'; \theta)\, dt' . \tag{2}$$

To satisfy the self-similarity requirement we will additionally assume that the random time $T(\theta)$ is a β-stable process with the self-similarity parameter $\beta, 0 < \beta < 1$, and p.d.f. $f_\beta(t; \theta)$ with the Laplace transform

$$\hat{f}_\beta(s; \theta) = e^{-\theta s^\beta} .$$

The self-similarity of $T(\theta)$ implies the following scaling property of the random time:

$$f_\beta(t; \theta) = \frac{1}{\theta^{1/\beta}}\, f_\beta\left(\frac{t}{\theta^{1/\beta}}\right) , \tag{3}$$

where, following the convention established above, $f_\beta(t) \equiv f_\beta(t, 1)$. Substituting (3) into (2) immediately yields the following self-similarity property for the p.d.f. $g_\beta(\varrho; t)$ of the inverse random time $\Theta(t)$:

$$g_\beta(\theta; t) = \frac{1}{t^\beta} g_\beta\left(\frac{\theta}{t^\beta}\right) , \tag{4}$$

where

$$g_\beta(\theta) = \frac{1}{\beta \theta^{1+1/\beta}}\, f_\beta\left(\frac{1}{\theta^{1/\beta}}\right) . \tag{5}$$

The above formula for $g_\beta(\theta)$, and the quoted above form of the Laplace transform of the β-stable p.d.f. $f_\beta(t)$, implies together the following useful equality:

$$\exp(-\theta\, s^\beta) = \beta \int_0^{\infty} e^{-st}\, \frac{\theta}{t^{\beta+1}}\, g_\beta\left(\frac{\theta}{t^\beta}\right) dt . \tag{6}$$

Moments $\langle T^\nu \rangle$ of the β-stable random time with p.d.f. $f_\beta(t)$ are infinite for $\nu \geq \beta$. However, moments $\langle \Theta^\nu \rangle$ of the inverse random time Θ with p.d.f.

$g_\beta(\theta)$ are finite for all $\nu > 0$, and we can find them by multiplying both sides of (6) by $\theta^{\nu-1}$, and integrating the resulting equality with respect to θ. This procedure gives

$$\frac{\Gamma(\nu)}{s^{\nu\beta}} = \beta\langle\Theta^\nu\rangle \int_0^\infty e^{-st}\, t^{\nu\beta-1}\, dt = \beta\langle\Theta^\nu\rangle\frac{\Gamma(\nu\beta)}{s^{\nu\beta}},$$

so that

$$\langle\Theta^\nu\rangle = \int_0^\infty \theta^\nu\, g_\beta(\theta)\, d\theta = \frac{\Gamma(\nu)}{\beta\Gamma(\nu\beta)} = \frac{\Gamma(\nu+1)}{\Gamma(\nu\beta+1)}. \tag{7}$$

Consequently, the Laplace transform $\hat{g}_\beta(s)$ of the p.d.f. $g_\beta(\theta)$ has the Taylor expansion

$$\hat{g}_\beta(s) = \langle e^{-s\Theta}\rangle = \sum_{n=0}^\infty \frac{(-s)^n}{\Gamma(n\beta+1)} = E_\beta(-s), \tag{8}$$

where

$$E_\beta(z) = \sum_{n=0}^\infty \frac{z^n}{\Gamma(n\beta+1)} \tag{9}$$

is the Mittag-Leffler function with parameter β (see, e.g., Volume 1, Section 6.10). As a result, the Laplace transform $\hat{g}_\beta(s;t) = \langle e^{-s\Theta(t)}\rangle$ of p.d.f. $g_\beta(\theta;t)$ (4) can be written in the form

$$\hat{g}_\beta(s;t) = E_\beta(-s\,t^\beta), \tag{10}$$

and it is easy to see (see, again, Volume 1, Section 6.10)) that $\hat{g}_\beta(s;t)$ from (10) is a causal solution of the nonhomogeneous linear *fractional differential equation*

$$\frac{d^\beta \hat{g}_\beta(s;t)}{dt^\beta} + s\,\hat{g}_\beta(s;t) = \frac{t^{-\beta}}{\Gamma(1-\beta)}\,\chi(t), \tag{11}$$

where d^β/dT^β is the Riemann–Liouville fractional differentiation operator defined via the formula

$$\frac{d^\beta u(t)}{dt^\beta} = \frac{1}{\Gamma(-\beta)} \int_0^t (t-s)^{-\beta-1} u(s)\, ds.$$

Notation $\chi(t)$ stands here for the Heaviside unit step function.

Remark 1. A couple of observations are in order here. We use the term "fractional differential equation" although fractional differentiation is really a nonlocal integral operator; this usage is, however, traditional. In this context, instead of solving an initial-value problem it is more natural to consider the inhomogeneous equation (11) whose source term on the right-hand side automatically enforces the causality condition and guarantees the uniqueness of the solutions. Notice that, as $\beta \to 1-$, the source term converges weakly to the Dirac delta $\delta(t)$. Thus, in the limit case $\beta = 1$, equation (11) becomes a classical differential equation $\hat{g}_1'(s;t) + s\hat{g}_1(s;t) = \delta(t)$, which is equivalent to the initial-value problem $\hat{g}' + s\hat{g} = 0$, $\hat{g}(0) = 1$.[5]

Recall that the Mittag-Leffler function $E_\beta(z)$ has an integral representation

$$E_\beta(z) = \frac{1}{2\pi i} \int_{\mathcal{H}} \frac{e^y \, y^{\beta-1}}{y^\beta - z} \, dy, \tag{12}$$

where the integration is performed along the Hankel loop depicted in Fig. 21.4.1. The loop envelops the negative part of the real axis and all singular points of the integrand.

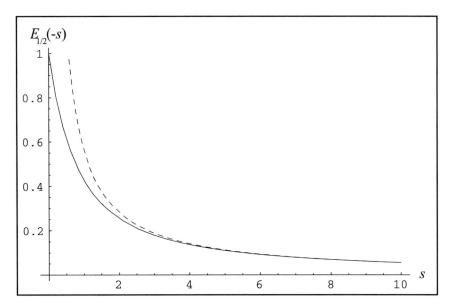

FIGURE 22.3.1.
The Mittag-Leffler function $E_{1/2}(-s)$ (solid line) and its asymptotic behavior $E_{1/2}(-s) \sim 1/\sqrt{\pi}s$, $(s \to \infty)$(dashed line). Notice the accuracy approximation of $E_{1/2}(-s)$ by $1/\sqrt{\pi}s$ for $s > 3$.

If $z = -s$ is a negative real number, then the integral in (12) is reduced to the definite integral

$$E_\beta(-s) = \frac{s}{\pi} \sin(\pi\beta) \int_0^\infty \frac{x^{\beta-1} \, e^{-x} \, dx}{x^{2\beta} + s^2 + 2sx^\beta \cos(\pi\beta)}, \qquad (s > 0, \quad 0 < \beta < 1). \tag{13}$$

---

[5]For the benefit of the reader we would also like to cite the following standard references on fractional calculus: Samko, Gilbas and Marichev [12], Gorenflo and Mainardi [13], Podlubny [14], and West, Bologna and Grigolini [15].

As $s \to \infty$, the terms in the denominator of the integrand are dominated by s^2 which gives the asymptotic formula

$$E_\beta(-s) \sim \frac{1}{s\Gamma(1-\beta)}, \qquad (s \to \infty). \tag{14}$$

Of course, for $\beta = 1/2$, integral (13) boils down to the familiar complementary error function:

$$E_{1/2}(-s) = \exp(s^2)\,\mathrm{erfc}\,(s),$$

where

$$\mathrm{erfc}\,(z) = \frac{2}{\sqrt{\pi}} \int_z^\infty e^{-z^2}\,dz\,.$$

The plot of $E_{1/2}(-s)$, indicating its asymptotics

$$E_{1/2}(-s) \sim 1/\sqrt{\pi}s, \ (s \to \infty),$$

is shown in Fig. 22.3.1.

Finally, note that in the limiting case $\beta = 1$, $E_1(-s)$ is simply the exponential function e^{-s}, with asymptotics dramatically different than the asymptotics of $E_\beta(-s)$, for $0 < \beta < 1$.

22.4 Distribution of the inverse random time

In this section we provide detailed information about p.d.f. $g_\beta(\theta)$ of the inverse random time $\Theta(1)$. Observe that, in view of (5), $g_\beta(\theta)$ is the p.d.f. of the random variable $T^{-\beta}$, where T has the β-stable p.d.f. $f_\beta(t)$. In particular, in view of the asymptotic relation

$$f_\beta(t) \sim \frac{\beta}{\Gamma(1-\beta)t^{1+\beta}} \qquad (t \to \infty),$$

we have

$$g_\beta(\theta = 0) = \frac{1}{\Gamma(1-\beta)}\,.$$

The characteristic function (Fourier transform) $\tilde{g}_\beta(u) = \langle e^{iu\Theta(1)}\rangle$ of p.d.f. $g_\beta(\theta)$ is of the form

$$\tilde{g}_\beta(u) = \hat{g}_\beta(s = -iu) = E_\beta(iu)\,,$$

so that the function $g_\beta(\theta)$ itself is equal to the inverse Fourier transform

$$g_\beta(\theta) = \frac{1}{2\pi} \int E_\beta(iu)\,e^{-iu\theta}\,du\,. \tag{1}$$

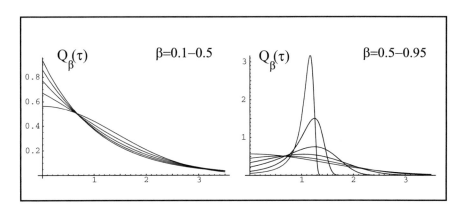

FIGURE 22.4.1.
Plots of p.d.f.s $g_\beta(\theta)$ of the inverse random time for different values of β; the values $g_\beta(\theta = 0)$ decrease as β increases. In contrast to the corresponding β-stable p.d.f.s $f_\beta(t)$, $g_\beta(\theta)$ decay exponentially as $\tau \to \infty$, ensuring that their moments **(7)** are finite. As $\beta \to 1$, we have $g_\beta(\theta) \to \delta(\theta - 1)$.

Using this equality and the integral representation **(12)** of the Mittag-Leffler function it is easy to show that

$$g_\beta(\theta) = \frac{1}{\pi\beta} \operatorname{Im} e^{i\pi\beta} \int_0^\infty \exp\left(-x^{1/\beta} - x\theta\, e^{i\pi\beta}\right) dx, \quad \text{for } 0 < \beta \le 1/2, \quad (2)$$

and

$$g_\beta(\theta) = \frac{1}{\pi\beta} \operatorname{Re} \int_0^\infty \exp\left(ix\theta + x^{1/\beta}\, e^{-i\pi/2\beta}\right) dx, \quad \text{for } 1/2 \le \beta \le 1. \quad (3)$$

Plots of p.d.f.s $g_\beta(\theta)$, for different values of β, are shown in Fig. 22.4.1.

For some special values of parameter β, the integral **(2)** can be explored analytically in more detail. For example, consider $\beta = 1/n$, $n = 2, 3, \ldots$, and introduce an auxiliary complex function

$$w_\beta(\theta) = \frac{1}{\pi\beta} e^{i\pi\beta} \int_0^\infty \exp\left(-x^{1/\beta} - x\theta\, e^{i\pi\beta}\right) dx.$$

Clearly

$$\operatorname{Im} w_\beta(\theta) = g_\beta(\theta), \qquad 0 < \beta \le 1/2.$$

Finding the $(n-1)$-st derivative of $w_{1/n}(\theta)$ with respect to θ we obtain that

$$\frac{d^{n-1} w_{1/n}(\theta)}{d\theta^{n-1}} = (-1)^n \frac{n}{\pi} \int_0^\infty x^{n-1} \exp\left(-x^n - x\theta\, e^{i\pi/n}\right) dx.$$

Integrating by parts we get the equation

$$\frac{d^{n-1}w_{1/n}(\theta)}{d\theta^{n-1}} + \frac{(-1)^n}{n}\,\theta\,w_{1/n}(\theta) = \frac{(-1)^n}{\pi}, \qquad \theta > 0.$$

Separating its imaginary part we arrive at the following differential equation for p.d.f. $g_{1/n}(\theta)$:

$$\frac{d^{n-1}g_{1/n}(\theta)}{d\theta^{n-1}} + \frac{(-1)^n}{n}\,\theta\,g_{1/n}(\theta) = 0.$$

The change of variables $z = \theta/\sqrt[n]{n}$ reduces the above equation to the generalized Airy equations

$$g^{(n-1)}(z) + (-1)^n\, z\, g(z) = 0, \qquad z > 0, n = 2, 3, \ldots.$$

The initial conditions for these equations can be determined by evaluation, for $\theta = 0$, of the integral (2) and its derivatives with respect to θ. For $n = 2$, the Airy equation has a solution $g(z) = e^{-z^2/2}$. As a result, we obtain that

$$g_{1/2}(\theta) = \frac{1}{\sqrt{\pi}}\exp\left(-\frac{\theta^2}{4}\right), \qquad \theta \geq 0. \tag{4}$$

In other words, the inverse random time, for $\beta = 1/2$, has a one-sided Gaussian distribution. As a matter of fact, the corresponding random time's $1/2$-stable density (in our case $\lambda = 1/\sqrt{2}$)

$$f_{1/2}(t) = \frac{\lambda}{\sqrt{2\pi}}t^{-3/2}e^{-\lambda^2/2t}, \tag{5}$$

is traditionally called the inverse Gaussian distribution (see, e.g., Seshadri [16]) and its history goes back to 1915 when Schrödinger and Smoluchowski independently derived it as the p.d.f. of the first passage time of level λ for the standard Brownian motion.

In the case $\beta = 1/3$,

$$g_{1/3}(\theta) = \sqrt[3]{9}\,\text{Ai}\left(\frac{\theta}{\sqrt[3]{3}}\right), \tag{6}$$

where $\text{Ai}\,(z)$ is the well-known Airy function; that is, the vanishing at infinity solution of the equation $g''(z) - zg(z) = 0$ (see *Mathematica*, for an efficient numerical and symbolic implementation). Taking into account relations (5) one obtains

$$f_{1/3}(t) = \frac{1}{\sqrt{3}}t^{-7/2}\text{Ai}\left(\frac{1}{\sqrt[3]{3t}}\right). \tag{7}$$

22.5　1-D anomalous diffusion

In this section we will study statistical properties of the anomalous diffusion

$$\mathcal{X}(t) = X(\Theta(t)),\tag{1}$$

where $X(\theta), \theta \geq 0$, is a random function (stochastic process) and $\Theta(t)$ is the inverse random time introduced in Section 22.3. The rationale for introduction of such anomalous diffusion was given in the introduction and an additional physical justification can be found in Section 8 which discusses anomalous random walks.

The above transformation is known as the *subordination*, and it was introduced in Bochner [17]. Both, Bertoin [8] and Sato [9], devote chapters to this concept and the related concept of the *subordinator*, i.e., $\Theta(t)$ in (1).

We shall assume that processes $X(\theta)$ and $\Theta(t)$ are statistically independent and that they have the one-point p.d.f.s $w(x; \theta)$ and $g(\theta; t)$, respectively. Then the p.d.f. $f(x; t)$ of the composite random function $\mathcal{X}(t)$, obtained with the help of the total probability formula, is given by the expression

$$f(x; t) = \int_0^\infty w(x; \theta)\, g(\theta; t)\, d\theta,\tag{2}$$

and the mean square of process $\mathcal{X}(t)$ is equal to

$$\langle \mathcal{X}^2(t) \rangle = \int_0^\infty \langle X^2(\theta) \rangle g(\theta; t)\, d\theta.\tag{3}$$

Also, the characteristic function $\tilde{f}(u; t) = \langle \exp(iu\mathcal{X}) \rangle$ can be written in the form

$$\tilde{f}(u; t) = \int_0^\infty \tilde{w}(u; \theta) g(\theta; t)\, d\theta,\tag{4}$$

where $\tilde{w}(u; \theta)$ is the characteristic function of the random function $X(\theta)$.

Let us now assume that $X(\theta), X(0) = 0$, is a Lévy process (see Chapter 21), that is a process with time-homogeneous (stationary) and independent increments and with an infinitely divisible one point p.d.f. with characteristic function

$$\tilde{w}(u; \theta) = \tilde{w}^{\gamma\theta}(u) = \exp[\gamma\theta\, \Psi(u)],$$

where $\Psi(u)$ is the logarithm of the characteristic function of the underlying "mother" infinitely divisible p.d.f. Substituting this expression into (4) gives

$$\tilde{f}(u; t) = \int_0^\infty e^{\gamma\Psi(u)\theta}\, g(\theta; t)\, d\theta = \hat{g}(-\gamma\, \Psi(u); t).\tag{5}$$

Furthermore, we shall assume that $T(\theta)$ is a β-stable random time introduced in Section 22.3 so that the inverse random time $\Theta(t)$ has the p.d.f. $g_\beta(\theta; t)$ given by the formula (4). Then, taking into account (10), the formula (5) yields

$$\tilde{f}(u; t) = E_\beta(\gamma t^\beta \Psi(u)).\tag{6}$$

In turn, it follows from (11) that the characteristic function of the anomalous diffusion $\mathcal{X}(t)$ is a solution of the fractional differential equation

$$\frac{d^\beta \tilde{f}(u;t)}{dt^\beta} = \gamma \Psi(u)\, \tilde{f}(u;t) + \frac{t^{-\beta}}{\Gamma(1-\beta)}\, \chi(t)\,. \tag{7}$$

In the special case when $X(t)$ is a symmetric α-stable process, $0 < \alpha \leq 2$, with the logarithm of the mother characteristic function $\Psi(u) = -\sigma^\alpha |u|^\alpha$, taking the inverse Fourier transform, (7) yields the following fractional diffusion equation for the p.d.f. $f(x;t)$ of the process $\mathcal{X}(t)$:

$$\frac{\partial^\beta f}{\partial \tau^\beta} = \sigma^\alpha \frac{\partial^\alpha f}{\partial |x|^\alpha} + \frac{\tau^{-\beta}}{\Gamma(1-\beta)}\, \chi(\tau)\, \delta(x) \qquad (\tau = \gamma^{1/\beta} t)\,; \tag{8}$$

the fractional derivative with respect to spatial variable x is understood here and thereafter in the Weyl sense, see Appendix A. The solution of this equation can be found by substituting into (2) a self-similar p.d.f. $w(x;\theta) = f_\alpha(x;\theta)$ of a symmetric α-stable p.d.f., and the β-stable p.d.f $g(\theta;t) = g_\beta(\theta;t)$ given in (4). As a result we obtain

$$f(x;t) = \frac{1}{\sigma \tau^\mu} f_{\alpha,\beta}\left(\frac{x}{\sigma \tau^\mu}\right)\,, \tag{9}$$

where

$$f_{\alpha,\beta}(y) = \int_0^\infty g_\beta(z)\, f_\alpha\left(\frac{y}{z^{1/\alpha}}\right) \frac{dz}{z^{1/\alpha}}\,, \qquad \mu = \frac{\beta}{\alpha}\,. \tag{10}$$

Let us find the moments of the p.d.f. (9). The self-similarity of $f(x;t)$ and (10) implies that

$$\langle |\mathcal{X}(t)|^\kappa \rangle = \langle |X|^\kappa \rangle_\alpha \langle \Theta^{\kappa/\alpha} \rangle_\beta \sigma^\kappa \tau^{\kappa\beta/\alpha}\,, \tag{11}$$

where the angled brackets' subscripts mean averaging, respectively, with help of the symmetric stable p.d.f. $f_\alpha(x)$, and the p.d.f. $g_\beta(\tau)$. In view of (7), we get

$$\langle |X|^\kappa \rangle_\alpha = \frac{2}{\pi} \Gamma(\kappa) \Gamma\left(1 - \frac{\kappa}{\alpha}\right) \sin\left(\frac{\pi\kappa}{2}\right)\,, \qquad \langle \Theta^{\kappa/\alpha} \rangle_\beta = \frac{\Gamma\left(\frac{\kappa}{\alpha} + 1\right)}{\Gamma\left(\frac{\kappa\beta}{\alpha} + 1\right)}\,.$$

Substituting these expressions into (11), and utilizing the symmetrization formula of the theory of the Gamma function, we obtain finally that

$$\langle |\mathcal{X}(t)|^\kappa \rangle = \frac{2}{\beta} \frac{\sin\left(\frac{\pi\kappa}{2}\right)}{\sin\left(\frac{\pi\kappa}{\alpha}\right)} \frac{\Gamma(\kappa)}{\Gamma\left(\frac{\kappa\beta}{\alpha}\right)} \sigma^\kappa \tau^{\kappa\beta/\alpha} \qquad (\kappa < \alpha)\,. \tag{12}$$

Similarly, for $\alpha > 1$, we can find the value of the p.d.f. $f(x;t)$ of the anomalous diffusion $\mathcal{X}(t)$, see (9), at $x = 0$:

$$\sigma f(0;t) = \frac{1}{\tau^\mu} f_\alpha(0) \langle \Theta^{-1/\alpha} \rangle_\beta = \frac{\tau^{-\mu}}{\alpha \Gamma\left(1 - \frac{\beta}{\alpha}\right) \sin\left(\frac{\pi}{\alpha}\right)} \qquad (\alpha > 1)\,.$$

Its asymptotics, for $x/\sigma\tau^\mu \to \infty$, is described by the formula

$$\sigma f(x;t) \sim \frac{1}{\pi} \frac{\tau^\beta}{|x|^{\alpha+1}} \frac{\Gamma(1+\alpha)}{\Gamma(1+\beta)} \sin\left(\frac{\pi\alpha}{2}\right) \qquad (|x| \gg \sigma\tau^\mu).$$

Observe that, for $\alpha < 1$ $(0 < \beta < 1)$, the value of the p.d.f. $f(x;t)$ diverges to ∞ as $x \to 0$.

For certain values of α and β, explicit solutions of equation (4.8) can be found, see, e.g., Saichev and Zaslavsky [3]. We shall list them below.

In the case $\alpha = \beta$, equation (8) has the solution of the form

$$f(x;t) = \frac{1}{\pi|y|\tau} \frac{\sin\left(\frac{\pi\beta}{2}\right)}{|y|^\beta + |y|^{-\beta} + 2\cos\left(\frac{\pi\beta}{2}\right)} \qquad \left(y = \frac{x}{\sigma\tau}\right).$$

For $\beta \to 1$, the above solution converges, as expected, to the Cauchy p.d.f.

For $\alpha = 1$, $\beta = 1/2$, the p.d.f. $f(x;t)$ can be expressed in terms of the integral exponential function $\mathrm{Ei}(z)$ since

$$f_{\alpha,\beta}(y) = -\frac{1}{2\pi\sqrt\pi} \exp\left(\frac{y^2}{4}\right) \mathrm{Ei}\left(-\frac{y^2}{4}\right).$$

So far we have only considered symmetric anomalous diffusions, and the corresponding equation (8) for the p.d.f. $f(x;t)$ contained only symmetric fractional derivatives with respect to x. However, modeling of various physical phenomena, such as the accumulation processes, requires consideration of processes $X(\theta)$ which are nondecreasing and have nonnegative values. In such cases, the p.d.f.s $f(x;t)$ of the anomalous diffusions $\mathcal{X}(t)$ satisfy an analogue of the fractional differential equation (8) in which the symmetric fractional derivative in x is replaced by a one-sided fractional derivative, either right-sided or left-sided. Here, a good example is the equation

$$\frac{\sqrt\partial f}{\partial\sqrt\tau} = \frac{\sqrt\partial f}{\partial\sqrt x} + \frac{1}{\sqrt{\pi\tau}}\chi(\tau)\delta(x).$$

Its solution is also described by the general formulas (9–10). Substituting into (10) $f_{\alpha=1/2}(x)$, and $g_{\beta=1/2}(\tau)$ given by (4), we obtain in this case that

$$f(x;t) = \frac{1}{\pi}\sqrt{\frac{t}{x}} \frac{\chi(x)}{x+t}, \qquad x > 0, t > 0.$$

Remark 1. It is worth mentioning that the case of coupled anomalous diffusion $\mathcal{X}(t) = X(\Theta(t))$, where the processes $X(\theta), \theta \geq 0$, and $\Theta(t)$ are not statistically independent, has been recently discussed by Meerschaert, Benson, Scheffler and Becker-Kern [18].

22.6 Anomalous drift process

In this section we will discuss statistical properties of the *fractional pure drift process*, a simple example of the anomalous diffusion introduced in Section 22.5 which corresponds to the case $X(\theta) = \gamma\theta$, $\gamma > 0$. In this case,

$$X(t) = \gamma\,\Theta(t)\,. \tag{1}$$

The simplicity of this case will permit us to elucidate the analysis of the general case, and the results obtained herein will prove useful in Section 22.7 where anomalous subdiffusion processes are going to be introduced.

For the drift process the logarithm of the characteristic function of the underlying "mother" infinitely divisible distribution has a degenerate form $\Psi(u) = iu$, and equation (7) assumes the form

$$\frac{\partial^\beta \tilde{f}}{\partial \tau^\beta} = iu\tilde{f} + \frac{\tau^{-\beta}}{\Gamma(1-\beta)}\,\chi(\tau) \qquad (\tau = \gamma^{1/\beta}t)\,. \tag{2}$$

Taking the inverse Fourier transform in variable u, we obtain the following equation for the p.d.f. $f(x;t)$ of the anomalous drift process $X(t)$:

$$\frac{\partial^\beta f}{\partial \tau^\beta} = \frac{\partial f}{\partial x} + \frac{\tau^{-\beta}}{\Gamma(1-\beta)}\,\chi(\tau)\,\delta(x)\,. \tag{3}$$

Substituting $\Psi(u) = iu$ into (6), applying the inverse Fourier transform, and taking into account formula (1), we arrive at the following solution of equation (3)

$$f(x;t) = \frac{1}{\tau^\beta}g_\beta\left(\frac{x}{\tau^\beta}\right)\,. \tag{4}$$

For different values of β, the shapes of $f(x;t)$, as a function of x, are the same as the shapes of the plots of $g_\beta(\theta)$ shown in Fig. 22.4.1. As $\beta \to 1$, the p.d.f. (4) weakly converges to the Dirac delta:

$$\lim_{\beta\to 1-} f(x;t) = \delta(x-\tau)\,.$$

This behavior can be viewed as a sort of law of large numbers for $\beta \to 1$.

The properties of the function $g_\beta(\theta)$ immediately give the following rates of growth of the moments of the fractional anomalous drift process (1):

$$\langle X^n(t)\rangle = \langle \Theta^n\rangle\,\tau^{n\beta}\,, \qquad \langle \Theta^n\rangle = \frac{n!}{\Gamma(n\beta+1)}\,.$$

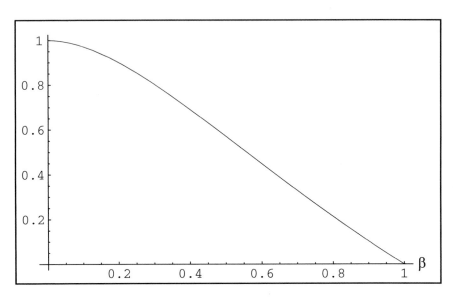

FIGURE 22.6.1.
Plot of the relative variance of the fractional anomalous drift process
as a function of the order β of the fractional derivative in equation (3).

In particular, for $\beta \to 1-$, the relative variance of the fractional drift,

$$\frac{\sigma_{\mathcal{X}}^2(t)}{\langle \mathcal{X}(t) \rangle^2} = 2 \frac{\Gamma^2(\beta+1)}{\Gamma(2\beta+1)} - 1 \tag{5}$$

converges to zero confirming the law of large numbers behavior as $\beta < 1$
approaches 1. A plot of the relative variance (5) as a function of β is shown
in Fig. 22.6.1.

22.7   Fractional anomalous subdiffusion

In this section the underlying Lévy infinitely divisible process $X(\theta)$ will be
taken to be the Brownian motion (Wiener) process $B(\theta)$ with the logarithm
of the characteristic function of the "mother" infinitely divisible distribution
of the form $\Psi(u) = -u^2$. Note that with this definition $\langle B^2(t) \rangle = 2t$ (and not
t as in the standard definition). The random time $T(\theta)$ will be assumed to
be a β-stable process with p.d.f. (3). The corresponding anomalous diffusion

$$\mathcal{X}(t) = \sqrt{\gamma}\, B(\Theta(t)) \tag{1}$$

will be called *fractional anomalous subdiffusion*.

Our first task is to find an equation for the one-point p.d.f. $f(x;t)$ of $\mathcal{X}(t)$. We will start with an equation for its characteristic function. The latter is obtained immediately from equation (7) by substituting $\Psi(u) = -u^2$ which yields

$$\frac{d^\beta \tilde{f}}{d\tau^\beta} + u^2 \tilde{f} = \frac{\tau^{-\beta}}{\Gamma(1-\beta)} \chi(\tau) \qquad (\tau = \gamma^{1/\beta} t).$$

Taking the inverse Fourier transform in variable u we obtain a *subdiffusion equation* for the p.d.f. $f(x;t)$ of the process (1):

$$\frac{\partial^\beta f}{\partial\tau^\beta} = \frac{\partial^2 f}{\partial x^2} + \frac{\tau^{-\beta}}{\Gamma(1-\beta)} \chi(\tau) \delta(x). \qquad (2)$$

This equation can be solved with the help of the solution (4) of the fractional drift equation (3).

Indeed, let us split the characteristic function of the fractional drift into the even and odd (in Fourier variable u) components:

$$\tilde{f} = \tilde{f}_{\text{even}} + \tilde{f}_{\text{odd}}$$

and substitute it into equation (2). As a result we obtain two equations:

$$\frac{\partial^\beta \tilde{f}_{\text{even}}}{\partial\tau^\beta} = iu\tilde{f}_{\text{odd}} + \frac{\tau^{-\beta}}{\Gamma(1-\beta)} \chi(\tau), \qquad \text{and} \qquad \frac{\partial^\beta \tilde{f}_{\text{odd}}}{\partial\tau^\beta} = iu\tilde{f}_{\text{even}}.$$

Applying fractional derivative of order β with respect to τ to the first equation, taking into account the standard fractional calculus formula

$$\frac{\partial^\beta \tau^\delta \chi(\tau)}{\partial\tau^\beta} = \frac{\Gamma(\delta+1)}{\Gamma(1+\delta-\beta)} \tau^{\delta-\beta} \chi(\tau),$$

see, e.g., Volume 1 of this monograph series, and eliminating the odd component $\tilde{f}_{\text{odd}}$, we obtain a closed equation for $\tilde{f}_{\text{even}}(u;t)$:

$$\frac{\partial^{2\beta} \tilde{f}_{\text{even}}}{\partial\tau^{2\beta}} + u^2 \tilde{f}_{\text{even}} = \frac{\tau^{-2\beta}}{\Gamma(1-2\beta)} \chi(\tau).$$

As a result, the even component

$$f_{\text{even}}(x;t) = \frac{1}{2}[f(x;t) + f(-x;t)] = \frac{1}{2}f(|x|;t) \qquad (3)$$

of the p.d.f. $f(x;t)$ of the fractional drift satisfies equation

$$\frac{\partial^{2\beta} f}{\partial\tau^{2\beta}} = \frac{\partial^2 f}{\partial x^2} + \frac{\tau^{-2\beta}}{\Gamma(1-2\beta)} \chi(\tau) \delta(x).$$

A substitution $\beta \mapsto \beta/2$ reduces this equation to the equation (2). Thus, the sought solution of the fractional subdiffusion equation (2) can be obtained from (3) and (4) by replacing β by $\beta/2$ which gives

$$f(x;t) = \frac{1}{2\,\tau^{\beta/2}}\, g_{\beta/2}\left(\frac{|x|}{\tau^{\beta/2}}\right).\tag{4}$$

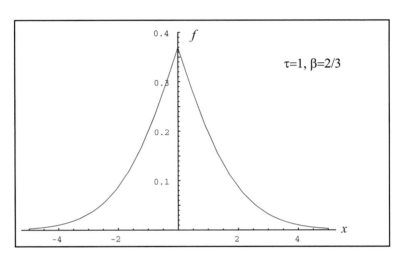

FIGURE 22.7.1.
The solution of the subdiffusion equation (2), for $\beta = 2/3$.

In particular, for $\beta = 2/3$, recalling (7), we obtain that

$$f(x;t) = \frac{1}{2}\,\sqrt[3]{\frac{9}{\tau}}\,\mathrm{Ai}\left(\frac{|x|}{\sqrt[3]{3\tau}}\right), \qquad \beta = \frac{2}{3}.$$

A plot of this solution, for $\tau = 1$, is given in Fig. 22.7.1.

The sub-Fickian behavior of the anomalous diffusion (1), justifying the term *subdiffusion*, is obtained by substituting $\alpha = 2$ into the general formula (12), and finding the limit as $\kappa \to 2-$, which gives

$$\langle \mathcal{X}^2(t)\rangle = \frac{2}{\Gamma(\beta+1)}\,\tau^\beta.\tag{5}$$

In other words, the variance of $\mathcal{X}^2(t)$ grows sublinearly as t increases.

The solution (4) of the subdiffusive equation (2) can be used immediately to calculate, for instance, the p.d.f. $f(t;\ell)$ of the first crossing time T of the level ℓ by a subdiffusion process (1) starting at 0:

$$T = \min\{t : \mathcal{X}(t) \geq \ell\}.$$

Indeed,

$$\mathbf{P}(T > t) = \int_{-\infty}^{\ell} f(x; t|\ell) \, dx \, , \qquad (6)$$

where $f(x; t|\ell)$ is a solution of the equation (2) satisfying the boundary condition

$$f(x = \ell; t) = 0 \, .$$

Solving the above boundary-value problem by the standard reflection method we obtain

$$f(x; \tau|\ell) = \frac{1}{2\tau^{\beta/2}} \left[g_{\beta/2}\left(\frac{|x|}{\tau^{\beta/2}}\right) - g_{\beta/2}\left(\frac{|x - 2\ell|}{\tau^{\beta/2}}\right) \right] . \qquad (7)$$

Substituting this expression in (6) we get

$$\mathbf{P}(T > t) = \int_{0}^{\ell/t^{\beta/2}} g_{\beta/2}(z) \, dz \, , \qquad (8)$$

which, together with (5), gives

$$f(\tau; \ell) = -\frac{d\mathbf{P}(T > t)}{dt} = \frac{1}{2\sqrt{\tau}\ell^{1/\beta}} f_{\beta/2}\left(\frac{\sqrt{\tau}}{\ell^{1/\beta}}\right) . \qquad (9)$$

22.8 Multidimensional fractional subdiffusion

The above analysis of the one-dimensional subdiffusion can be easily extended to the multidimensional case. Let us consider, for example, a d-dimensional vector random process

$$\vec{\mathcal{X}}(t) = \{\mathcal{X}_1(t), \mathcal{X}_2(t), \dots, \mathcal{X}_d(t)\} \, , \qquad (1)$$

with components which are independent Wiener processes of random argument $\Theta(t)$:

$$\mathcal{X}_k(t) = \sqrt{\gamma} \, B_k(\Theta(t)) \qquad (k = 1, 2, \dots, d) \, .$$

The d-dimensional characteristic function of the random vector $\vec{\mathcal{X}}(t)$

$$\tilde{f}_d(\vec{u}; t) = \langle e^{i(\vec{u} \cdot \vec{\mathcal{X}}(t))} \rangle = \int_0^{\infty} \tilde{w}_d(\vec{u}; \theta) \, g(\theta; t) \, d\theta \, , \qquad (2)$$

where

$$\tilde{w}_d(\vec{u}; \theta) = e^{-\gamma(\vec{u} \cdot \vec{u}) \theta}$$

is the d-dimensional characteristic function of the vector Gaussian diffusion process $\vec{X}(\theta) = \sqrt{\gamma} \, \vec{B}(\theta)$.

If $T(\theta)$ is a β-stable random time, then an evaluation of the integral in (2) gives $\tilde{f}_d$ in terms of the Mittag-Leffler function:

$$\tilde{f}_d(\vec{u};t) = E_\beta(-\tau^\beta\,(\vec{u}\cdot\vec{u}))\,, \qquad (\tau = \gamma^{1/\beta}t).$$

Also, the characteristic function $\tilde{f}_d$ satisfies equation

$$\frac{d^\beta \tilde{f}_d}{d\tau^\beta} + (\vec{u}\cdot\vec{u})\,\tilde{f}_d = \frac{\tau^{-\beta}}{\Gamma(1-\beta)}\,\chi(\tau)\,.$$

This equation, in turn, is equivalent to the d-dimensional subdiffusion equation for the p.d.f. $f_d(\vec{x};t)$ of the random vector $\vec{\mathcal{X}}(t)$ (1):

$$\frac{\partial^\beta f_d}{\partial\tau^\beta} = \Delta\,f_d + \frac{\tau^{-\beta}}{\Gamma(1-\beta)}\,\chi(\tau)\,\delta(\vec{x})\,, \qquad (3)$$

where $\Delta = \partial^2/\partial x_1^2 + \cdots + \partial^2/\partial x_d^2$ is the usual d-dimensional Laplacian.

A solution of this equation can be found substituting into (2) the p.d.f.

$$w_d(\vec{x};\theta) = \left(\frac{1}{2\pi\theta}\right)^{d/2}\exp\left(-\frac{r^2}{4\theta}\right)\,, \qquad r = \sqrt{x_1^2 + \cdots + x_d^2}\,,$$

of random process $\vec{X}(\theta)$, and the p.d.f. $g_\beta(\theta;t)$ given in (4). As a result, we obtain a self-similar solution

$$f_d(\vec{x};t) = \frac{1}{\tau^{n\beta/2}}\,h_d\left(\frac{r}{\tau^{\beta/2}}\right)\,, \qquad \tau = \gamma^{1/\beta}t\,, \qquad (4)$$

where

$$h_d(y) = \frac{1}{(4\pi)^{d/2}}\int_0^\infty g_\beta(z)\exp\left(-\frac{y^2}{4z}\right)\frac{dz}{z^{d/2}}\,.$$

In the three-dimensional space, a comparison of (4) and (4) gives

$$h_3(y) = -\frac{1}{4\pi\,y}\,\frac{d\,g_{\beta/2}(y)}{dy}\,. \qquad (5)$$

In particular, for $\beta = 2/3$,

$$h_3(y) = -\frac{3^{1/3}}{4\pi y}\mathrm{Ai}'\left(\frac{y}{3^{1/3}}\right)\,,$$

where $\mathrm{Ai}'(\theta)$ is the derivative of the Airy function introduced in (6).

22.9 Tempered anomalous subdiffusion

Introduced in Section 22.7 fractional subdiffusion $\mathcal{X}(t) = \sqrt{\gamma}\, B(\Theta(t))$, $t \geq 0$, where B is the Brownian motion process and random time $\Theta(t)$ has the inverse β-stable p.d.f. , $0 < \beta < 1$, owed its subdiffusive, for all times t, behavior $\langle \mathcal{X}^2(t) \rangle = \frac{2}{\Gamma(\beta+1)} \tau^\beta$, with $\tau = \gamma^{1/\beta} t$, see (5), to the fact that, for the β-stable random time $T(\theta)$, the mean, and thus the second moment as well, were infinite. From the physical perspective it is desirable to also have a model that circumvents the infinite-moment difficulty while preserving the subdiffusive behavior, at least for small times.

For this purpose we will consider in this section an anomalous diffusion of the form

$$\mathcal{X}(t) = \sqrt{\gamma}\, B(\Theta(t)), \tag{1}$$

where $\Theta(t)$ is the inverse *tempered β-stable* random time, with the tempered β-stable random time $T(\theta)$ defined by the Laplace transform

$$\hat{f}_{\beta,\mathrm{tmp}}(s; \delta) = \langle e^{-sT(\theta)} \rangle = e^{\theta \Phi(s)} = \exp\left(\delta^\beta - (s+\delta)^\beta \right), \qquad 0 < \beta < 1, \tag{2}$$

where

$$\Phi(s) = \int_0^\infty (e^{-sz} - 1)\phi(z)\, dz, \qquad \phi(z) = \frac{\beta}{\Gamma(1-\beta)}\, z^{-\beta-1} e^{-\delta z} \chi(z),$$

for some $\delta > 0$.[6] The form of the Lévy intensity function $\phi(z)$ guarantees its β-stable behavior for small jump sizes z; the tempering influence of the exponential term gives the existence of all moments of $T(\theta)$. In that sense tempered stable processes occupy an intermediate place between pure stable processes and the classical Brownian motion diffusion processes.

Using formulas (3) and (2) for the p.d.f. of the inverse random time, we can express the mean square of the process $\mathcal{X}(t)$ (1) via the cumulative distribution function (c.d.f.) $F_{\beta,\mathrm{tmp}}(t; \theta)$ of the random time $T(\theta)$:

$$(\sigma^2)(t) = \langle \mathcal{X}^2(t) \rangle = \langle \langle W^2(\Theta(t)) \rangle_B \rangle_\Theta = \langle 2\Theta(t) \rangle = 2\gamma \int_0^\infty F_{\beta,\mathrm{tmp}}(t; \theta)\, d\theta.$$

The last expression was obtained by integration by parts and application of the fact that, for each t, the values of $F_{\beta,\mathrm{tmp}}(t; \theta) \to 0$ as $\theta \to \infty$.

The Laplace image of $\langle \mathcal{X}^2(t) \rangle$ in variable t is

$$\widehat{(\sigma^2)}(s) = \int_0^\infty e^{-st} \langle \mathcal{X}^2(t) \rangle\, dt = \frac{2\gamma}{s} \int_0^\infty (\hat{f}_{\beta,\mathrm{tmp}})^\theta(s)\, d\theta = -\frac{2\gamma}{s \ln \hat{f}_{\beta,\mathrm{tmp}}(s)}, \tag{3}$$

---

[6]For a comprehensive analysis of tempered stable diffusions, see Rosiński [19].

where $\hat{f}_{\beta,\mathrm{tmp}}(s) = \hat{f}_{\beta,\mathrm{tmp}}(s; 1)$ is the mother Laplace transform of the tempered β-stable random time $T(\theta)$. Substituting into (3) the Laplace image of $f_\beta(s; \delta)$ from (2), we obtain

$$\widehat{(\sigma^2)}(s) = \frac{2\gamma}{s[(s + \delta)^\beta - \delta^\beta]} \, .$$

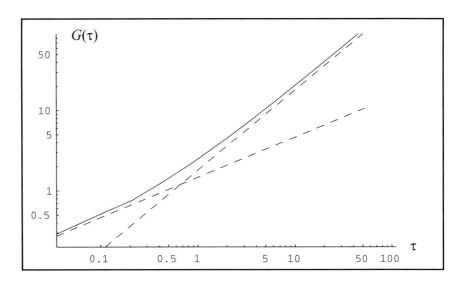

FIGURE 22.9.1.
Plot of function $k(\tau; \beta = 1/2)$, drawn in loglog scale (solid line). The dashed lines show the subdiffusive ($\sqrt{\tau}$) and linear (τ) asymptotics of this function.

In a dimensionless variable $p = s/\delta$ the above expression takes the form

$$\widehat{(\sigma^2)}(s) = \frac{2\gamma}{\delta^{1+\beta}} \, \hat{k}(p; \beta) , \quad \text{where} \quad \hat{k}(p; \beta) = \frac{1}{p[(p + 1)^\beta - 1]} , \qquad p = \frac{s}{\delta} .$$

So the mean square

$$\langle \mathcal{X}^2(t) \rangle = \frac{2\gamma}{\delta^\beta} \, k(\tau; \beta) , \qquad \tau = \delta \, t ,$$

where $k(\tau; \beta)$ is the inverse Laplace transform of $\hat{k}(p; \beta)$. The asymptotics of the function $k(\tau; \beta)$ is then determined by the asymptotics of $\hat{k}(p; \beta)$ which is

$$\hat{k}(p; \beta) \sim \frac{1}{p^{\beta+1}} \quad (p \to \infty); \qquad \hat{k}(p; \beta) \sim \frac{1}{\beta \, p^2} \quad (p \to 0) .$$

The corresponding asymptotics of $k(\tau; \beta)$ is as follows:

$$k(\tau; \beta) \sim \frac{\tau^\beta}{\Gamma(1 + \beta)} \quad (\tau \to 0); \qquad k(\tau; \beta) \sim \frac{\tau}{\beta} \quad (\tau \to \infty) .$$

Thus, for small τ, the mean square $\langle \mathcal{X}^2(t) \rangle$ grows according to the sub-diffusive law τ^β, while for large τ, the subdiffusive law is replaced by the classical Fickian law of the Gaussian linear diffusion.

We illustrate the above dichotomy of the behavior of $\langle \mathcal{X}^2(t) \rangle$ in Fig. 22.9.1 in the case of $\beta = 1/2$, where the explicit expression

$$k(\tau; \beta = 1/2) = \sqrt{\frac{\tau}{\pi}}\, e^{-\tau} + \tau + \left(\frac{1}{2} + \tau \right) \operatorname{erf}\left(\sqrt{\tau} \right),$$

is available.

22.10 Anomalous random walks

Anomalous diffusions discussed in the preceding sections can be viewed as scaling limits of anomalous random walks which change their positions by independent random jumps occurring at independent random intervals which are, as an ensemble, independent of the jumps. These anomalous random walks are not Markovian, except in the special case when the random intervals widths have an exponential p.d.f.

22.10.1. Anomalous diffusion as a limit of anomalous random walks. Anomalous random walk will be introduced here in the spirit of the so-called *renewal theory*. The starting point will be a sequence,

$$\tau(n), \quad n = 1, 2, \ldots,$$

of statistically independent, positive, infinite mean (and thus infinite variance) random quantities representing duration of time intervals between consecutive events, which in our case are the jumps of the anomalous random walk $X(t), t \geq 0$, under consideration. The jump sizes $h(n), n = 1, 2, \ldots$, are themselves random quantities. So, the jumps occur at times

$$T(n) = \sum_{k=1}^{n} \tau(k), \quad n = 1, 2, \ldots,$$

with

$$\tau(n) = T(n) - T(n-1).$$

Then the *anomalous random walk*, $X(t)$, itself can be thought of as a solution of the stochastic differential equation,

$$\frac{dX(t)}{dt} = \sum_{k} \xi(k)\, \delta(t - T(k)), \qquad X(t = 0) = 0, \tag{1}$$

where $X(t)$ is the description of the coordinate of a particle in the 1-D space which starts at the origin and then jumps the distance $\xi(n)$ at times $T(n)$. However, many other interpretations are possible, including an application to modeling evolution of prices in financial markets.

The general assumption is that jumps $\{\xi(n)\}$ are assumed to be independent from each other, and, as an ensemble, independent of the sequence $\tau(n)$, $n = 1, 2, \ldots$, which are statistically independent random quantities themselves. The jumps $\xi(n)$ have identical p.d.f.s, $w(x)$, and the time intervals $\tau(n)$ have identical p.d.f.s, $f(t)$.

The time intervals, $T(m)$, which, for any $n \geq m$, have the same probability distributions as the time intervals

$$T(n) - T(n-m) = \tau(n) + \tau(n-1) + \cdots + \tau(n-m+1), \qquad (2)$$

have the following p.d.f., and c.d.f., respectively,

$$f(t;m) = \underbrace{f(t) * \cdots * f(t)}_{m-\text{times}}, \qquad \text{and} \qquad F(t;m) = F(t) * \underbrace{f(t) * \cdots * f(t)}_{(m-1)-\text{times}},$$
$$(3)$$

where

$$F(t) = \mathbf{P}(\tau(n) \leq t) = \langle \chi(t - \tau(n)) \rangle$$

is the common c.d.f. of the random intervals $\tau(n)$.

The solution of equation (1), which is the anomalous random walk in question, is of the form

$$X(t) = \sum_{k=1}^{M(t)} \xi(k), \qquad (4)$$

where $M(t)$ is the random number of jumps taken by the anomalous random walk up to time t. The process $M(t)$ is often called the *counting process* (or the *renewal process*) associated with the random walk $X(t)$. The probability $R(m;t)$ that the interval $(0, t]$ contains m jumps is

$$R(m;t) = \mathbf{P}(M(t) = m) = \mathbf{P}(m \leq M(t) < m+1) =$$

$$\mathbf{P}(M(t) < m+1) - \mathbf{P}(M(t) < m).$$

On the other hand,

$$\mathbf{P}(M(t) < m) = \mathbf{P}(T(m) > t) = 1 - F(t;m), \qquad (5)$$

so that

$$R(m;t) = F(t;m) - F(t;m+1), \quad \text{for} \quad m \geq 1, \qquad (6)$$

and

$$R(0;t) = 1 - F(t;1).$$

According to the total probability formula, the p.d.f. of $X(t)$ is equal to

$$f(x;t) = \sum_{m=0}^{\infty} R(m;t)\, w(x;m)\,, \tag{7}$$

where the probabilities $R(m;t)$ are defined by equalities (6), while

$$w(x;m) = \underbrace{w(x) * \cdots * w(x)}_{m-\text{times}}\,, \qquad w(x;0) = \delta(x)\,.$$

Our final assumption is that the common p.d.f. $f(t)$ of the random durations $\tau(n)$ of the inter-jump time intervals is infinitely divisible. In this case there is an obvious relationship between formulas (2) and (6). Indeed, the infinitely divisible time $T(\theta)$ introduced in (1) coincides, for integer values $\theta = n$, with the values of the random time $T(n)$ introduced at the beginning of this subsection. Similarly, the c.d.f.s $F(t;m)$, and $F(t;m+1)$, appearing in the right-hand side of the equality (6), can be viewed as the values of the c.d.f., $F(t;\theta)$, of the infinitely divisible time $T(\theta)$, for integer values of its variable θ.

The inverse random time $\Theta(t)$ of Section 22.3 is then related to the counting process $M(t)$ via the following relations:

$$M(t) \le \Theta(t), \qquad M(t = T(m)) = \Theta(T(m)),$$

while the probability that the time interval $(0, t]$ contains m jumps is

$$R(m;t) = \int_{m}^{m+1} g(\theta;t)\, d\theta, \tag{8}$$

where, we recall, $g(\theta;t)$ is the p.d.f. (2) of the inverse random time $\Theta(t)$. Integrating both sides of (2) with respect to θ over the interval $[m, m+1]$, and taking into account (8), brings us back to the formula (6).

Our final goal in this subsection is to find the asymptotic behavior of the p.d.f. $f(x;t)$ of the anomalous random walk $X(t)$ at large temporal and spatial scales. We begin by observing that if the p.d.f. $g(\theta;t)$ varies slowly (smoothly) with θ, then the equality (8) can be replaced by an approximate equality

$$R(m;t) \simeq g(m;t)\,.$$

Correspondingly, the sum (7) takes the form

$$f(x;t) \approx R(0;t)\, \delta(x) + \sum_{m=1}^{\infty} g(m;t)\, w(x;m)\,. \tag{9}$$

Now, if we assume that the p.d.f. $w(\xi)$ of the random jump sizes is infinitely divisible and generates a one-parameter semigroup $w(x; \theta)$ of infinitely divisible p.d.f.s such that $w(x; m) = w(x; \theta = m)$, and such that $w(x; \theta)$ vary slowly (smoothly) as θ varies over intervals of length one, then the sum in (9) can be approximated by the integral and we have

$$f(x; t) \approx R(0; t)\, \delta(x) + \int_{1}^{\infty} g(\theta; t) w(x; \theta)\, d\theta \,. \qquad (10)$$

If the probability of the absence of the jumps,

$$R(0; t) \approx \int_{0}^{1} g(\theta; t)\, d\theta, \qquad (11)$$

is small then the expression (10) can be replaced by

$$f(x; t) \approx \int_{0}^{\infty} g(\theta; t) w(x; \theta)\, d\theta \,. \qquad (12)$$

We recognize the right-hand side as the familiar integral expression (2) for the p.d.f. of the anomalous diffusion $\mathcal{X}(t)$ introduced in (1).

Taking the Laplace transform in variable t of both sides of the expression (2) for the p.d.f. $g(\theta; t)$, we obtain

$$\hat{g}(\theta; s) = -\frac{1}{s} \frac{\partial \hat{f}^{\,\theta}(s)}{\partial \theta} = -\frac{1}{s} \hat{f}^{\,\theta}(s)\, \Phi(s), \quad \text{where} \quad \Phi(s) = \ln \hat{f}(s)\,. \qquad (13)$$

Recall that if the logarithm of the Laplace transform of the p.d.f. of the inter-jump time intervals has a β-stable-like asymptotics

$$\Phi(s) \sim -\frac{1}{\gamma} s^{\beta} \qquad (\gamma > 0, \quad 0 < \beta < 1, \quad s \to \infty), \qquad (14)$$

then the mean $\langle \tau \rangle = \infty$. Replacing $\Phi(s)$ in (13) by its asymptotics (14), we obtain the asymptotics of the Laplace image $\hat{g}(\theta; t)$:

$$\hat{g}(\theta; s) \sim \frac{1}{\gamma} s^{\beta - 1} \exp\left(-\frac{\theta}{\gamma} s^{\beta}\right) = -\frac{1}{\beta\, \theta} \frac{\partial}{\partial s} \exp\left(-\frac{\theta}{\gamma} s^{\beta}\right) \qquad (s \to \infty).$$

Applying the inverse Laplace transform to both sides of the above equality, and taking into account the fact that the function $-\frac{\partial}{\partial s} \exp(-\mu s^{\beta})$ is the Laplace transform of $\frac{t}{\mu^{1/\beta}} f_{\beta}\left(\frac{t}{\mu^{1/\beta}}\right)$, with $f_{\beta}(t)$ being the β-stable p.d.f. introduced in Section 22.3, we finally obtain the asymptotics,

$$g(\theta; t) \sim \frac{\tau}{\beta\, \theta^{1 + 1/\beta}} f\left(\frac{\tau}{\theta^{1/\beta}}\right) = \frac{1}{\tau^{\beta}} g_{\beta}\left(\frac{\theta}{\tau^{\beta}}\right) \qquad (\tau = \gamma^{1/\beta}\, t \to \infty). \qquad (15)$$

Similarly, if the logarithm of the characteristic function of the p.d.f. $w(x;\theta)$ possesses the asymptotics,

$$\ln \tilde{w}(u;\theta) \sim -\sigma^\alpha |u|^\alpha \theta, \qquad (u \to \infty), \tag{16}$$

then the p.d.f. $w(x;\theta)$ converges, as $\theta \to \infty$, to a symmetric stable p.d.f. (with the substitution $\tau = \theta$, $\gamma = 1$). Substituting asymptotics (14), (16) into (12), we obtain the solution (9) of the fractional diffusion equation (8). Thus, the p.d.f. of the anomalous diffusion described by a fractional diffusion equation provides asymptotics of the p.d.f. of the anomalous random walk at large temporal and spatial scales.

22.10.2. The fractional Kolmogorov–Feller equations for anomalous random walks. In this subsection we will find an equation for the p.d.f. $f(x;t)$ of the anomalous random walk $X(t)$ which is explicitly given by formula (7). For this purpose let us take the Fourier transform of (7) with respect to x, which gives

$$\tilde{f}(u;t) = \sum_{m=0}^{\infty} R(m;t)\, \tilde{w}^m(u). \tag{17}$$

The above series is equal to the generating function $\mathcal{R}(t,z)$ of $R(m,t)$, evaluated at $z = \tilde{w}(u)$. Hence, the Laplace image of the series (17) is given by the expression

$$\hat{\tilde{f}}(u;s) = \frac{1 - \hat{f}(s)}{s[1 - \hat{f}(s)\,\tilde{w}(u)]}. \tag{18}$$

This equality can be rewritten in the form

$$\frac{1}{\hat{f}(s)}\,\hat{\tilde{f}}(u;s) - \tilde{w}(u)\,\hat{\tilde{f}}(u;s) = \frac{1 - \hat{f}(s)}{s\,\hat{f}(s)}. \tag{19}$$

After taking the inverse Fourier transform in space, and the inverse Laplace transform in time, equation (19) gives an integral equation for the p.d.f. $f(x;t)$ of the anomalous random walk which is given by the expression (7).

As an example consider first the Poissonian case of independent inter-jump intervals identically distributed with the exponential p.d.f. with mean $1/\nu$. Its Laplace image is

$$\hat{f}(s) = \frac{\nu}{\nu + s}. \tag{20}$$

Substituting this expression in (18) we obtain the following equation,

$$s\,\hat{\tilde{f}}(u;s) + \nu\,[1 - \tilde{w}(u)]\,\hat{\tilde{f}}(u;s) = 1.$$

Applying the inverse Laplace transform in s to both sides of this equality, and setting $\tilde{f}(u; t = 0) = 0$, we obtain a differential equation for the characteristic function of the anomalous random walk:

$$\frac{d\tilde{f}(u; t)}{dt} + \nu\,[1 - \tilde{w}(u)]\,\tilde{f}(u; t) = \delta(t - 0_+)\,.$$

Taking the inverse Fourier transform we find that the p.d.f. $f(x; t)$ of random walk $X(t)$ satisfies the integro-differential equation

$$\frac{\partial f(x; t)}{\partial t} + \nu\,[f(x; t) - f(x; t) * w(x)] = \delta(t - 0_+)\delta(x),$$

which is equivalent to the following initial-value problem for the homogeneous equation

$$\frac{\partial f(x; t)}{\partial t} + \nu\,[f(x; t) - f(x; t) * w(x)] = 0\,, \qquad f(x; t = 0) = \delta(x)\,, \qquad (21)$$

analogous to the classical Kolmogorov–Feller equation.

For small s and u the functions $\hat{f}(s)$ and $\tilde{w}(u)$ have the asymptotics of the form,

$$\frac{1}{\hat{f}(s)} \sim 1 + \frac{s^\beta}{\gamma} \quad (s \to 0, \gamma > 0)\,; \qquad \tilde{w}(u) \sim 1 - \sigma^\alpha |u|^\alpha \quad (u \to 0)\,.$$

Substituting them into (19), we find the asymptotic equation

$$s^\beta\,\hat{\tilde{f}} + \gamma\sigma^\alpha|u|^\alpha\,\hat{\tilde{f}} = s^{\beta-1} \qquad (s \to 0, \quad u \to 0)$$

for $\hat{\tilde{f}}(u; s)$, which is obviously equivalent to the fractional diffusion equation (8).

If the inter-jump intervals $\tau(n)$ have the fractional exponential p.d.f. with the Laplace transform $(1 + s^\beta)^{-1}$, $0 < \beta < 1$, then (18) yields the equation

$$s^\beta\,\hat{\tilde{f}} + [1 - \tilde{w}(u)]\,\hat{\tilde{f}} = s^{\beta-1}\,.$$

Applying the inverse Laplace transform to both sides of this equality we obtain an equation for the characteristic function:

$$\frac{\partial^\beta\,\tilde{f}(u; t)}{\partial t^\beta} + [1 - \tilde{w}(u)]\tilde{f}(u; t) = \frac{t^{-\beta}}{\Gamma(1 - \beta)}\,\chi(t)\,. \qquad (22)$$

Now, applying the inverse Fourier transform to (22) we obtain the *fractional Kolmogorov–Feller equation*

$$\frac{\partial^\beta\,f(x; t)}{\partial t^\beta} + f(x; t) - f(x; t) * w(x) = \frac{t^{-\beta}}{\Gamma(1 - \beta)}\,\delta(x)\,\chi(t)\,. \qquad (23)$$

An anomalous random walk with the p.d.f. satisfying the above equation will be called here *fractional wandering*.

Comparing (23) with (7), (6), we find an explicit expression for the characteristic function of the fractional wandering:

$$\tilde{f}(u;t) = E_\beta\left([\tilde{w}(u) - 1]\, t^\beta\right). \tag{24}$$

Formula (4) also gives an alternative form of this characteristic function

$$\tilde{f}(u;t) = \int_0^\infty g_\beta(\theta;t)\, e^{\theta[\tilde{w}(u)-1]}\, d\theta\,, \tag{25}$$

where $g_\beta(\theta;t)$ is given by equality (4).

It follows from (24) that the p.d.f. of the fractional wandering has the following structure:

$$f(x;t) = R(0;t)\, \delta(x) + f_c(x;t)\,, \tag{26}$$

where $f_c(x;t)$ is the continuous part of the p.d.f., while

$$R(0;t) = E_\beta(-t^\beta) \tag{27}$$

represents the probability of the event that, up to time t, the particle has not moved. For $\beta = 1$, when the fractional Kolmogorov–Feller equation reduces to the standard Kolmogorov–Feller equation (21), the probability of jumps' absence (27) decays exponentially to zero since

$$R(0;t) = e^{-t}\,.$$

However, for $0 < \beta < 1$, it decays much slower. More precisely, according to (14), we have the following asymptotic formula:

$$R(0;t) \sim \frac{1}{\Gamma(1 - \beta)\, t^\beta} \quad (t \to \infty).$$

Functions $R(0;t)$, for different values of β, are plotted in Fig. 22.10.1.

Formula (25) also gives an expressions for the probabilities that the fractional wandering particle makes m jumps by time t:

$$R(m;t) = \frac{t^{m\beta}}{m!} \int_0^\infty \theta^m\, e^{-\theta\, t^\beta}\, g_\beta(\theta)\, d\theta\,.$$

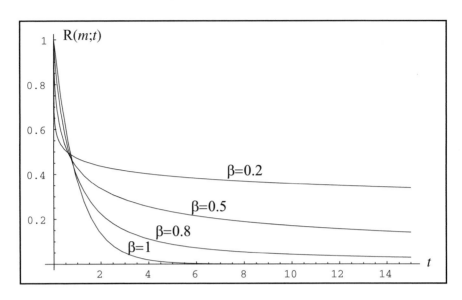

FIGURE 22.10.1.
Plots of functions $R(0; t)$, for different values of β, describing the
probabilities of the event that the fractional wandering particle has
not moved up to time t.

Using the steepest descent method it is easy to show that the functions
$R(m; t)$ satisfy the asymptotic relation

$$R(m; t) \sim \frac{1}{t^\beta} g_\beta \left(\frac{m}{t^\beta} \right), \qquad (t^\beta \to \infty).$$

22.10.3.   Subdiffusive Fickian laws for anomalous random walk.
Expressions (18) and (22) are convenient for analysis of the Fickian laws of
diffusion of anomalous random walks. Indeed, expansion of the right-hand
sides of the equalities (18) and (22) in the Taylor series in powers of u gives
the Laplace transform of the moments of the process $X(t)$. If, for instance,
the small u asymptotics of $\tilde{w}(u)$ is of the form

$$\tilde{w}(u) \sim 1 - \frac{1}{2} \sigma^2 u^2 \qquad (u \to 0),$$

then, substituting it into (18), and expanding the resulting expression into
the Taylor series in u, we obtain

$$\hat{\tilde{f}}(u; s) \sim \frac{1}{s} - \frac{1}{2} \sigma^2 u^2 \frac{\hat{f}(s)}{s[1 - \hat{f}(s)]}.$$

The first summand on the right-hand side is responsible for the normalization
condition. The second summand gives rise to the Fickian diffusion law for

the anomalous random walk. To see this, observe that the Laplace image of the mean square of the anomalous random walk is of the form,

$$\hat{\sigma}^2(s) = \int_0^\infty \langle X^2(t) \rangle e^{-st}\, dt = \sigma_\xi^2 \frac{\hat{f}(s)}{s[1 - \hat{f}(s)]}\,.$$

Substituting the asymptotics of the Laplace image of the one-sided β-stable p.d.f., $0 < \beta < 1$, we obtain the following subdiffusive Fickian law for the anomalous random walk:

$$\hat{\sigma}^2(s) \sim \frac{\sigma^2}{s^{\beta+1}} \quad (s \to 0) \qquad \Rightarrow \qquad \langle X^2(t) \rangle \sim \frac{\sigma^2}{\Gamma(1+\beta)} t^\beta \quad (t \to \infty)\,.$$

If the mean duration of the inter-jump intervals is finite ($\langle \tau \rangle < \infty$), then one obtains the Laplace image asymptotics

$$\hat{f}(s) \to 1 - \langle \tau \rangle s \quad (s \to 0) \qquad \Rightarrow \qquad \hat{\sigma}^2(s) \sim \frac{D}{s^2}\,, \qquad D = \frac{\sigma^2}{\langle \tau \rangle}\,,$$

which yields, asymptotically, the classical linear Fickian law of diffusion

$$\langle X^2(t) \rangle \sim D t, \qquad (t \to \infty).$$

22.10.4. Anomalous random walks: the stationary case. In this subsection we will briefly discuss anomalous random walks in the stationary case which corresponds, intuitively speaking, to the situation when the sequence of random jump times

$$\ldots, T(-2), T(-1), T(0), T(1), T(2), \ldots$$

extends from $-\infty$ to $+\infty$, with independent, and identically distributed inter-jump time intervals

$$\tau(n) = T(n) - T(n-1)$$

with a common p.d.f. $f(t)$.

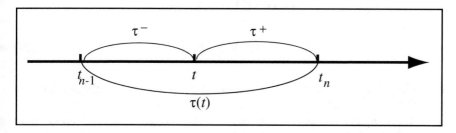

FIGURE 22.10.2.
A schematic illustration of different random times studied in the renewal theory.

For a certain random n_0, the time $t = 0$ is contained in the random interval

$$[T(n_0 - 1), T(n_0)),$$

and the p.d.f. of $T(n_0)$ is equal to the stationary p.d.f.

$$f^+(t) = \frac{d}{dt} \lim_{s \to \infty,\ s \in [T(n_s - 1), T(n_s))} \mathbf{P}(T(n_s) - s \le t) = \frac{1 - F(t)}{\langle \tau \rangle} \qquad (28)$$

of the so-called *excess life* $\tau^+ = T(n_s) - s$, where $T(n_s - 1) \le s < T(n_s)$, cf. Fig. 22.10.2. The term "excess life" is borrowed from the reliability theory.

More formally, the starting point of Subsection 1 will be here adjusted as follows: The stationary anomalous random walk $X_{st}(t), t \ge 0$, is defined as a solution of the stochastic differential equation

$$\frac{dX_{st}}{dt} = \sum_{k=1}^{\infty} \xi(k)\, \delta(t - T_k), \quad X(t = 0) = 0, \qquad (29)$$

where

$$T(1) = \tau^+, \qquad T(n) = \tau^+ + \tau(2) + \cdots + \tau(n), \quad n = 2, 3, \ldots, \qquad (30)$$

where $\tau^+, \tau(2), \tau(3), \ldots$, are independent, τ^+ has the p.d.f. $f^+(t)$ defined in (28), and $\tau(2), \tau(3), \ldots$, have the p.d.f. $f(t)$. In this case, for any $s > 0$, the excess life $T(n_s) - s$ has the same stationary p.d.f. $f^+(t)$, independent of time s. As before, the jump sizes $\xi(k)$ are independent, independent of $T(k)$'s, and have a common p.d.f. $w(x)$.

In the stationary case the discussion of Subsections 1–3 has to be adjusted to reflect assumptions (29–30). Thus, the p.d.f. and the c.d.f. of the time intervals $T(m)$ are, respectively,

$$f_{st}(t; m) = f^+(t) * \underbrace{f(t) \cdots * f(t)}_{(m-1)-\text{times}}, \quad \text{and} \quad F_{st}(t; m) = F^+(t) * \underbrace{f(t) * \cdots * f(t)}_{(m-1)-\text{times}},$$
$$(31)$$

and the solution of the equation (29) is of the form

$$X_{st}(t) = \sum_{k=1}^{N(t)} \xi(k), \qquad (32)$$

where the counting process $N(t)$ represents the random number of jumps taken by the anomalous random walk $X_{st}(t)$ up to time t.

With $P(m; t)$ denoting the probability that the interval $(0, t]$ contains m jumps of $X_{st}(t)$, the p.d.f. of $X_{st}(t)$ is

$$f_{st}(x; t) = \sum_{m=0}^{\infty} P(m; t) w(x; m), \qquad (33)$$

with $w(x; m)$ defined, as in (7), as the m-fold convolution of $w(x)$, the common p.d.f. of ξ_k. However, in the present stationary case, taking first the Fourier transform of (33) in x, and then the Laplace transform in t (denoted here by $\mathcal{L}$), gives the following result

$$\mathcal{L}\tilde{f}_{st}(u; s) = \frac{1}{s}\left[1 + \frac{\tilde{w}(u) - 1}{\langle\tau\rangle\, s}\, \frac{1 - \hat{f}(s)}{1 - \hat{f}(s)\,\tilde{w}(u)}\right]. \tag{34}$$

It differs from the analogous expressions (18) and from the equation (19) which follows from (18). Formula (34) can also can be written in the form

$$\frac{1}{\hat{f}}\mathcal{L}\tilde{f}_{st} - \tilde{w}\,\mathcal{L}\tilde{f}_{st} = \frac{1}{s}\left[1 - \hat{f}\tilde{w} + \frac{1}{\langle\tau\rangle\, s}(\tilde{w} - 1)(1 - \hat{f})\right], \tag{35}$$

analogous to (19), which is equivalent to the integral equation for the p.d.f. $f(x; t)$ of the random wandering. In particular, substituting here the Laplace image (20) of the exponential p.d.f., we again obtain the classical Kolmogorov–Feller equation.

To use formula (28), the p.d.f. $f(t)$ must have the finite mean so, in particular, that formula is not applicable to the β-stable inter-jump intervals with $\beta \leq 1$. However, it is perfectly applicable in the case of the tempered stable p.d.f. $f_{\beta,\text{tmp}}(\tau; \delta)$ discussed in Section 9. For $s \gg \delta$ that p.d.f. has the asymptotic behavior corresponding to the Laplace transform of one-sided β-stable p.d.f., while for $|s| \ll \delta$, it has the asymptotics

$$\hat{f}_{\beta,\text{tmp}}(s; \delta) \sim \exp\left(-\beta\,\delta^{\beta-1}\, s + \frac{1}{2}\,\delta^{\beta-2}\beta(1 - \beta)\, s^2\right). \tag{36}$$

Hence, in particular, in this case, both the mean and the variance of τ are finite:

$$\langle\tau\rangle = \beta\,\delta^{\beta-1}, \qquad \sigma_\tau^2 = \delta^{\beta-2}\beta(1 - \beta). \tag{37}$$

We should emphasize here that in contrast to the expression (18), formula (34) implies the exact law of the linear diffusion. Indeed, the main asymptotics of (34), for $u \to 0$, is

$$\mathcal{L}\tilde{f}_{st}(u; s) \sim \frac{1}{s} + \frac{\tilde{w}(u) - 1}{\langle\tau\rangle\, s^2} \qquad (u \to 0),$$

so that

$$g(s) = \frac{D}{s^2} \qquad \Rightarrow \qquad \langle X^2(t)\rangle \equiv D\,t.$$

Remark 1. Observe that if $\langle\tau\rangle = \infty$ (as in the case of the β-stable inter-jump time intervals) then the stationary excess time τ^+ is infinite with probability 1, since, for each $t \geq 0$,

$$F^+(t) = \mathbf{P}(\tau^+ \leq x) = \frac{1}{\langle\tau\rangle}\int_0^x (1 - F(s))\, ds = 0.$$

The same picture emerges from (34) as

$$\mathcal{L}\tilde{f}_{st}(u;s) = \frac{1}{s} \quad \Rightarrow \quad f_{st}(x;t) \equiv \delta(x).$$

In other words, if $\langle \tau \rangle = \infty$, then the stationary case is trivial with the wandering particle always remaining motionless $(X(t) \equiv 0)$.

22.11   The Langevin equation

In this section we will discuss statistical properties of stationary solutions of the Langevin equation driven by anomalous random walks and anomalous diffusion processes. Initially, suppose that process $Z(t)$ satisfies the Langevin equation driven by an anomalous random walk, that is

$$\frac{dZ(t)}{dt} + \gamma Z(t) = \sum_{n=1}^{\infty} \xi(n)\delta(t - T(n)), \tag{1}$$

where $\{\xi(n)\}$ are independent Gaussian random variables with zero mean and identical variances σ_ξ^2, while $\{T(n)\}$ are the random jump times.

In the case of the stationary anomalous random walk discussed in Subsection 22.10.4, the random quantity

$$Z_{\text{st}} = \sum_{n=1}^{\infty} \xi(n)e^{-2\gamma T(n)} \tag{2}$$

has the statistical properties equivalent to the statistical properties of the stationary solution of the stochastic equation (1). Such solutions are often called the *Ornstein–Uhlenbeck processes*. Here $T(n)$ are given by equalities (30) and, to avoid the trivial case, we assume that $\langle \tau^+ \rangle$ is finite. From (2) it is easy to see that the characteristic function of the stationary solution of the Langevin equation (1) is of the form

$$\left\langle \exp\left[-\frac{1}{2}\sigma_\xi^2 u^2 \sum_{n=1}^{\infty} e^{-2\gamma T(n)} \right] \right\rangle, \tag{3}$$

where the angled brackets denote averaging over the ensemble of independent random variables

$$\{\tau^+, \tau(2), \ldots, \tau(k) \ldots\}. \tag{4}$$

In particular, it follows from (3) that the mean square of the stationary solution is

$$\langle Z_{\text{st}}^2 \rangle = \sigma_\xi^2 \sum_{n=1}^{\infty} \langle e^{-2\gamma T(n)} \rangle. \tag{5}$$

A calculation of the means on the right-hand side, with help of the p.d.f. $f_{st,n}(t; n)$ in (31), gives

$$\langle e^{-2\gamma T(n)} \rangle = \frac{1 - \hat{f}(2\gamma)}{2\gamma \langle \tau \rangle} \, \hat{f}^n(2\gamma) \, .$$

Substituting these expressions into (5), and summing the resulting geometric progression, yields

$$\langle Z_{\text{st}}^2 \rangle = \frac{\sigma_\xi^2}{2\gamma \langle \tau \rangle} \, . \tag{6}$$

In the particular case of the Ornstein–Uhlenbeck process driven by an anomalous random walk with tempered β-stable inter-jump times, in view of (37), we have

$$\langle Z_{\text{st}}^2 \rangle = \frac{\sigma_\xi^2 \delta^{1-\beta}}{2\gamma\beta} \, .$$

In the case of nonstationary anomalous random walk discussed in Subsections 10.1-3 the situation changes qualitatively. The characteristic function of the random variable Z_{st} is then described by by the expression

$$\left\langle \exp\left[-\frac{1}{2}\sigma_\xi^2 u^2 \sum_{m=1}^{\infty} e^{-2\gamma T(m)} \right] \right\rangle , \tag{7}$$

where the random time $T(m)$ is described by the equality given at the beginning of Subsection 9.1, and is equal to the sum of m independent random variables with identical p.d.f. $f(t)$. Proceeding as in (5–6), it follows from (3) that

$$\langle Z_{\text{st}}^2 \rangle = \sigma_\xi^2 \frac{\hat{f}(2\gamma)}{1 - \hat{f}(2\gamma)} \, . \tag{8}$$

If the mean duration of the inter-jump intervals $\tau(m) = T(m) - T(m-1)$ is finite, then the following relation is true:

$$\hat{f}(s) \sim 1 - \langle \tau \rangle s \qquad (s \to 0) \, ,$$

while the asymptotic behavior of the expression (8), for $\gamma \to 0$, is described by the formula (6).

If the asymptotics of the corresponding characteristic function of the inter-jump time intervals $\tau(m)$ is of the form

$$\hat{f}(s) \sim 1 - \varepsilon s^\beta, \qquad (s \to 0, \quad 0 < \beta < 1) \, , \tag{9}$$

then, for $\gamma \to 0$, the formula (8) implies the following fractional asymptotics:

$$\langle Z_{\text{st}}^2 \rangle \sim \frac{\sigma_\xi^2}{\varepsilon (2\gamma)^\beta} \qquad (0 < \beta < 1) \, . \tag{10}$$

Finally, let us now explore in more detail the limiting case of a continuous infinitely divisible time, putting

$$\sigma_\xi^2 = D\,\varepsilon$$

and replacing the p.d.f. in (3) by its continuous time analogue with the Laplace transform $\hat{f}^\varepsilon(s)$, where $\hat{f}(s)$ is the Laplace transform of an infinitely divisible mother p.d.f. $f(t)$. Letting $\varepsilon \to 0$ in (7) we see that the limit characteristic function of the corresponding Ornstein–Uhlenbeck process is of the form

$$\left\langle \exp\left[-\frac{D}{2}\,u^2 \int_0^\infty e^{-2\gamma\,T(\theta)}\,d\theta\right]\right\rangle. \tag{11}$$

Here, the averaging is over the ensemble of the infinitely divisible time $T(\theta)$, see (1). In particular, it follows from (11) that

$$\langle Z_{st}^2\rangle = D\int_0^\infty \langle e^{-2\gamma\,T(\theta)}\rangle\,d\varrho = D\int_0^\infty \hat{f}^\theta(2\gamma)\,d\theta = \frac{D}{\ln\hat{f}(2\gamma)}. \tag{12}$$

Substituting here the Laplace transform $\hat{f}_\beta(s)$ of the β-stable p.d.f. we return to the formula (10):

$$\langle X_{st}^2\rangle = \frac{D}{(2\gamma)^\beta}. \tag{13}$$

In conclusion we shall calculate the fourth moment of the random variable Z_{st} which is given by the expression

$$\langle Z_{st}^4\rangle = 3D^2\int_0^\infty d\theta_1 \int_0^\infty d\theta_2 \langle e^{-2\gamma\,[T(\theta_1)+T(\theta_2)]}\rangle.$$

Recalling that $T(\theta)$ is a process with statistically independent increments, we can rewrite the last equality in the form

$$\langle Z_{st}^4\rangle = 6D^2\int_0^\infty d\varrho_1\,\hat{f}^{\theta_1}(2\gamma)\int_0^{\theta_1} d\theta_2 \left(\frac{\hat{f}(4\gamma)}{\hat{f}(2\gamma)}\right)^{\theta_2}.$$

Evaluating the integrals we finally obtain that

$$\langle Z_{st}^4\rangle = \frac{6D^2}{\ln\hat{f}(2\gamma)\,\ln\hat{f}(4\gamma)}.$$

In the case of a β-stable time, the above formula implies that

$$\langle Z_{st}^4\rangle = \frac{6D^2}{(2\gamma)^\beta\,.(4\gamma)^\beta}. \tag{14}$$

It follows from (13–14) that, for $\beta < 1$, the p.d.f. of the stationary solutions of the Langevin equation is not Gaussian. Indeed, the kurtosis excess

$$\kappa = \frac{\langle Z_{st}^4\rangle - 3\langle Z_{st}^2\rangle^2}{\langle Z_{st}^2\rangle^2} = 3\left[2^{1-\beta} - 1\right] > 0.$$

22.12 Exercises

The general tempered α-stable distribution on $\mathbf{R}^d$ is defined as a Lévy infinitely divisible distribution with the Lévy measure

$$M(A) = \int_{\mathbf{R}^d} \int_0^\infty I_A(tx) \frac{e^{-t}}{t^{\alpha+1}} dt R(dt),$$

where R is a unique measure (called the Rosiński measure) on $\mathbf{R}^d$ satisfying the condition,

$$\int_{\mathbf{R}^d} \|x\|^\alpha R(dx) < \infty,$$

and such that $R(\{0\}) = 0$.

1. Prove that the tempered distribution has all the moments finite.

2. Show that for the tempered α-stable Lévy process $X(t)$, with the Rosiński measure satisfying the condition, $\int_{\mathbf{R}^d} \|x\|^\alpha R(dx) < \infty$, the short time behavior is described by the asymptotics

$$h^{-1/\alpha} X(th) \longrightarrow_d Y(t)$$

where $\{Y(t) : t \geq 0\}$ is a strictly α-stable Lévy process.

3. Show that for the tempered α-stable Lévy process $X(t)$, with the Rosiński measure satisfying the condition $\int_{\mathbf{R}^d} \|x\|^2 R(dx) < \infty$, the long time behavior is described by the asymptotics

$$h^{-1/2} X(th) \longrightarrow_d B(t)$$

where $\{B(t) : t \geq 0\}$ is a d-dimensional Brownian motion (see, Rosiński [19], and Terdik and Woyczyński [20].)

Chapter 23

Nonlinear and Multiscale Anomalous Fractional Dynamics in Continuous Media

Nonlocal nonlinear evolution equations of the form, $u_t + \mathcal{L}u + \nabla \cdot f(u) = 0$, where $-\mathcal{L}$ is the generator of a Lévy semigroup on $L^1(\mathbf{R}^n)$, are encountered in continuum mechanics as model equations with anomalous diffusion. In the case when $\mathcal{L}$ is the Laplacian and the nonlinear term, $f(u)$ is quadratic, the equation boils down to the classical Burgers equation that has been studied in Volume 2, and Chapter 19 of the present volume where it has been analyzed in the context of passive tracer transport in Burgers turbulence.

We shall start this chapter with two examples covering the physical situations from the micro- to macro-worlds (see Mann and Woyczynski [1]). Then we will consider the critical case when the diffusion and nonlinear terms are balanced. The results include decay rates of solutions and their genuinely nonlinear asymptotic behavior as time t tends to infinity, determined by self-similar source solutions with the Dirac delta as the initial condition (see Biler, Karch, and Woyczynski [3].)

In the case of supercritical nonlinearities considered in the following sections, conservation laws driven by Lévy processes have solutions which have an asymptotic behavior dictated by the solutions of the linearized equations. Thus the explicit representation of the latter is of interest in the nonlinear theory, and in the last part of this chapter we concentrate on the case where the driving Lévy process is a multiscale stable (anomalous) diffusion, which corresponds to the case of multifractal conservation laws considered in Biler, Funaki and Woyczyński [2–3], and Biler, Karch and Woyczyński [4–6]. The explicit representations are developed in terms of the special functions (such

© Springer International Publishing AG, part of Springer Nature 2018

A. I. Saichev and W. A. Woyczynski, *Distributions in the Physical and Engineering Sciences, Volume 3*, Applied and Numerical Harmonic Analysis, https://doi.org/10.1007/978-3-319-92586-8_23

as Meijer G functions) and are amenable to direct numerical evaluations of relevant probabilities (see Górska nad Woyczyński [7]).

23.1   Anomalous nonlinear dynamics, from micro- to macro-worlds

We begin by presenting an analytic model (see Mann and Woyczynski [1]) for growing interfaces created by deposition of microparticles in the presence of Brownian diffusion and hopping transport. The model is based on a continuum formulation of mass conservation at the interface, including reactions. The Burgers-KPZ equation for the rate of elevation change emerges after a number of approximations are invoked. We add to the model the possibility that surface transport may be by a hopping mechanism of a Lévy flight, which leads to the (multi)fractal Burgers-KPZ model. The issue how to incorporate experimental data on the jump length distribution in our model is discussed, and controlled algorithms for numerical solutions of such fractal Burgers-KPZ equations are provided.

23.1.1. Modeling chemical vapor deposition.   Consider the problem of modeling the chemical vapor deposition (CVD) of thin diamond films where the presence of substitutional impurity atoms influences surface self-diffusion. Long jumps corresponding to Lévy flights are then possible between trap sites; this in contrast to the nearest neighbor hopping which leads to the Brownian motion in the continuum approximation. So the goal here is to extend the classical KPZ model, originally derived from a thermodynamic-kinetics viewpoint, to the situation where nonlocal long-range anomalous diffusion is permitted.

Recall that the KPZ evolution equation is of the form

$$h_t = \nu \Delta h + (\lambda/2)(\nabla h)^2 (+\eta),$$

see, e.g., Kardar, Parisi, and Zhang [8][1], where $h = h(x,t)$ is the interface elevation function, ν is identified by KPZ as a "surface tension" or "hight diffusion coefficient," Δ and ∇ stand, respectively, for the usual Laplacian and gradient differential operators in spacial variables, and λ scales the intensity of the ballistic rain of molecular fragments onto the surface (the noise η reflects the random fluctuations in the ballistic rain). The simple nonlinear governing KPZ equation, complemented by (random) initial data reflecting the roughness of the original substrate, gives rise to rich mathematics because it is easily translated into the Burgers turbulence model (for an exposition of the area, see, e.g., Woyczynski [10]).

To take account of trapping effects that lead to anomalous surface diffusion on the growing interface the above equation has to be generalized by

---

[1]See, also, Fritz and Hairer [9], based chiefly on the 2014 Fields-Medal-winning work of Martin Hairer.

including the term corresponding to the infinitesimal general of the Lévy flights. The resulting fractional KPZ equation

$$h_t = \nu(\Delta_\alpha h) + \frac{\lambda}{2}(\nabla h)^2 + \eta(x, t)$$

where $\Delta_\alpha \equiv -(-\Delta)^{\alpha/2}$ is the *fractional (power of the) Laplacian* defined, for any $0 < \alpha \le 2$, via the Fourier transform $\mathcal{F}$:

$$\mathcal{F}(\Delta_\alpha f)(\omega) \equiv (|\omega|^2)^{\alpha/2}(\mathcal{F}f)(\omega), \qquad \omega \in \mathbf{R}^d.$$

The equation includes the ballistic deposition term $\frac{\lambda}{2}(\nabla h)^2$ and satisfies the required symmetries as suggested by Barabási and Stanley

(*i*) *Invariance under translation in time;*

(*ii*) *Translation invariance along the growth direction;*

(*iii*) *Translation invariance in the direction perpendicular to the growth direction;*

(*iv*) *Rotation and inversion symmetry about the growth direction.*

A rigorous study of the evolution of the rough random surface leads to a fairly complex picture even in the case of the classical Burgers-KPZ model. In particular, its structure depends strongly on the spectral properties of the initial random substrate $h(x, 0)$ and of the additive noise η. For certain initial conditions, the density of paraboloid "bumps" can decay as $t^{-d\gamma/(\gamma-d)}$, for a parameter $\gamma > d$ which characterizes initial data, while for others it can decay as $t^{-d}(\log t)^{d(\alpha-1)/\alpha}$, for an $\alpha > 1$ (for details, see Molchanov, Surgailis, Woyczyński [12], Woyczyński [10]).

23.1.2. Experimental evidence of anomalous diffusion at the molecular level.

Long jumps in surface diffusion at the molecular level have been observed in a number of experimental situations, and in this subsection we would like to mention two such cases.

The first, described in Senft and Ehrlich [13], is the jump processes in the one-dimensional diffusion of palladium (Pd) on the tungsten W(211) lattice. Here, due probably to the low diffusion barrier of palladium on W(211), the jumps by two and three lattice sites participate significantly in the diffusion process. The distribution of displacements obtained by observing individual adatoms via the field ion microscopy is shown in Fig. 23.1.1, borrowed from Senft and Ehrlich [13]. Comparison with the distribution for diffusion by nearest neighbor jumps only does not produce a good fit. However the model which permits significant contributions from double jumps (at rate β) and triple jumps (at rate γ) fits well, and in the above experiment the best fit of the data produced the ratios

$$\frac{\beta}{\alpha} = 0.21 \qquad \frac{\gamma}{\alpha} = 0.14$$

where α was the rate of the nearest neighbor jumps, which means that at least 10 % of the jumps were of size equal to three lattice spacings and at least 16 % of the jumps were of size equal to two lattice spacings.

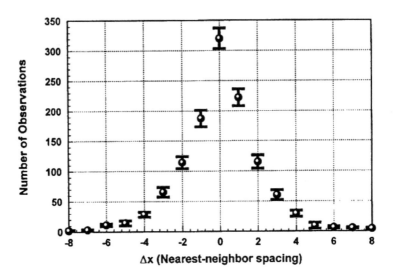

FIGURE 23.1.1.
Distribution of displacements for a single Pd atom diffusing on a one-dimensional W(211) lattice (from Senft and Ehrlich [13]). Superimposed is our least squares fit of a α-stable density.

Now we will return to the experimental data on jump statistics obtained by Senft and Ehrlich (1995) and will find the best-fitting Lévy α-stable distribution for them. Its Fourier transform $\Phi(s, \alpha, u) = e^{-|su|^{\alpha}}$ has two parameters: scale parameter s and the fractality parameter α and the two-parameter least squares fitting of the data from Fig. 23.1.1 yielded parameter values $\alpha = 1.251$ and $s = 7.4$.(see Fig. 23.1.2)

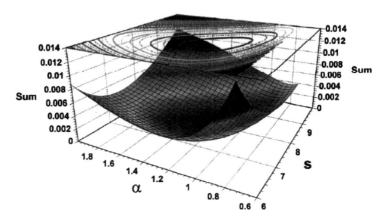

FIGURE 23.1.2.
The two-parameter dependence of the sum-of-squares residuals $\sum(p(s, \alpha, n) - p(n))^2$ versus α and s, based on the data (from Senft and Ehrlich [13] of Pd atoms diffusing on W(221).

Actually the minimum of least squares residuals in the above fit is relatively sharp; clearly the best fit is quite far from being Gaussian.

The second experiment, described by Linderoth et al. [14], measured the one-dimensional self-diffusion of Pt adatoms in the missing row troughs of the Pt(110) surface via atomically resolved time-lapsed scanning tunneling microscopy images. This example is significant because for metal on metal self-diffusion long jumps are expected to be less likely as the coupling of the adatoms to the lattice is strong due to the equal masses of adsorbate and substrate atoms which causes the energy transfer in collisions to be at a maximum. In this experiments, the hopping rates α and β for single and double jumps respectively were obtained as functions of temperature and both exhibited an Arrhenius dependence. At the temperature T=375 K, the ratio β/α was found to be 9.5 %. All these data were obtained at the very low coverages so that correlations between adatoms could be neglected.

The fact that the above experimental data were obtained by two totally different techniques, the field ion microscopy and the scanning tunneling microscopy, only reinforces the strength of the evidence for the existence of long jumps and their contribution to the surface diffusivity.

On the theoretical side, molecular dynamics simulations (Tully, Gilmer and Shugard [15], Mruzik and Pound [16]) have demonstrated that, in self-diffusion on Lennard-Jones crystals, jumps by more than one lattice site become increasingly important as the temperature rises. Other, more realistic potentials have been considered in, e.g., De Loranzi and Jacucci [17] and Sanders and DePristo [18], with a similar result.

In a more recent paper, Beenakker and Krylov [19] derived a kinetic theory-based formula for the jump length distribution in molecule-on-substrate diffusion. Their basic assumption is that the particle moves in a periodic potential with wells that the probability of trapping a particle passing over the well is α and that "once the molecule has escaped from the well the probability p_k to show a jump of the 'length' k is given by the product of the probabilities to pass over the $(k-1)$ wells without trapping and the probability of being trapped by the kth well," that is $p_k = (1-\alpha)^{k-1}\alpha$. Then they calculate the dependence of α on other parameters of the system such as characteristic energies, temperature. The above assumption, perhaps physically justified in some cases, automatically gives an exponential decay of probabilities of jump length and excludes self-similarity property which is of essence in our later considerations.

Remark 1. Multifractal KPZ equation. Self-similar α-stable Lévy distributions and the associated fractional Laplacians Δ_α are useful in what follows because of their attractive scaling properties. However, their fit to experimental data, although better in some cases than the Gaussian fit, is not perfect. A more ambitious program (see material in the following sections) would be to utilize the infinitesimal operator of a general Lévy process with its Lévy measure fitted "perfectly" to the distribution of jumps sizes from the experimental data such as those considered above. A partial step in this direction could involve treating the experimental data as multifractal. In this case the fractional Laplacian Δ_α of the fractal KPZ equation would have to be replaced by the multifractal Laplacian

$$\mathcal{L} = \nu_1 \Delta_{\alpha_1} + \cdots + \nu_K \Delta_{\alpha_K}$$

involving several scaling indices $\alpha_1, \ldots, \alpha_K$ (see also, Biler, Karch, and Woyczynski [20], Gunaratnam and Woyczynski [21], where the asymptotic behavior of multifractal conservation laws was considered).

In the next two subsections we briefly mention anomalous dynamics appearing in the biological and astrophysical context.

23.1.3. Detection of anomalous diffusion in the biological cell membranes. There is a great variety of diffusive phenomena in biological systems that are not fitting the classical Brownian motion model and the Fickian law of linear dependence of the mean square displacement on time. A good review of some of those situations can be found in Ritchie et al. [22] which describes efforts to detect a non-Brownian diffusion in the cell membrane via single molecule tracking.

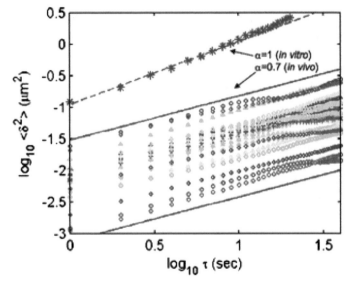

FIGURE 23.1.3.
Comparison of diffusive behavior of RNA molecules in the cell *in vitro*, where it is close to the classical Brownian motion with $\alpha \approx 1.04$, and *in vivo* where the anomalous character of the diffusion is obvious with $MSD = \tau^\alpha$ with $\alpha \approx 0.7$ (from Golding and Cox [26]).

The fact that membrane molecules perform anomalous diffusion has been known for some time, see, e.g., Kusumi et al. [23], Smith et al. [24], and Fujiwara et al. [25]. The technique that permitted direct assessment of the

type of diffusion in the cell membranes relies now on single molecule tracking through single fluorescent-molecule video imaging. The experimental
evidence suggests that the cytoplasmic portion of transmembrane proteins
collides with membrane skeleton causing temporary trapping of the diffusing protein by the membrane skeleton meshwork, and the result is what the
biologists call the *hop diffusion*.

Fig. 23.1.3, borrowed from Golding and Cox [26], shows the dependence on time τ of the mean square displacements (MSD) of different RNA
molecules in the cell. *In vitro* (top line) diffusion was close to the standard
Brownian diffusion with $MSD = \tau^\alpha$ with $\alpha \approx 1.04$, while *in vivo* (bottom
lines) the subdiffusive nature of the movements was clear with $\alpha \approx 0.7$.

**23.1.4. Fractional cosmogony in the large-scale structure
of mass distribution of the Universe.**   It is a well-known observational fact,
confirmed by the Sloan Digital Sky Survey (SDSS), that matter in the Universe
is distributed in cellular "pancake" structures, clusters and sheets of galaxies,
with giant voids between them. Since 1982, starting with the work of Zeldovich,
Einasto, and Shandarin [27] a major effort was undertaken to provide a mathematical model of the Universe evolution that would explain the present structure
starting with an essentially uniform mass distribution of matter following the Big
Bang, with perhaps small random quantum fluctuations.

Assuming the expanding Universe with the accelerating rate of expansion
$a(t) = t^{2/3}$, and mean density $\bar\rho \sim a^{-3}$, the evolution of the density of matter, $\rho = \rho(t, \vec{x}), \vec{x} \in \mathbf{R}^3$, is usually described by the following system of three
coupled partial differential equations; see, e.g., Peebles [28], and Kofman and
Raga [29]):

$$\frac{\partial \rho}{\partial t} + 3H\rho + a^{-1}\nabla(\rho\vec{w}) = 0$$

$$\frac{D\vec{w}}{Dt} + H\vec{w} = -a^{-1}\nabla\varphi,$$

$$\Delta\varphi = 4\pi G a^2(\rho - \bar\rho),$$

where $\vec{w}$ is the local velocity, φ is the gravitational potential, D/Dt stands for
the usual Eulerian derivative, and H and G are, respectively, the Hubble and
the gravitational constants. The three equations are, of course, the *continuity
equation*, the *Euler equation*, and the *Poisson equation*.

This system is not easy to analyze rigorously, and several attempts have
been made to construct simplified models which would preserve the predictive
ability of the above system of three equations while permitting their rigorous mathematical analysis. One of such approaches has been developed in
Molchanov, Surgailis, and Woyczynski [30], and it relies on consideration of
the structure of shock fronts (corresponding to the clusters of matter) in the
inviscid nonhomogeneous Burgers equation in $\mathbf{R}^d$ in the presence of random
forcing due to a degenerate potential:

$$\frac{\partial \vec{v}}{\partial t} + (\vec{v}\nabla)\vec{v} = \frac{1}{2}\mu\Delta\vec{v} - \vec{F}, \qquad \vec{v}(0, x) = -\nabla S_0(x), \qquad (1)$$

where $\vec{v} = \vec{v}(t,x) = d\vec{x}/da$ represents the velocity in the coordinates comoving with the expanding universe, $S_0(x)$ is the initial velocity potential, and the "viscosity" μ is supposed to mimic the gravitational "adhesion" which is supposed to be small enough so that the "viscosity" effects do not affect the motion of matter outside clusters. This *adhesion model* of the large-scale structure of the universe demonstrates self-organization at large times and has the ability to reproduce the formation of cellular structures in mass distribution.

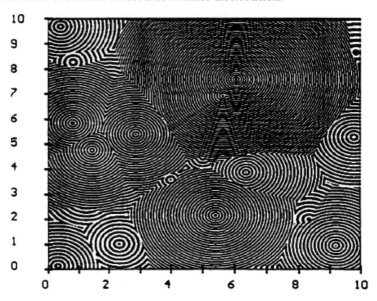

FIGURE 23.1.4.
Shock fronts for the velocity field $\vec{v}(t,x)$ in the zero viscosity limit form a quasi-Voronoi tessellation of the space. The boundaries between the white and black areas are level curves of the action potential (6) (from Molachanov, Surgailis, and Woyczyński [30]).

The main results from the above-mentioned work are as follows: Under some regularity conditions on Φ, in the zero viscosity limit, $\mu \to 0$, the velocity potential $S(t,x)$ of the solution $\vec{v}(t,s) = -\nabla S(t,x)$ is given by the formula,

$$S(t,x) = \sup_{\gamma \in \Gamma_{x,t}} S(t,x;\gamma), \qquad (2)$$

where the action functional

$$S(t,x;\gamma) = \int_0^t \left(\Phi(\gamma(s)) - \frac{1}{2}|\dot{\gamma}(s)|^2 \right) ds + S_0(\gamma(t)), \qquad (3)$$

is the difference of the potential and kinetic energy, and the supremum is taken over the class $\Gamma_{x,t}$ of all paths, $\gamma : [0,t] \to \mathbf{R}^d, \gamma(0) = x$, which are absolutely continuous and satisfy the condition $\int_0^t |\dot{\gamma}(s)|^2 ds < \infty$.

As a consequence, the zero viscosity limit solution of (1),

$$\vec{v}(t, x) = \lim_{\mu \to 0} \vec{v}(t, x; \mu) \tag{4}$$

is given by the formula

$$\vec{v}(t, x) = -\int_0^t \nabla\Phi(\gamma^*(s))\, ds - \nabla S(\gamma^*(t)) = -\gamma^*(0), \tag{4a}$$

where γ^* is the unique solution of the variational problem (2–3).

In the particular case of the degenerate discrete point potential,

$$\Phi(x) = \sum_{j \in I} h_j \delta(x - x_j), \tag{5}$$

the maximum action functional

$$S(t, x) = \max\{\sup_j (th_j - \sqrt{2h_j}|x - x_j|), 0\} \tag{6}$$

is the upper envelope (that is, supremum) of the cones

$$c_j(t, x) = \max\{(th_j - \sqrt{2h_j}|x - x_j|, 0\} \tag{7}$$

of height th_j and centered at x_j.

Thus the shock fronts form a quasi-Voronoi tessellation of the space $\mathbf{R}^d$ and the situation is pictured in Fig. 23.1.4 in the two-dimensional case.

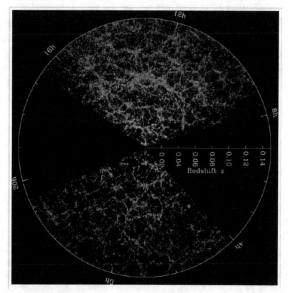

FIGURE 23.1.5.
A plot of sky coordinates vs. distance for galaxies in the Sloan Digital Sky Survey.

Even though the paper [30] also studies the case of a random singular potential $\Phi(x)$ with the Poissonian distribution, the above model is a great simplification of the real situation as pictured in Fig. 23.1.5 which is based on the results of the Sloan Digital Sky Survey, and there were numerous efforts to demonstrate that the aggregation of galaxies display obvious fractal properties (see, e.g., Slobodrian [31], Joyce, Gabrielli and Labini [32], Baryshev and Teerikorpi [33], and the references therein).

Thus the remainder of this chapter will develop tools to analyze the general fractional conservation laws (including as special cases the Burgers, Hamilton-Jacobi, and KPZ equations) in which the usual Laplacian is replaced by the fractional Laplacian introduced earlier in Chapter 22. These tools can then be applied to the physical problems described above[2].

23.2 Nonlinear and nonlocal evolution equations in which anomalous diffusion and nonlinearity balance each other

23.2.1. Classical Burgers equation vs. fractal Burgers equation. In this section[3] we shall describe the behavior of the solutions of the *Lévy conservation laws*,

$$u_t + \mathcal{L}u + \nabla \cdot \boldsymbol{f}(u) = 0, \tag{1}$$

where $\boldsymbol{x} \in \mathbf{R}^n$, $t \geq 0$, $u : \mathbf{R}^n \times \mathbf{R}^+ \to \mathbf{R}$, $\boldsymbol{f} : \mathbf{R} \to \mathbf{R}^n$ is a nonlinear term, and $-\mathcal{L}$ is the generator of a symmetric, positivity-preserving, Lévy operator semigroup $e^{-t\mathcal{L}}$, $t > 0$, on $L^1(\mathbf{R}^n)$, in the critical case when the diffusive part $\mathcal{L}$, and the nonlinear part $\nabla \cdot \boldsymbol{f}(u)$ balance each other. The asymptotic behavior of solutions of the Cauchy problem in the noncritical cases, when the diffusive term dominates, will be studied in Section 23.3.

The initial condition

$$u_0(\boldsymbol{x}) = u(\boldsymbol{x}, 0), \tag{2}$$

which supplements (1), is assumed to be an $L^1(\mathbf{R}^n)$ function. If $u_0(x) \geq 0$, with $\int_{\mathbf{R}^n} u_0(x)\,dx < \infty$, then (1) can model evolution of the *densities* u, satisfying the conditions, $u(\boldsymbol{x}, t) \geq 0$, and $\int_{\mathbf{R}^n} u(\boldsymbol{x}, t)\,d\boldsymbol{x} = \int_{\mathbf{R}^n} u_0(\boldsymbol{x})\,d\boldsymbol{x} < \infty$, for all $t > 0$.

---

[2]Work in progress. A description of a great variety of physical phenomena that require Lévy flight tools for their analysis can be found in the Proceedings volume of the International Workshop "Lévy Flights and Related Topics in Physics" (see Shlesinger, Zaslavsky and Frisch [34]) held in Nice, France, in 1994. More recently, another conference on the same topic was held at the Wroclaw University of Technology, Wroclaw, Poland, in 2016.

[3]Based on Biler, Karch, and Woyczyński [4].

The operator $\mathcal{L}$ is a pseudodifferential operator (nonlocal, in general) defined in the Fourier domain by the symbol (Fourier multiplier) $a = a(\xi) \geq 0$, via the formula

$$\widehat{\mathcal{L}v}(\xi) = a(\xi)\hat{v}(\xi). \tag{3}$$

The function $e^{-ta(\xi)}$ is positive-definite, so the symbol $a(\xi)$ can be represented, by the Lévy–Khintchine formula in the Fourier variables (see, e.g., Bertoin [35]),

$$a(\xi) = ib\xi + q(\xi) + \int_{\mathbf{R}^n} \left(1 - e^{-i\eta\xi} - i\eta\xi \, \mathbb{1}_{\{|\eta|<1\}}(\eta)\right) \Pi(d\eta). \tag{4}$$

We assume (with no loss of generality) that $b = 0$, i.e., there is no drift; indeed, a shift of the x variable removes the drift term b. The function $q(\xi) = \sum_{j,k=1}^n q_{jk}\xi_j\xi_k$ in (4) is a positive-definite (in the wide sense) quadratic form on $\mathbf{R}^n$. The case when $q(\xi) = |\xi|^2$ corresponds to the usual Laplacian $-\Delta$ on $\mathbf{R}^n$ as the Gaussian part of $\mathcal{L}$. Finally, Π is a Borel measure such that $\Pi(\{0\}) = 0$ and $\int_{\mathbf{R}^n} \min(1, |\eta|^2)\,\Pi(d\eta) < \infty$.

The idea of balanced diffusion and nonlinearity in conservation laws goes back to the classical Burgers equation

$$u_t - u_{xx} + (u^2)_x = 0, \tag{5}$$

with $x \in \mathbf{R}$ and $t > 0$, which, of course, is the special case of (1). In this case, with the initial condition in $L^1(\mathbf{R})$, the solutions become asymptotically self-similar as $t \to \infty$. Indeed, for each $1 \leq p \leq \infty$

$$t^{(1-1/p)/2}\|u(t) - U(t)\|_p \to 0, \qquad \text{as} \qquad t \to \infty, \tag{6}$$

where the function $U = U_M(x,t)$ has an explicit form

$$U(x,t) = \frac{1}{\sqrt{t}} \exp(-x^2/(4t)) \left(K - \int_0^{x/\sqrt{t}} \exp(-z^2/4)\,dz\right)^{-1}.$$

The function U is the, so-called, *source solution* of (5) such that

$$\int_{\mathbf{R}} U(x,1)\,dx = \int_{\mathbf{R}} u_0(x)\,dx \equiv M$$

with the constant K depending on the total mass M, and it satisfies the singular initial condition

$$\lim_{t\to 0} U(t,x) = M\delta(x), \tag{7}$$

in the sense that

$$\lim_{t\to 0} \int_{\mathbf{R}} U(x,t)\varphi(x)\,dx = M\varphi(0)$$

for each bounded $\varphi \in C(\mathbf{R})$.

This particular solution of (5) is self-similar because it satisfies the identity

$$U(x,t) = \frac{1}{\sqrt{t}}U\left(\frac{x}{\sqrt{t}},1\right).$$

for all t and x. In other words, U is invariant under the parabolic space-time scaling of functions $u \mapsto u_\lambda$ defined, for $\lambda > 0$, by

$$u_\lambda = \lambda u(\lambda x, \lambda^2 t),$$

that is, $U \equiv U_\lambda$ for each $\lambda > 0$. Note that this scaling preserves the integrals: $\int_{\mathbf{R}} u_\lambda \, dx = \int_{\mathbf{R}} u \, dx = M$. Moreover, the convergence property (6) can be restated as

$$\|u_\lambda(t) - U(t)\|_p \to 0 \text{ as} \lambda \to \infty$$

for each fixed $t > 0$. All these properties can be established using the Hopf–Cole substitution $u = -(\log v)_x$ which reduces (5) to the linear heat equation (see Volume 2).

Equations (1–2) also generalize the *fractal Burgers* equation

$$u_t + (-\Delta)^{\alpha/2}u + b \cdot \nabla(u|u|^{r-1}) = 0 \tag{8}$$

with $r > 1$, $b \in \mathbf{R}^n$, studied in Biler, Funaki, and Woyczyński [2], as well as the one-dimensional *multifractal conservation laws*

$$u_t - u_{xx} + \sum_{j=1}^{k} a_j(-\partial^2/\partial x^2)^{\alpha_j/2}u + f(u)_x = 0, \tag{9}$$

with $0 < \alpha_j < 2$, $a_j > 0$, and a polynomial nonlinearity f, considered Biler, Karch, and Woyczyński [20]. Here, the fractional power of order $\alpha/2$, $0 < \alpha < 2$, of the Laplacian in $\mathbf{R}^n$ (or of the second derivative $-\partial^2/\partial x^2$ in $\mathbf{R}$) is the pseudodifferential operator with the symbol $|\xi|^\alpha$.

The balanced case for the equation (1) corresponds to the case when $1 < \alpha < 2$ and

$$r = 1 + (\alpha - 1)/n. \tag{10}$$

for the fractal Burgers equation (8) when the linear and nonlinear terms are of the same importance over the entire time scale $t > 0$. Indeed, if the relation (10) holds, then equation (8) written for the rescaled solution

$$u_\lambda(\boldsymbol{x},t) = \lambda^n u(\lambda\boldsymbol{x}, \lambda^\alpha t)$$

is again the same fractal Burgers equation (10).

In this context, *one can think about the equation (8) with critical nonlinearity as a true fractal analog of the classical Burgers equation (5).*

23.2.2. Self-similar solutions of the fractal Burgers equation with critical nonlinearity.

Our first goal is to prove the existence and uniqueness of the source solution to the fractal Burgers equation with a critical nonlinearity and $1 < \alpha < 2$.

THEOREM 1. *The Cauchy problem*

$$u_t + (-\Delta)^{\alpha/2} u + b \cdot \nabla(u|u|^{(\alpha-1)/n}) = 0, \qquad u(\boldsymbol{x}, 0) = M\delta(\boldsymbol{x}), \quad M > 0, \, (11)$$

for the n-dimensional fractal Burgers equation with $1 < \alpha < 2$, and critical power nonlinearity with the exponent $r = 1 + (\alpha - 1)/n$, has a unique solution U which is positive and has the self-similar form

$$U(\boldsymbol{x}, t) = t^{-n/\alpha} U(\boldsymbol{x}t^{-1/\alpha}, 1).$$

The proof's idea relies on an approximation of equation (11) by its parabolic regularization

$$u_t + \mathcal{L}u + b \cdot \nabla(u|u|^{(\alpha-1)/n}) = 0,$$

where $\mathcal{L} = -\varepsilon\Delta + (-\Delta)^{\alpha/2}$ for small $\varepsilon > 0$. At the same time the singular initial data are also approximated by smooth positive functions u_0^ε with compact supports shrinking to $\{0\}$, $u_0^\varepsilon \in L^1(\mathbf{R}^n) \cap L^\infty(\mathbf{R}^n)$, and $\int_{\mathbf{R}^n} u_0^\varepsilon(x)\varphi(\boldsymbol{x})\,d\boldsymbol{x} \to M\varphi(0)$ as $\varepsilon \to 0$. For the full proof, depending on a sequence of ten technical lemmas, see Biler, Karch, and Woyczynski [4], where the theorem is also verified for solutions of arbitrary sign.

23.2.3. Self-similar asymptotics of solutions of general Lévy conservation laws with critical nonlinearity.

It turns out that the *unique* self-similar solution in Theorem 1 determines the long time behavior of solutions to a large class of Cauchy problems

$$u_t + \mathcal{L}u + b \cdot \nabla(u|u|^{(\alpha-1)/n}) = 0, \qquad u(\boldsymbol{x}, 0) = u_0(\boldsymbol{x}), \tag{12}$$

for which the symbol $a(\xi)$ of the Lévy infinitesimal generator $\mathcal{L}$ satisfies the following condition:

$$a(\xi) = \ell|\xi|^\alpha + k(\xi), \tag{13}$$

where $\ell > 0$, $1 < \alpha < 2$, and $k(\xi)$ is a symbol of another Lévy operator $\mathcal{K}$ such that

$$\lim_{\xi \to 0} \frac{k(\xi)}{|\xi|^\alpha} = 0. \tag{14}$$

Without loss of generality (changing the scale of the spatial variable x) we can assume that $\ell = 1$.

Remark 1. It is well known that $a(\xi)$, as a symbol of an operator generating a Lévy semigroup, satisfies the bound $0 \le a(\xi) \le C_a(1+|\xi|^2)$, for all $\xi \in \mathbf{R}^n$, and a constant C_a. This fact, combined with the assumptions (13) and (14), gives the inequality

$$0 \le a(\xi) \le C(|\xi|^\alpha + |\xi|^2), \tag{15}$$

for all $\xi \in \mathbf{R}^n$, and another constant C. Similarly,

$$0 \le k(\xi) \le \varepsilon|\xi|^\alpha + C(\varepsilon)|\xi|^2 \tag{16}$$

holds for each $\varepsilon > 0$ and a constant $C(\varepsilon)$.

Example 1. Multifractal diffusions.   Note that the assumptions (13) and (14) are fulfilled by *multifractal diffusion operators*

$$\mathcal{L} = -a_0\Delta + \sum_{j=1}^{k} a_j(-\Delta)^{\alpha_j/2}$$

with $a_0 \ge 0$, $a_j > 0$, $1 < \alpha_j < 2$, and $\alpha = \min_{1 \le j \le k} \alpha_j$.

Our second main result of this section provides asymptotics of general solutions of the Lévy conservation law with balanced nonlinearity.

THEOREM 2. *Let u be a solution of the Cauchy problem (12) for the n-dimensional Lévy conservation law with the Lévy diffusion operator $\mathcal{L}$ satisfying assumptions (13–14), and the initial data $u_0 \in L^1(\mathbf{R}^n)$ such that $\int_{\mathbf{R}^n} u_0(x)\,dx = M$. Then, for each $p \in [1,\infty]$,*

$$t^{n(1-1/p)/\alpha}\|u(t) - U(t)\|_p \to 0 \quad as \quad t \to \infty, \tag{17}$$

where $U = U_M$ is the unique self-similar solution of the problem (11) with the initial data $M\delta_0$ constructed in Theorem 1.

The complete proof, again, can be found in Biler, Karch, and Woyczyński [4], but we provide a sketch below.

The crucial observation is that the investigation of the asymptotic behavior of a solution u can be reduced to studying the convergence of the family $\{u_\lambda\}_{\lambda>0}$ as $\lambda \to \infty$. Indeed, if we note that

$$u_\lambda(x, 1) - U(x, 1) = \lambda\left(u(\lambda x, \lambda^\alpha) - U(\lambda x, \lambda^\alpha)\right),$$

then choosing $\lambda^\alpha = t$ we have

$$\begin{aligned}
\|u_\lambda(1) - U(1)\|_p &= \lambda^{n(1-1/p)}\|u(\lambda^\alpha) - U(\lambda^\alpha)\|_p \\
&= t^{n(1-1/p)/\alpha}\|u(t) - U(t)\|_p.
\end{aligned}$$

Thus, the convergence in the L^p-norm of $u_\lambda(x, 1)$ to $U(\boldsymbol{x}, 1)$ as $\lambda \to \infty$ is equivalent to (17). Of course, the same is true, if we replace $t = 1$ by any fixed $t_0 > 0$.

From this point on, the proof is based on several observations.

(i) First, using estimates of the family $\{u_\lambda\}_{\lambda \geq 1}$ uniform with respect to λ, and compactness arguments, we find $\bar{u}$ which satisfies the equation (11) in the sense of distributions.

(ii) Next, we verify that $\bar{u}(\boldsymbol{x}, 0) = M\delta_0$. This implies, by the uniqueness of $U = U_M$ proved in Theorem 1, that $\bar{u}(\boldsymbol{x}, t) = U(\boldsymbol{x}, t)$ and $u_\lambda \to U$ as $\lambda \to \infty$.

(iii) Finally, we demonstrate that the convergence of $u_\lambda(t)$ toward $U(t)$ takes place not only in the local or weak sense but actually in $L^p(\mathbf{R}^n)$ for each $1 \leq p < \infty$.

The proof of (17) for $p = \infty$ requires another argument involving the integral equation

$$u(t) = e^{-t\mathcal{L}}u_0 - \int_0^t \nabla e^{-(t-\tau)\mathcal{L}} \cdot f(u(\tau))d\tau \tag{18}$$

which is an version of the Duhamel formula familiar from the classical theory of partial differential equations. Solution of (18) is called *mild solutions* of (1). Recall that $\int_{\mathbf{R}^n} u(x, t)\,dx = M$ and $U = U_M$ is the corresponding source solution of (11). A calculation involving (18) yields the following identity

$$u(t+1) - U(t+1) = e^{-t\mathcal{L}}(u(1) - U(1)) \tag{19}$$

$$- \int_0^t b \cdot \nabla e^{-(t-\tau)\mathcal{L}} \left(u|u|^{(\alpha-1)/n} - U|U|^{(\alpha-1)/n} \right) (\tau + 1)\, d\tau.$$

Since $\int_{\mathbf{R}^n} (u(x, 1) - U(x, 1))\, dx = 0$, it follows that

$$t^{n/\alpha} \|e^{-t\mathcal{L}}(u(1) - U(1))\|_\infty \to 0 \text{ as } t \to \infty.$$

We split the integration range with respect to τ in the second term on the right-hand side of (19) into $[0, t/2]$ and $[t/2, t]$. For $\tau \in [0, t/2]$ the L^∞-norm of the integrand is bounded by

$$\|b \cdot \nabla(e^{-(t-\tau)\mathcal{L}})\|_\infty \|u(\tau + 1) - U(\tau + 1)\|_1$$

$$\times \left(\|u(\tau + 1)\|_\infty^{(\alpha-1)/n} + \|U(\tau + 1)\|_\infty^{(\alpha-1)/n} \right)$$

$$\leq C(t - \tau)^{-n/\alpha - 1/\alpha} \tau^{-1+1/\alpha} \|u(\tau + 1) - U(\tau + 1)\|_1.$$

Hence, by the Lebesgue Dominated Convergence Theorem we have, as $t \to \infty$,

$$t^{\alpha/n} \left\| \int_0^{t/2} b \cdot \nabla e^{-(t-\tau)\mathcal{L}} \left(u|u|^{(\alpha-1)/n} - U|U|^{(\alpha-1)/n} \right) (\tau + 1)\, d\tau \right\|_\infty$$

$$\leq C t^{n/\alpha} \int_0^{t/2} (t - \tau)^{-n/\alpha - 1/\alpha} \tau^{-1+1/\alpha} \|u(\tau + 1) - U(\tau + 1)\|_1\, d\tau$$

$$\leq C \int_0^{1/2} (1 - s)^{-n/\alpha - 1/\alpha} s^{-1+1/\alpha} \|u(ts + 1) - U(ts + 1)\|_1\, ds \to 0.$$

344   23. *Nonlinear anomalous dynamics in continuous media*

We proceed analogously with the integral on the right-hand side of (19) for $\tau \in [t/2, t]$. By the Hölder inequality, for every $p, q \in [1, \infty]$ satisfying $1/p + 1/q = 1$, we have the L^∞-norm of the integrand in (19) bounded from above by

$$\|b \cdot \nabla(e^{-(t-\tau)\mathcal{L}})\|_q \|u(\tau+1) - U(\tau+1)\|_p$$
$$\times \left(\|u(\tau+1)\|_\infty^{(\alpha-1)/n} + \|U(\tau+1)\|_\infty^{(\alpha-1)/n} \right)$$
$$\leq C(t-\tau)^{-n/(p\alpha)-1/\alpha}(\tau+1)^{-1+1/\alpha-n(1-1/p)/\alpha}$$
$$\times \left((\tau+1)^{n(1-1/p)/\alpha} \|u(\tau+1) - U(\tau+1)\|_p \right).$$

From now on, we repeat the reasoning in the case $\tau \in [0, t/2]$ provided $q \in (1, \infty)$ is chosen so that $-n/(p\alpha) - 1/\alpha > -1$, i.e., $p > n/(\alpha - 1)$. This concludes the sketch of the proof of Theorem 2.

Remark 2. The assumptions (13–14) on the operator $\mathcal{L}$ can be replaced by a weaker assumptions

$$\lim_{\xi \to 0} \frac{a(\xi)}{|\xi|^\alpha} \in (0, \infty), \qquad \text{and} \qquad \inf_{\xi \in \mathbf{R}^n} \frac{a(\xi)}{|\xi|^\alpha} > 0.$$

In this case the proof necessitates a supplementary smoothness assumption on a for $\xi \neq 0$.

Remark 3. The result in Theorem 2 remains true if the nonlinear term $f(s)$ in equation (1) just enjoys the correct critical asymptotics at 0, that is, if it satisfies the condition $\lim_{s \to 0} f(s)/(s|s|^{(\alpha-1)/n}) \in (0, \infty)$.

23.3   Nonlocal anomalous conservation laws with supercritical nonlinearity

In this section[4] we shall describe the asymptotic behavior of the solutions of the *Lévy conservation laws*,

$$u_t + \mathcal{L}u + \nabla \cdot f(u) = 0, \qquad u(\boldsymbol{x}, 0) = u_0(\boldsymbol{x}), \tag{1}$$

where, as in Section 23.2, $x \in \mathbf{R}^n$, $t \geq 0$, $u : \mathbf{R}^n \times \mathbf{R}^+ \to \mathbf{R}$, $-\mathcal{L}$ is the generator of a symmetric, positivity-preserving, Lévy operator semigroup $e^{-t\mathcal{L}}$, $t > 0$, on $L^1(\mathbf{R}^n)$, with the symbol $a(\xi)$ asymptotically behaving like $|\xi|^\alpha$ near the origin, but the nonlinearity is supercritical, that is, roughly speaking, $f(u) \sim |u|^r$, $u \to 0$, with

$$r > 1 + \frac{\alpha - 1}{n}. \tag{2}$$

---

[4]Based on Biler, Karch, and Woyczyński [5], where complete details of the proofs can be found.

Note that in the one-dimensional case this simply means that $r > \alpha$. The main result is that in the supercritical case the solutions of the nonlinear problem (1) have the asymptotics of the linearized equation

$$u_t + \mathcal{L}u = 0. \tag{3}$$

Remark 1. Observe that equation (1) can, in the general case, be formally interpreted as a "Fokker–Planck–Kolmogorov equation" for a "nonlinear" diffusion process in McKean's sense (see Funaki and Woyczyński [36], and Biler, Funaki and Woyczyński [37], for more details on this subject and related issues of the interacting particles approximation). Indeed, consider a Markov process $X(t)$, $t \geq 0$, which is a solution of the stochastic differential equation

$$
\begin{aligned}
dX(t) &= dS(t) - u^{-1}f(u(X(t),t))\,dt, \\
X(0) &\sim u_0(\boldsymbol{x})\,d\boldsymbol{x} \quad \text{in law,}
\end{aligned}
$$

where $S(t)$ is the Lévy process with generator $-\mathcal{L}$. Assuming that $X(t)$ is a unique solution of the above stochastic differential equation, we see that the measure-valued function $v(dx, t) = P(X(t) \in dx)$ satisfies the weak forward equation

$$
\begin{aligned}
\frac{d}{dt}\langle v(t), \eta \rangle &= \langle v(t), \widetilde{\mathcal{L}}_{u(t)}\eta \rangle, \quad \eta \in \mathcal{S}(\mathbf{R^n}), \\
v(0) &= u(x,0)\,dx
\end{aligned}
$$

with $\widetilde{\mathcal{L}}_u = -\mathcal{L} + u^{-1}f(u) \cdot \nabla$. On the other hand $u(dx, t) = u(x,t)\,dx$ also solves the above weak forward equation since

$$\frac{d}{dt}\langle u(t), \eta \rangle = \langle -\mathcal{L}u - \nabla \cdot f(u), \eta \rangle = \langle u, (-\mathcal{L} + u^{-1}f(u) \cdot \nabla)\eta \rangle$$

so that $v(d\boldsymbol{x}, t) = u(d\boldsymbol{x}, t)$ and, by uniqueness, u is the density of the solution of the stochastic differential equation.

THEOREM 1. *Assume that u is a solution of the Cauchy problem (1) with $u_0 \in L^1(\mathbf{R}^n) \cap L^\infty(\mathbf{R}^n)$, and that $-\mathcal{L}$ generates the semigroup $e^{-t\mathcal{L}}$ satisfying*

$$0 < \liminf_{\xi \to 0} \frac{a(\xi)}{|\xi|^\alpha} \leq \limsup_{\xi \to 0} \frac{a(\xi)}{|\xi|^\alpha} < \infty, \tag{4}$$

$$0 < \inf_\xi \frac{a(\xi)}{|\xi|^2}, \tag{5}$$

for some $0 < \alpha < 2$, and sufficiently smooth semigroup symbol $a(\xi)$. Furthermore, suppose that $f \in C^1(\mathbf{R}, \mathbf{R}^n)$ and $|f(s)| \leq c(R)|s|^r$, for some $r > \max\big((\alpha - 1)/n + 1, 1\big)$, a continuous nondecreasing function $c(\cdot)$ on $[0, \infty)$, and $|s| \leq R$. Then, for every $p \in [1, \infty]$,

$$t^{n(1-1/p)/\alpha}\|u(t) - e^{-t\mathcal{L}}u_0\|_p \to 0, \quad \text{as} \quad \to \infty, \tag{6}$$

where $e^{-t\mathcal{L}}u_0$ is the solution of the linear equation (3).

The proof (see Biler, Karch, and Woyczynski [5] for complete details) depends on a number of observations about the solution of equation (1) which will be summarized below.

(i) Assume that $f \in C^1(\mathbf{R}, \mathbf{R}^d)$ and $\mathcal{L}$ is the infinitesimal generator of a Lévy process with the symbol $a(\xi)$ satisfying the condition

$$\limsup_{|\omega| \to \infty} \frac{a(\xi) - q_0 |\xi|^2}{|\xi|^\alpha} < \infty \quad \text{for some } 0 < \alpha < 2, \text{and } f_0 > 0. \tag{7}$$

Given $u_0 \in L^1(\mathbf{R}) \cap L^\infty(\mathbf{R})$, there exists a unique solution $u \in \mathcal{C}([0, \infty); L^1(\mathbf{R}) \cap L^\infty(\mathbf{R}))$ of the problem (1). This solution is regular[5],

$$u : C((0, \infty); W^{2,2}(\mathbf{R})) \cap C^1((0, \infty); L^2(\mathbf{R})),$$

satisfies the conservation of "mass" property,

$$\int u(\boldsymbol{x}, t) \, d\boldsymbol{x} = \int u_0(x) \, dx,$$

and the contraction property in the $L_p(\mathbf{R})$ space,

$$\|u(t)\|_p \le \|u_0\|_p, \tag{8}$$

for each $p \in [1, \infty]$, and all $t > 0$.

(ii) Moreover, the *maximum and minimum principles* hold, that is,

$$\operatorname{ess \ inf} u_0(\boldsymbol{x}) \le u(\boldsymbol{x}, t) \le \operatorname{ess \ sup} u_0(\boldsymbol{x}), \quad \text{a.e.} \boldsymbol{x}, t, \tag{9}$$

and the *comparison principle* is valid, which means that if $u_0 \le v_0 \in L^1(\mathbf{R})$, then

$$u(x, t) \le v(x, t) \quad \text{a.e.} \quad x, t, \text{and} \quad \|u(t) - v(t)\|_1 \le \|u_0 - v_0\|_1. \tag{10}$$

(iii) Under the following additional conditions on the symbol of $\mathcal{L}$,

$$0 < \liminf_{\xi \to 0} \frac{a(\xi)}{|\xi|^\alpha} \le \limsup_{\xi \to 0} \frac{a(\xi)}{|\xi|^\alpha} < \infty, \qquad 0 < \inf_\xi \frac{a(\xi)}{|\xi|^2}, \tag{11}$$

for some $0 < \alpha < 2$, the more precise bound,

$$\|u(t)\|_p \le C_p \min(t^{-(1-1/p)/2}, t^{-(1-1/p)/\alpha})\|u_0\|_1$$

---

[5]Recall that the Sobolev space $W^{2,2}(\mathbf{R})$ is the space of twice differentiable functions with square integrable derivatives of order one and two.

holds for all $1 \leq p \leq \infty$. Moreover, if $u_0 \in L^1(\mathbf{R}) \cap L^\infty(\mathbf{R})$, then

$$\|u(t)\|_p \leq C(1+t)^{-(1-1/p)/\alpha} \tag{12}$$

with a constant C which depends only on $\|u_0\|_1$ and $\|u_0\|_p$.

Remark 2. The asymptotics of the solution of the linear Cauchy problem (3) is well known: There exists a nonnegative function $\eta \in L^\infty(0, \infty)$ satisfying $\lim_{t \to \infty} \eta(t) = 0$ such that

$$\left\| e^{-t\mathcal{A}} * u_0 - \int_{\mathbf{R}} u_0(\boldsymbol{x})\, d\boldsymbol{x} \cdot p_{\mathcal{A}}(t) \right\|_p \leq t^{-(1-1/p)/\alpha} \eta(t), \tag{13}$$

where $p_{\mathcal{L}}(t)$ is the kernel of the operator $\mathcal{L}$. Higher-order asymptotics is also available in the paper cited in Footnote 4.

The above general results have direct consequences for (not necessarily symmetric) *multifractal conservation laws* driven by *multiscale anomalous diffusions*[6] (Lévy processes) X_t with the characteristic functions of the form,

$$\mathbf{E} e^{i\omega X_t} = \prod_{j=1}^{n} \tilde{v}_{\alpha_j, \beta_j, \gamma_j}(\omega, t), \tag{14}$$

where, for each $j = 1, 2, \ldots, n$,

$$\tilde{v}_{\alpha_j, \beta_j, \gamma_j}(t, \omega) = \mathcal{F}[v_{\alpha_j, \beta_j, \gamma_j}(t, x); \omega] = \exp\left[-t\gamma_j |\omega|^{\alpha_j} e^{\frac{i\pi}{2} \beta_j \operatorname{sgn}(\omega)} \right] \tag{15}$$

The symbol $\operatorname{sgn}(\omega)$ denotes the sign of the parameter ω, and $\mathcal{F}$ denotes the usual characteristic function.

The multiparameter $(\vec{\alpha}; \vec{\beta}; \vec{\gamma}) = (\alpha_1, \ldots, \alpha_n; \beta_1, \ldots \beta_n; \gamma_1, \ldots, \gamma_n)$ has to satisfy the following conditions: If $0 < \alpha_j < 1$ then $|\beta_j| \leq \alpha_j$, and if $1 < \alpha_j \leq 2$ then $|\beta_j| \leq 2 - \alpha_j$; for all j, we assume that $\gamma_j > 0$. Thus the Fourier multiplier describing the infinitesimal generator of X_t is of the form,

$$a_{(\vec{\alpha}; \vec{\beta}; \vec{\gamma})}(\omega) = \sum_{j=1}^{n} -\gamma_j |\omega|^{\alpha_j} e^{\frac{i\pi}{2} \beta_j \operatorname{sgn}(\omega)}. \tag{16}$$

The generator itself will be denoted $\mathcal{L}_{(\vec{\alpha}; \vec{\beta}; \vec{\gamma})}$. For the sake of convenience, and without loss of generality, we will assume that

$$\alpha_1 < \alpha_2 < \cdots < \alpha_n.$$

---

[6]Note the parabolic regularization included in the operator $\mathcal{L}$ because of the conditions (5), and (7).

The densities $v_{\alpha_j,\beta_j}(x,t)$ appearing in (15) are unimodal (see, e.g., Bertoin [35]). The skewness parameter, β_j, measures the degree of asymmetry of $v_{\alpha_j,\beta_j}(x,t)$: For $\beta_j = 0$ they are just the previously mentioned symmetric α_j-stable densities with fractional Laplacians as the corresponding infinitesimal generators. Moreover, all of those densities are self-similar, since, for any $x \in \mathbf{R}$, and $t > 0$,

$$v_{\alpha_j,\beta_j,\gamma_j}(t,x) = \frac{1}{t^{1/\alpha_j}} v_{\alpha_j,\beta_j,\gamma_j}\left(1, \frac{x}{t^{1/\alpha_j}}\right), \tag{17}$$

and

$$v_{\alpha_j,\beta_j,\gamma_j}(-x,t) = v_{\alpha_j,-\beta_j,\gamma_j}(t,x). \tag{18}$$

Equation (18) is a consequence of the identity $\tilde{v}_{\alpha_j,\beta_j,\gamma_j}(t,-\omega) = \tilde{v}_{\alpha_j,-\beta_j,\gamma_j}(t,\omega)$ satisfied by $\tilde{v}_{\alpha_j,\beta_j,\gamma_j}(t,\omega)$ given in (15).

COROLLARY 1. *All the statements of Theorem 1 are valid for the one-dimensional conservation laws*

$$\frac{\partial u}{\partial t} + \mathcal{L}_{(\vec{\alpha};\vec{\beta},\vec{\gamma})}u + \frac{\partial}{\partial x}f(u) = 0, \qquad u(x,0) = u_0(x), \tag{19}$$

with

$$\alpha = \alpha_1 < \alpha_2 < \cdots < \alpha_n = 2.$$

In particular, if u is a solution of the Cauchy problem (19) with $u_0 \in L^1(\mathbf{R}) \cap L^\infty(\mathbf{R})$, and the nonlinearity $g \in C^1$ is supercritical, i.e., $\limsup_{s\to 0}|f(s)|/|s|^r < \infty$, for $r > \max(\alpha,1)$, then the relation

$$\lim_{t\to\infty} t^{(1-1/p)/\alpha}\|u(t) - e^{-t\mathcal{L}}u_0\|_p = 0 \tag{20}$$

holds for every $1 \leq p \leq \infty$. Moreover,

$$\left\|e^{t\mathcal{L}_{(\vec{\alpha};\vec{\beta},\vec{\gamma})}} * u_0 - \int_{\mathbf{R}} u_0(x)\,dx \cdot p_{\mathcal{A}_{(\vec{\alpha};\vec{\beta},\vec{\gamma})}}(t)\right\|_p \leq t^{-(1-1/p)/\alpha}\eta(t), \tag{21}$$

where $p_{\mathcal{L}_{(\vec{\alpha};\vec{\beta},\vec{\gamma})}}(t)$ is the kernel of the operator $\mathcal{L}_{(\vec{\alpha};\vec{\beta},\vec{\gamma})}$ in (19).

To prove Corollary 1 it suffices to show that the assumptions of Theorem 1 are satisfied. Indeed, for the multiscale Lévy process with symbol (16), we have

$$\limsup_{|\omega|\to\infty} \frac{a_{(\vec{\alpha};\vec{\beta},\vec{\gamma})}(\omega) - \gamma_n|\omega|^2}{|\omega|^{\alpha^*}} = a_{j^*} < \infty$$

with $\alpha_{j^*} = \alpha^*$ where $\alpha^* = \max(\alpha_1,\ldots,\alpha_{(n-1)})$. Also,

$$0 < \lim_{\omega\to 0} \frac{a_{(\vec{\alpha};\vec{\beta},\vec{\gamma})}(\omega)}{|\omega|^{\alpha_*}} = \alpha_{j_*} < \infty,$$

with $\alpha_{j_*} = \alpha_*$ where $\alpha_* = \min(\alpha_1, \ldots, \alpha_n) = \alpha_1$; and

$$\inf_\omega \frac{a_{(\vec{\alpha};\vec{\beta},\vec{\gamma})}(\omega)}{|\omega|^2} \geq \gamma_n > 0.$$

The above results depended on the subcritical behavior

$$\limsup_{s \to 0} \frac{|g(s)|}{|s|^r} < \infty, \qquad \text{for} \qquad r > \max(\alpha, 1),$$

of the nonlinearity in the conservation laws discussed above.

23.4 Explicit formulas for multiscale fractional anomalous dynamics

23.4.1. Preliminaries. In this section we provide[7] explicit representations of the kernels for multiscale one-dimensional Lévy processes (multifractal anomalous diffusions). In view of the results of the previous section they also give explicit asymptotic behavior of solutions of conservation laws driven by multiscale, $(\alpha_1, \ldots, \alpha_k)$-stable diffusions with supercritical nonlinearities. The starting point here is to obtain exact representation, via known special functions such as Meijer G functions, of the solutions of linear multiscale evolution equations, that is, for the evolution equations for the densities of the multiscale Lévy processes themselves.

The explicit representations using a fairly standard set of special functions produce a framework that permits a straightforward calculation of probabilities related to the multiscale diffusions using a symbolic manipulation platform such as *Mathematica*.

Let us begin by establishing the notation for the standard integral transforms (Fourier, Laplace, and Mellin), their inverses, and special functions we are going to work with in this section.

(i) The Fourier transform (and its inverse) of an integrable function $f(x)$ is defined for $x \in \mathbf{R}$, and real ω, is defined by the formula

$$\tilde{f}(\omega) = \mathcal{F}[f(x); \omega] = \int_{-\infty}^{\infty} e^{i\omega x} f(x) dx, \tag{1}$$

$$f(x) = \mathcal{F}^{-1}[\tilde{f}(\omega); x] = \frac{1}{2\pi} \int_{-\infty}^{\infty} e^{-i\omega x} \tilde{f}(\omega) d\omega. \tag{2}$$

---

[7]Based on results in Górska and Woyczyński [7].

(ii) For a function $f(x) \equiv 0$, for $x < 0$, such that $e^{-cx} f(x)$ is integrable on the positive half-line for some fixed number $c > 0$, the Laplace transform

$$f^\star(p) = \mathcal{L}[f(x); p] = \int_0^\infty e^{-px} f(x) dx, \tag{3}$$

$$f(x) = \mathcal{L}^{-1}[f^\star(p); x] = \frac{1}{2\pi i} \int_{c-i\infty}^{c+i\infty} e^{px} f^\star(p) dp, \tag{4}$$

where $p = c + i\omega$. There is an obvious relationship between the Fourier transform of $f_1(x) = e^{-cx} f(x)$ and the Laplace transform of $f(x)$.

(iii) The Mellin transform of $f(x)$ is here defined as follows:

$$\hat{f}(s) = \mathcal{M}[f(x); s] = \int_0^\infty x^{s-1} f(x) dx, \tag{5}$$

$$f(x) = \mathcal{M}^{-1}[\hat{f}(s); x] = \frac{1}{2\pi i} \int_L x^{-s} \hat{f}(s) ds, \tag{6}$$

where s is a complex variable. The contour of integration L is determined by the domain of analyticity of $\hat{f}(s)$, and, usually, it is an infinite strip parallel to the imaginary axis.

(iv) The Meijer G function is defined as the inverse Mellin transform of products and ratios of the classical Euler's gamma functions. More precisely,

$$G_{p,q}^{m,n}\left(z \left| \begin{matrix} A_1 \ldots A_p \\ B_1 \ldots B_q \end{matrix} \right.\right) = \mathcal{M}^{-1}\left[\frac{\prod_{j=1}^m \Gamma(B_j + s) \prod_{j=1}^n \Gamma(1 - A_j - s)}{\prod_{j=m+1}^q \Gamma(1 - B_j - s) \prod_{j=n+1}^p \Gamma(A_j + s)}; z\right], \tag{7}$$

where empty products in (7) are taken to be equal to 1, and the following assumptions are satisfied:

$$z \neq 0, \quad 0 \leq m \leq q, \quad 0 \leq n \leq p, A_j \in \mathbb{C}, \quad j = 1, \ldots, p; \quad B_j \in \mathbb{C}, \quad j = 1, \ldots, q. \tag{8}$$

A description of the integration contours in (7), and the general properties and special cases of the Meijer G functions can be found in, e.g., Prudnikov, Brychkov, and Marichev [38]. If the integral in (7) converges and if no confluent poles appear among $\Gamma(1 - A_j - s)$ or $\Gamma(1 - B_j - s)$, then the Meijer G function can be expressed as a finite sum of the generalized hypergeometric functions. Recall that a generalized hypergeometric function can be represented in terms of the following series,

$$_pF_q\left(\begin{matrix} a_1, \ldots, a_p \\ b_1, \ldots, b_q \end{matrix}; x\right) = \sum_{n=0}^\infty \frac{x^n \prod_{j=1}^p (a_j)_n}{n! \prod_{j=1}^q (b_j)_n}, \tag{9}$$

where the upper and lower lists of parameters are denoted by (a_p) and (b_q), respectively, and $(a)_n = \Gamma(a + n)/\Gamma(a)$ is the so-called *Pochhammer symbol*.

Finally, we introduce the special notation for a specific uniform partition of the unit interval,

$$\Delta(n, a) = \left\{ \frac{a}{n}, \frac{a+1}{n}, \ldots, \frac{a+n-1}{n} \right\}. \tag{10}$$

For later reference, we also quote the Euler's reflection formula,

$$\Gamma(z)\Gamma(1-z) = \frac{\pi}{\sin(\pi z)}, \tag{11}$$

and the Gauss–Legendre multiplication formula,

$$\Gamma(na) = (2\pi)^{\frac{1-n}{2}} n^{na-\frac{1}{2}} \prod_{j=0}^{n-1} \Gamma\left(a + \frac{j}{n}\right), \tag{12}$$

23.4.2. Explicit representation of the kernels of the two-scale, two-sided Lévy generators, $0 < \alpha \le 2$. In this subsection we will describe explicit representations for kernels $v_{(\vec{\alpha};\vec{\beta},\vec{\gamma})}$ of the infinitesimal generators $\mathcal{A}_{(\vec{\alpha};\vec{\beta},\vec{\gamma})}$ which dictate the long time behavior of the nonlinear supercritical conservation laws discussed in Section 23.3. To make our presentation more straightforward we will assume that the scaling parameter $\vec{\gamma} = (1, \ldots, 1)$ and simplify our notation: $v_{(\vec{\alpha};\vec{\beta},\vec{\gamma})} \equiv v_{(\vec{\alpha};\vec{\beta})}$.

Our goal is to find an explicit expression for the Fourier convolution of $v_{\alpha_j,\beta_j}(t, x)$, $j = 1, 2$, $x \in \mathbf{R}$, and $t > 0$:

$$H(t, x) = \int_{-\infty}^{\infty} v_{\alpha_1,\beta_1}(t, y) v_{\alpha_2,\beta_2}(t, x - y) dy, \tag{13}$$

where $H(t, x) = H(\vec{\alpha}; \vec{\beta}; t, x)$, $\vec{\alpha} = (\alpha_1, \alpha_2)$, and $\vec{\beta} = (\beta_1, \beta_2)$.

The functions $v_{\alpha_j,\beta_j}(t, x)$, $j = 1, 2$, represent the unimodal probability density functions of two-sided Lévy stable distributions, and the series representation of two-sided Lévy stable PDFs for $0 < \alpha_j < 1$, and $|\beta_j| \le \alpha_j$, can be found in, e.g., Lukacs [39], and Feller [40]. Those series expansions, different for $0 < \alpha_j < 1$, and $1 < \alpha_j \le 2$, were calculated for rational values of parameter α_j and β_j. We quote some solutions which will be used later on.

Example 1. Gaussian and Lévy–Smirnov PDFs. The basic Gaussian distribution

$$v_{2,0}(t, x) = \frac{\exp(-\frac{x^2}{4t})}{2\sqrt{\pi t}} \tag{14}$$

corresponds to $\alpha = 2$ and $\beta = 0$, and the Lévy–Smirnov distribution

$$v_{\frac{1}{2},-\frac{1}{2}}(t, x) = \frac{t \exp(-\frac{t^2}{4x})}{2\sqrt{\pi} x^{3/2}}, \qquad \text{for} \qquad x > 0 \tag{15}$$

and $= 0$, for $x \leq 0$, corresponds to $\alpha = 1/2$ and $\beta = -1/2$. Additionally, for $\alpha = 3/2$ and $\beta = -1/2$, we have[8]

$$v_{\frac{3}{2},-\frac{1}{2}}(t,x) = \frac{Re}{2\pi} \int_{-\infty}^{\infty} e^{-i\omega x} e^{-t|\omega|^{3/2} \exp(-\frac{i\pi}{4}\mathrm{sgn}(\omega))} d\omega \tag{16}$$

$$= \frac{(2/t)^{2/3}}{3\sqrt{\pi}} \frac{\Gamma(\frac{5}{6})}{\Gamma(\frac{2}{3})} {}_1F_1\left(\begin{matrix}5/6\\2/3\end{matrix}; -\frac{4x^3}{27t^2}\right) + \frac{(2/t)^{4/3}}{9\sqrt{\pi}} \frac{\Gamma(\frac{7}{6})}{\Gamma(\frac{4}{3})} x\, {}_1F_1\left(\begin{matrix}7/6\\4/3\end{matrix}; -\frac{4x^3}{27t^2}\right)$$

The symbol ${}_1F_1$ stands for the hypergeometric function introduced above.

Next we will find the explicit form of $H(t,x)$ given in (13). Applying the property (18), Section 23.3, to (13) we can rewrite H in the form,

$$H(t,x) = \mathcal{F}^{-1}[\tilde{v}_{\alpha_1,\beta_1}(t,\omega)\,\tilde{v}_{\alpha_2,\beta_2}(t,\omega); x] = H_-(t,-x)\Theta(-x) + H_+(t,x)\Theta(x), \tag{17}$$

where

$$H_+(t,x) = H_+(\alpha_1,\beta_1,\alpha_2,\beta_2; t,x) = \frac{Re}{\pi} \int_0^{\infty} e^{-ix\omega} \exp\left(-t\omega^{\alpha_1} e^{\frac{i\pi}{2}\beta_1} - t\omega^{\alpha_2} e^{\frac{i\pi}{2}\beta_2}\right) d\omega, \tag{18}$$

and

$$H_-(t,x) = H_+(\alpha_1, -\beta_1, \alpha_2, -\beta_2; t,x). \tag{19}$$

The function $\Theta(x)$ is here the usual Heaviside step function. Note, that in the case $\alpha_1 = \alpha_2$, and $\beta_1 = \beta_2$, $H(x,t) = v_{\alpha_1,\beta_1}(x,2t)$.

In what follows, without loss of generality, we will consider only the case of $H_+(x,t)$. For rational α_j and β_j, $j = 1,2$, such that $\alpha_1 = \frac{l}{k}$, $\beta_1 = \frac{l-2a}{k}$, $\alpha_2 = \frac{p}{q}$, and $\beta_2 = \frac{p-2b}{q}$, where l, k, p, q, a, and b are integers, we have

$$\hat{H}_+(s,t) = \frac{1}{M\pi} \sum_{j=0}^{M_1-1} \frac{(-1)^j}{j!} \frac{\Gamma(s)\Gamma\left(\frac{1-s}{M} + \frac{m}{M}j\right)}{t^{\frac{1-s}{M}+(\frac{m}{M}-1)j}} Re\left\{e^{-i\pi\left[\frac{1}{2}-u\frac{1-s}{M}+\left(v-u\frac{m}{M}\right)j\right]}\right.$$

$$\times {}_{1+m_1}F_{M_1}\left(\begin{matrix}1, \Delta(m_1, \frac{1-s}{M} + \frac{m}{M}j)\\ \Delta(M_1, 1+j)\end{matrix}; \left(-\frac{te^{-i\pi v}}{M_1}\right)^{M_1}\left(\frac{m_1}{te^{-i\pi u}}\right)^{m_1}\right)\right\}$$

$$= \frac{1}{M\pi} \sum_{j=0}^{M_1-1} \sum_{r=0}^{\infty} \frac{(-1)^{j+rM_1}}{(j+rM_1)!} \frac{\Gamma(s)\Gamma[\frac{1-s}{M} + \frac{m}{M}(j+rM_1)]}{t^{\frac{1-s}{M}+(\frac{m}{M}-1)(j+rM_1)}} \tag{20}$$

$$\times \sin\left[\pi u \frac{1-s}{M} - \pi(j+rM_1)\left(v - u\frac{m}{M}\right)\right]$$

$$= \frac{1}{M\pi} \sum_{r=0}^{\infty} \frac{(-1)^r}{r!} \frac{t^{-\frac{1-s}{M}-r(\frac{m}{M}-1)}\Gamma(s)\Gamma(\frac{1-s}{M} + \frac{m}{M}r)}{\Gamma\left[1 - u\frac{1-s}{M} + (v - u\frac{m}{M})r\right]\Gamma\left[u\frac{1-s}{M} - (v - u\frac{m}{M})r\right]},$$

---

[8]See Górska and Penson [41].

where m, M, m_1, and M_1, are as follows:

$$m = \min(\frac{l}{k}, \frac{p}{q}), \quad M = \max(\frac{l}{k}, \frac{p}{q}), \quad m_1 = \min(kp, lq), \quad \text{and} \quad M_1 = \max(kp, lq).$$

The parameters u and v are determined by the equalities,

$$u = \frac{a}{k}, \quad v = \frac{b}{q}, \quad \text{for } \alpha_1 > \alpha_2, \quad \text{and} \quad u = \frac{b}{q}, \quad v = \frac{a}{k} \quad \text{for } \alpha_1 < \alpha_2.$$

The next step requires inverting the Mellin transform in (20). To accomplish this task we will introduce the new variable of integration, $\tilde{s} = (1-s)/(lp)$, and use (11) and (12). Putting all of these terms together, we get, for $x > 0$,

$$H_+(x,t) = \frac{m_1\sqrt{M}}{x(2\pi)^{\frac{lp+m_1}{2}-um_1}} \sum_{r=0}^{\infty} \frac{(-t)^r}{r!} \left(\frac{m_1}{t}\right)^{\frac{m}{M}r} \tag{21}$$

$$\times G_{lp+um_1,m_1+um_1}^{m_1,lp}\left(\frac{(lp)^{lp}\, t^{m_1}}{x^{lp}\, m_1^{m_1}}\,\middle|\, \begin{array}{l} \Delta(lp,0), \Delta(um_1, -(v - u\frac{m}{M}r)) \\ \Delta(m_1, \frac{m}{M}r), \Delta(um_1, -(v - u\frac{m}{M}r)) \end{array}\right).$$

The Meijer G functions in (21) can be expressed in terms of a generalized hypergeometric function. With respect to the values of α_1 and α_2, we can consider two different cases:

A. *The case of $0 < \alpha_i < 1$, $i = 1, 2$.* Then

$$H_+(x,t) = -\frac{1}{\pi} \sum_{r=0}^{\infty} \sum_{j=0}^{m_1-1} \frac{(-t)^{r+j}}{r!\,j!} \frac{\Gamma(1+Mj+mr)}{x^{1+Mj+rm}} \sin(r\pi v + j\pi u) \tag{22}$$

$$\times \,_{1+lp}F_{m_1}\left(\begin{array}{l} 1, \Delta(lp, 1+Mj+mr) \\ \Delta(m_1, 1+j) \end{array}; (-1)^{m_1u-m_1}\frac{t^{m_1}(lp)^{lp}}{m_1^{m_1}x^{lp}}\right),$$

where u and v were defined above. Moreover, using the series expansion of the function $_{1+lp}F_{m_1}$, the function H_+ can be expressed as follows:

$$H_+(x,t) = -\frac{1}{\pi} \sum_{r,n=0}^{\infty} \frac{(-t)^{n+r}}{n!\,r!} \frac{\Gamma(1+\alpha_1 r + \alpha_2 n)}{x^{1+\alpha_1 r+\alpha_2 n}} \sin\left(\pi r \frac{\alpha_1-\beta_1}{2} + \pi n \frac{\alpha_2-\beta_2}{2}\right), \tag{23}$$

which for $t = 1$, $x > 0$, and $r = 0$ (or $n = 0$), is identical with the series expression for two-sided Lévy stable distribution given in, e.g., Lukacs [39].

B: *The case of $1 < \alpha_i \le 2$, $i = 1, 2$.* Then, in view of Eq. (8.2.2.4) on p. 520 of Prudnikov, Brychkov and Marichev [38],

$$H_+(x,t) = \frac{1}{\pi M} \sum_{r=0}^{\infty} \sum_{j=0}^{lp-1} \frac{(-1)^{r+j}}{r!\,j!} \frac{x^j}{t^{\frac{1+j}{M}+(\frac{m}{M}-1)r}} \Gamma\left(\frac{1+j}{M} + \frac{m}{M}r\right) \tag{24}$$

$$\times \sin\left[\pi u \frac{1+j}{M} - \pi r\left(v - u\frac{m}{M}\right)\right]$$

$$\times \, _{1+m_1}F_{lp}\left(\frac{1, \Delta(m_1, \frac{1+j}{M} + \frac{m}{M}r), }{\Delta(lp, 1+j)}; (-1)^{m_1 u + lp} \frac{m_1^{m_1} x^{lp}}{t^{m_1}(lp)^{lp}}\right),$$

which can be rewritten as

$$H_+(x,t) = \frac{1}{\pi M} \sum_{r,n=0}^{\infty} \frac{(-1)^{r+n}}{r!\,n!} \frac{x^n}{t^{\frac{1+n}{M} + (\frac{m}{M}-1)r}} \Gamma\left(\frac{1+n}{M} + \frac{m}{M}r\right) \tag{25}$$

$$\times \sin\left[\pi u\frac{1+n}{M} - \pi r\left(v - u\frac{m}{M}\right)\right].$$

23.4.3.   Special cases when $H(x,t)$ can be expressed via simpler special functions. In this subsection we provide three examples of the densities H with relatively simple analytical representation. Two-scale asymmetric one-sided Lévy generators are discussed in the Exercise Section.

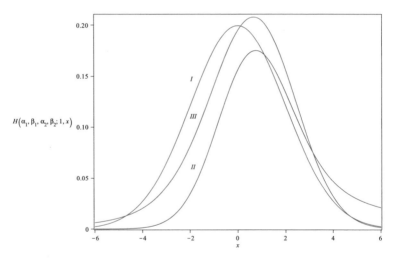

FIGURE 23.4.1.
Multiscale densities $H(\alpha_1, \beta_1, \alpha_2, \beta_2; t, x)$, for $t = 1$, and given values of α_1, β_1, α_2, and β_2. In plot I (Example 1), $\alpha_1 = \alpha_2 = 2$, and $\beta_1 = \beta_2 = 0$; in plot II (Example 2), $\alpha_1 = 2$, $\beta_1 = 0$, $\alpha_2 = \frac{1}{2}$, and $\beta_2 = -\frac{1}{2}$; in plot III (Example 3), $\alpha_1 = 2$, $\beta_1 = 0$, $\alpha_2 = \frac{3}{2}$, and $\beta_2 = -\frac{1}{2}$.

Example 1. The bi-Gaussian case. The elementary case $\alpha_1 = \alpha_2 = 2$, $\beta_1 = \beta_2 = 0$, is straightforward, and we include it here only for verification's sake:

$$H(2,0,2,0;t,x) = \frac{\exp(-\frac{x^2}{4t})}{4\pi t}\int_{-\infty}^{\infty} e^{-\frac{y^2}{2t} + \frac{xy}{2t}}\,dy = \frac{\exp(-\frac{x^2}{8t})}{2\sqrt{2\pi t}} = v_{2,0}(x,2t). \tag{26}$$

Its plot is presented in Fig. 23.4.1 as curve I.

Example 2. The Gaussian–Lévy case. Here $\alpha_1 = 2$, $\alpha_2 = 1/2$, $\beta_1 = 0$, and $\beta_2 = -1/2$, and

$$
H(2,0,\tfrac{1}{2},-\tfrac{1}{2};t,x) = \frac{1}{4\pi\sqrt{t}} e^{-\frac{x^2}{4t}} \int_0^\infty y^{-\frac{3}{2}} e^{-\frac{t^2}{4y}} e^{-\frac{y^2}{4t}+\frac{xy}{2t}} \, dy
$$

$$
= \frac{\sqrt{t}}{4\pi} e^{-\frac{x^2}{4t}} \sum_{r=0}^\infty \frac{(-t^2/4)^r}{r!} \int_0^\infty y^{-\frac{3}{2}-r} e^{-\frac{y^2}{4t}+\frac{xy}{2t}} \, dy
$$

$$
= \frac{1}{2} e^{-\frac{x^2}{8t}} \sum_{r=0}^\infty \frac{(-1)^{2r+1}}{r!\Gamma(\frac{3}{2}+r)} \frac{t^{\frac{1}{4}+\frac{3}{4}r}}{2^{\frac{5}{4}+\frac{5}{2}r}} D_{\frac{1}{2}+r}\left(-\frac{x}{\sqrt{2t}}\right)
$$

$$
= \frac{1}{2\sqrt{\pi t}} e^{-\frac{x^2}{8t}} \sum_{r=0}^\infty \frac{(-1)^{2r+1}}{(2r+1)!} \left(\frac{t^3}{2}\right)^{\frac{1}{4}(2r+1)} D_{\frac{2r+1}{2}}\left(-\frac{x}{\sqrt{2t}}\right)
$$

$$
= \frac{1}{2\sqrt{\pi t}} e^{-\frac{x^2}{8t}} \sum_{n=0}^\infty \frac{(-1)^n}{n!} \left(\frac{t^3}{2}\right)^{\frac{n}{4}} D_{\frac{n}{2}}\left(-\frac{x}{\sqrt{2t}}\right),
$$

where $D_\nu(z)$ is the parabolic cylinder function. The plot of $H(2,0,\frac{1}{2},-\frac{1}{2};t,x)$ for $t = 1$ is illustrated in Fig. 23.4.1 as curve II.

Example 3. For $\alpha_1 = 2$, $\alpha_2 = 3/2$, $\beta_1 = 0$, and $\beta_2 = -1/2$,

$$
H(2,0,\tfrac{3}{2},-\tfrac{1}{2};t,x) = \int_{-\infty}^\infty \frac{\exp(-\frac{y^2}{4t})}{2\sqrt{\pi t}} \frac{\mathrm{Re}}{2\pi} \left\{ \int_{-\infty}^\infty e^{-i\omega(x-y)} e^{-t|\omega|^{3/2} \exp[-\frac{i\pi}{4}\mathrm{sgn}(\omega)]} d\omega \right\} dy
$$

$$
= \frac{\mathrm{Re}}{2\pi} \int_{-\infty}^\infty e^{-i\omega x} e^{-t|\omega|^{3/2} \exp[-\frac{i\pi}{4}\mathrm{sgn}(\omega)]} \left[\int_{-\infty}^\infty e^{i\omega y} \frac{\exp(-\frac{y^2}{4t})}{2\sqrt{\pi t}} dy \right] d\omega
$$

$$
= \frac{\mathrm{Re}}{2\pi} \int_{-\infty}^\infty e^{-i\omega x} e^{-t|\omega|^{3/2} \exp[-\frac{i\pi}{4}\mathrm{sgn}(\omega)]-t\omega^2} d\omega.
$$

Thus we can repeat all the steps from the Subsection 23.4.2 which gives

$$
H(2,0,\tfrac{3}{2},-\tfrac{1}{2};t,x) = \frac{1}{2\pi} \sum_{r=0}^\infty \sum_{j=0}^5 \frac{(-1)^{r+j}}{r!j!} \frac{x^j}{t^{\frac{1+j}{2}-\frac{r}{4}}} \Gamma\left(\frac{1+j}{2}+\frac{3}{4}r\right) \cos\left(\frac{\pi}{2}j - \frac{\pi}{4}r\right)
$$

$$
\times\, _4F_6\left(\begin{array}{c} 1,\Delta(3,\frac{1+j}{2}+\frac{3}{4}r) \\ \Delta(6,1+j) \end{array}; -\frac{x^6}{1728t^3}\right).
$$

For $t = 1$, the function is plotted as curve III in Fig. 23.4.1.

23.5   Exercises

1. Consider the linear equation,

$$u_t - \mathcal{L}u = 0, \tag{1}$$

where $\mathcal{L} = -\partial^2/\partial x^2 + (-\partial^2/\partial x^2)^{\alpha/2}$. Denote

$$p_\alpha(x,t) = (2\pi)^{-1/2} \int_{\mathbf{R}} e^{-t|\xi|^\alpha + ix\xi} d\xi, \quad \text{for} \quad 0 < \alpha < 2,$$

and $p_2(x,t) = (4\pi t)^{-1/2} \exp(-|x|^2/(4t))$. (a) Prove, using the Fourier transform technique, that

$$u(x,t) = p_\alpha(t) * p_2(t) * u_0(x),$$

is the solution of (1) with the initial condition $u(x,0) = u_0(x)$.

(b) Verify that for every $p \in [1,\infty]$ there exists a positive constant C independent of t such that

$$\|p_2(t) * p_\alpha(t)\|_p \le C \min\{t^{(1-1/p)/2}, t^{-(1-1/p)/\alpha}\},$$

and

$$\|\partial_x(p_2(t) * p_\alpha(t))\|_p \le C \min\{t^{(1-1/p)/2-1/2}, t^{-(1-1/p)/\alpha-1/\alpha}\},$$

for every $t > 0$. *Hint:* Use the classical Young inequality : $\|h * g\|_p \le \|h\|_q\|g\|_r$, where $1 + 1/p = 1/q + 1/r$.

(c) Verify that for every $p \in [1,\infty]$ there exists a positive constant C independent of t such that

$$\|p_2(t) * p_\alpha(t) - p_\alpha(t)\|_p \le Ct^{-(1-1/p)/\alpha+1/2-1/\alpha},$$

and

$$\|\partial_x(p_2(t) * p_\alpha(t) - p_\alpha(t))\|_p \le Ct^{-(1-1/p)/\alpha+1/2-2/\alpha},$$

for every $t > 0$. (d) Show that the solution of the Cauchy problem from (a) satisfies the following inequality

$$\|u(x,t) - Mp_\alpha(t)\|_p \le t^{-(1-1/p)/\alpha}\eta(t),$$

for all $t > 0$, $M = \int_{\mathbf{R}} u_0(x)\,dx$, and some nonnegative and bounded function η such that $\lim_{t\to\infty} \eta(t) = 0$. *Hint:* Consult Biler, Karch, and Woyczyński [20], if necessary.

2. Consider the two-scale, one-sided Lévy generators with $0 < \alpha_1 < \alpha_2 \le 1$ (here the tool is, of course, the Laplace transform) and two one-parameter families, $v_{\alpha_1}(t,x) \equiv v_{(\alpha_1,-\alpha_1,1)}(t,x)$, and $v_{\alpha_2}(t,x) \equiv v_{(\alpha_2,-\alpha_2,1)}(t,x)$, of one-sided stable Lévy densities whose Laplace convolution has the form

$$h_{\alpha_1,\alpha_2}(t,x) = \int_0^x v_{\alpha_1}(t,y)v_{\alpha_2}(t,x-y)dy = \int_0^x v_{\alpha_1}(t,x-y)v_{\alpha_2}(t,y)dy.$$

Functions $v_{\alpha_j}(t,x)$, $j = 1,2$, are given by the "stretched exponential" Laplace transform $\exp(-t\omega^{\alpha_j})$ and have the series representation

$$v_{\alpha_j}(t,x) = -\frac{1}{\pi}\sum_{r=0}^{\infty}\frac{(-t)^r}{r!}\frac{\Gamma(1+\alpha_j r)}{x^{1+\alpha_j r}}\sin(\pi r\alpha_j).$$

Show that for $\alpha_1 > \alpha_2$, with $\alpha_1 = \frac{l}{k}$ and $\alpha_2 = \frac{p}{q}$, where l, k, p, and q are integers,

$$h_{\frac{l}{k},\frac{p}{q}}(t,x) = \frac{x^{-1}m_1\sqrt{M}}{(2\pi)^{\frac{m_1-lp}{2}}}\sum_{r=0}^{\infty}\frac{(-t)^r}{r!}\left(\frac{m_1}{t}\right)^{\frac{m}{M}r}G_{lp,m_1}^{m_1,0}\left(\kappa\,\middle|\,\begin{array}{c}\Delta(lp,0)\\\Delta(m_1,\frac{m}{M}r)\end{array}\right),$$

where $G_{p,q}^{m,n}$ is the Meijer G function defined in Section 23.4. *Hint:* Consult Górska and Woyczyński [7], if necessary.

Appendix A

Basic facts about distributions

This appendix provides a compact version of the foundational material on distributions (generalized functions) contained in Volume 1 of this book series. It explains the basic concepts and applications needed for the development of the theory of random distributions (generalized stochastic processes), anomalous fractional dynamics, and its diverse applications in physics and engineering as presented in the present Volume 3. The goal is to make the book more self-contained if the reader does not have an easy access to the first volume. We aimed at the compression level that permits the reader to get sufficient (for our purposes) operational acquaintance with the distributional techniques while skipping more involved theoretical arguments. Obviously, browsing through it is no replacement for a thorough study of Volume 1. The structure of this appendix roughly mimics that of Volume 1, with chapters replaced by sections, sections by subsection, etc., while in some cases several units were merged into one.

A.1 Basic definitions and operations

A.1.1. **Conceptual framework.** We begin by providing a definition of *distributions*, also often called *generalized functions*. Given a function $f(x)$ consider the linear functional

$$\phi \mapsto T[\phi(x)] = \int f(x)\phi(x)dx \ , \tag{1}$$

mapping functions $\phi(x)$ into numbers $T[\phi(x)]$. Function $f(x)$ is called the *kernel* of the functional T. Most of the time, unless specified otherwise, we shall interpret the above-mentioned functions $f(x)$, and $\phi(x)$, as real func-

© Springer International Publishing AG, part of Springer Nature 2018   359
A. I. Saichev and W. A. Woyczynski, *Distributions in the Physical and Engineering Sciences, Volume 3*, Applied and Numerical Harmonic Analysis, https://doi.org/10.1007/978-3-319-92586-8_A

tions of real variable, mapping points $x \in \mathbf{R}$ into the points of the same axis. Functions $\phi(x)$, called *test functions*, are assumed to belong to a functional space $\mathcal{D}$ to be specified later on. So, the linear functional $T[\phi(x)]$ produces a mapping of the functional space $\mathcal{D}$ into the set of real numbers $\mathbf{R}$. Recall that in our notation, the integral sign without explicit limits is meant as the integral over the whole real line, that is, from $-\infty$, to $+\infty$.

Once the space $\mathcal{D}$ of test functions $\phi(x)$ is chosen, the set of linear continuous functionals on $\mathcal{D}$, called the *dual space* of $\mathcal{D}$, and denoted $\mathcal{D}'$, is automatically determined. There are a couple of natural demands on the space $\mathcal{D}$ of test functions. On the one hand, it has to be large enough so that the kernel $f(x)$ is uniquely identified via the values of the integrals (1). In other words, once the values of the functional $T[\phi]$ are known for all $\phi \in \mathcal{D}$, the kernel must be uniquely determined. On the other hand, the space $\mathcal{D}$ of the test functions cannot be too broad to assure existence of a large variety of kernels $f(x)$ for which the integral (1) is well defined. In other words, we want the set $\mathcal{D}'$ of linear continuous functionals $T[\phi]$ to be rich enough to be useful.

It turns out that the *family of all infinitely differentiable functions with compact support* is a good candidate for the space $\mathcal{D}$. From now on, we shall reserve the notation $\mathcal{D}$ for this particular space of test functions. Recall, that a function $\phi(x)$ is said to have compact support, if it is equal to zero outside a certain bounded interval on the x-axis. The support of $\phi(x)$ itself, denoted supp ϕ, is, by definition, the closure of the set of x's, such that $\phi(x) \neq 0$.

Definition: *A linear functional $T[\phi]$, which is continuous on the set $\mathcal{D}$ of infinitely differentiable functions with compact support, is called a distribution.*

The terms "linear," and "continuous," appearing in the above definition have to be clarified to make the latter rigorous:

1. A functional $T[\phi]$ is said to be *linear* if, for any test functions $\phi(x) \in \mathcal{D}$, and $\psi(x) \in \mathcal{D}$, it satisfies the equality

$$T[\alpha\phi + \beta\psi] = \alpha T[\phi] + \beta T[\phi] \ ,$$

where α and β are arbitrary real constants.

2. A functional T is said to be *continuous* on $\mathcal{D}$, if for any sequence $\{\phi_k(x)\}$ of test functions from $\mathcal{D}$, which converges, for $k \to \infty$, to some test function $\phi(x) \in \mathcal{D}$, the corresponding sequence of numbers $\{T_k[\phi]\}$ converges to the number $T[\phi]$. The convergence of the sequence $\{\phi_k(x)\}$ of test functions is understood here as follows:

a) The supports of the ϕ_k's are all contained in a fixed bounded set on the x-axis.

b) As $k \to \infty$, the functions $\phi_k(x)$ themselves, and all their derivatives $\phi_k^{(n)}(x)$, $n = 1, 2, \ldots$, converge *uniformly* to the corresponding derivatives of the limit test function $\phi(x)$.

If one can express a linear continuous functional $T[\phi(x)]$ via the integral (1), whose kernel $f(x)$ is an everywhere continuous function, then such a functional is called a *regular distribution*, and it is identified with its kernel $f(x)$. In this sense all usual continuous functions, for instance, $f(x) = \sin(x)$, are identified with the corresponding regular distributions. There exists, however, a large variety of linear continuous functionals $T[\phi]$ on $\mathcal{D}$, which cannot be identified with any continuous, or even locally integrable, kernel $f(x)$. They are called *singular distributions*.

The key example of a singular distribution is the functional

$$\delta[\phi(x)] = \phi(0) \ , \tag{2}$$

that assigns to each test function $\phi(x) \in \mathcal{D}$ its value at $x = 0$. This linear and continuous functional are usually called the *Dirac delta* distribution. Although it is impossible to represent Dirac delta via an integral like (1), in the physical and other applications it is convenient to write it in the symbolic integral form

$$\delta[\phi(x)] = \int \delta(x)\phi(x)dx = \phi(0) \ , \tag{3}$$

where $\delta(x)$ is the so-called *Dirac delta function*, which is defined solely by its *probing property*: the integral in (3) assigns to any test function $\phi(x)$ its value at $x = 0$, i.e., $\phi(0)$.

The biggest advantage of the Dirac delta symbolic notation lies in its heuristic transparency which facilitates actual calculations involving the Dirac delta distribution. Its convenience is evident if, for example, one introduces the shifted Dirac delta, $\delta_a[\phi(x)] = \phi(a)$, via the probing property of the Dirac delta $\delta(x - a)$ of a shifted argument:

$$\int \delta(x - a)\phi(x)dx = \phi(a) \ . \tag{4}$$

So, in what follows we shall routinely employ the symbolic integral notations like (3), and (4), even for singular distributions; their symbolic kernels will also be called generalized functions. The notation is somewhat mathematically imprecise, but very familiar to those working on various applied problems.

A.1.2. Singular distributions as limits of regular functions. Although Dirac delta distribution itself cannot be represented rigorously in the form of an integral (3) with respect to a function, it can be thought of as a limit of a sequence of regular integral functionals

$$T_k[\phi(x)] = \int f_k(x)\phi(x)dx \tag{5}$$

with respect to a sequence of kernels $f_k(x)$, which are locally integrable functions. The limit is understood in the following sense: for any $\phi(x) \in \mathcal{D}$,

$$T_k[\phi(x)] \to \phi(0) \qquad (k \to \infty) \ . \tag{6}$$

If this is the case, then the sequence $\{f_k(x)\}$ is said to be *weakly convergent* to the Dirac delta $\delta(x)$.

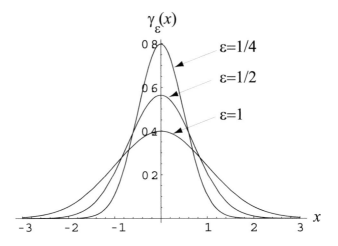

FIGURE A.1.1.
Graphs of three elements of a weakly convergent to the Dirac delta sequence of Gaussian functions (8) from Example 1. Here, $k = 1, 2, 4$ ($\varepsilon = 1, 1/2, 1/4$).

The choice of a sequence of regular distributions $\{T_k\}$, represented by kernels $\{f_k(x)\}$, and weakly convergent to $\delta(x)$, is clearly not unique. A few examples of such sequences described below will also provide a transparent illustration of some of the properties of Dirac delta.

Example 1. Gaussian kernels. Consider the family of Gaussian functions (see, Fig. A.1.1),

$$\gamma_\varepsilon(x) = \frac{1}{\sqrt{2\pi\varepsilon}} \exp\left(-\frac{x^2}{2\epsilon}\right), \tag{7}$$

parameterized by the parameter $\varepsilon > 0$, and taken as a weakly approximating sequence, $f_k(x) = \gamma_{1/k}(x)$, $k = 1, 2, \ldots$, of the Dirac delta. Notice that the factor in front of the exponent in (8) has been selected to satisfy the normalization condition

$$\int \gamma_\varepsilon(x)dx = 1, \tag{8}$$

which replicates the normalization condition (6) of the Dirac delta itself.

As $k \to \infty$, we have $\varepsilon \to 0$, and the approximating Gaussian functions are more and more concentrated around the origin $x = 0$, their peaks become higher and higher, while preserving the total area underneath them, thus satisfying the above normalization condition. ∎

Example 2. Complex oscillating kernels Consider a family of complex-valued oscillating functions

$$f_\varepsilon(x) = \sqrt{\frac{i}{2\pi\varepsilon}} \exp\left(-\frac{ix^2}{2\varepsilon}\right), \qquad (9)$$

parameterized by $\varepsilon > 0$. For a fixed ε, the modulus of each of these functions does not depend on x,

$$|f_\varepsilon(x)| = \frac{1}{\sqrt{2\pi\varepsilon}}, \qquad (10)$$

but, as $\varepsilon \to 0$, it diverges to ∞. Nevertheless, as $\varepsilon \to 0$, these functions converge weakly to Dirac delta. Physicists explain that the reason for this counterintuitive behavior is the higher and higher oscillation rate the smaller ε becomes. As a result, the integral of products of these kernels with any test function $\phi(x) \in \mathcal{D}$, converges to $\phi(0)$ as $\varepsilon \to 0$. A plot of the real part of function $f_{\varepsilon=1}(x)$ (9), demonstrating its oscillating nature, is shown in Fig. A.1.2. ∎

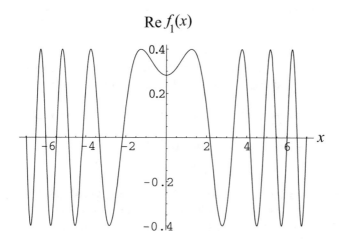

Re $f_1(x)$

FIGURE A.1.2.
Graph of the real part of the first function of the sequence $\{f_{1/k}(x)\}$ (10), weakly converging to the Dirac delta.

Example 3. Asymmetric kernels. Note that all of above examples of weakly approximating families $f_\varepsilon(x)$ for the Dirac delta have been constructed with the help of a single mother function $f(x)$, which was rescaled following the same rule,

$$f_\varepsilon(x) = \frac{1}{\sqrt{\varepsilon}} f\left(\frac{x}{\varepsilon}\right). \qquad (11)$$

The properties of the limiting Dirac delta do not depend on the particular form of the original function $f(x)$. Actually, any function, satisfying the normalization condition

$$\int f(x)dx = 1 \ , \tag{12}$$

will do. In particular, function $f(x)$ need not be symmetric (even).

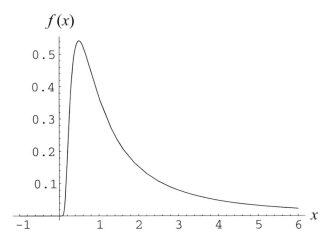

FIGURE A.1.3.
Graph of function $f(x)$ (A7), providing family of rescaling functions (A5), weakly converging, as $\varepsilon \to 0$, to the Dirac delta.

For instance, the family (11), produced by rescaling function

$$f(x) = \begin{cases} x^{-2}e^{-1/x} \ , & x > 0; \\ 0 \ , & x \leqslant 0, \end{cases} \tag{13}$$

whose plot is represented in Fig. A.1.3, will also approximate the delta function. In what follows, for the sake of convenience, we shall write expression (13), for functions which are identically equal to zero for any $x < 0$, in the more compact form

$$f(x) = x^{-2}e^{-1/x}\chi(x),$$

where $\chi(x)$ is the so-called *Heaviside* (or *unit step*) function, defined by equality

$$\chi(x) = \begin{cases} 1 \ , & x > 0 \ , \\ \frac{1}{2} \ , & x = 0 \ , \\ 0 \ , & x < 0 \ . \end{cases} \tag{14}$$

A plot of Heaviside function, which plays an important role in the following analysis, is shown in Fig. A.1.4.[1] ∎

---

[1]Note, that in this book Heaviside function is assumed to have the value $1/2$ at $x = 0$, which makes it more symmetric as compared with the usual definition.

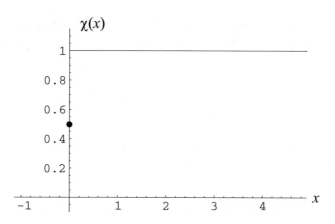

FIGURE A.1.4.
Graph of Heaviside function $\chi(x)$

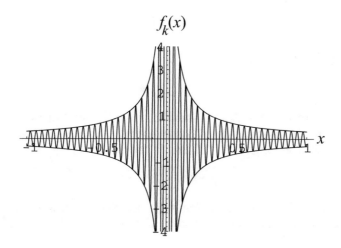

FIGURE A.1.5.
Graph of function $f_k(x)$ **(15), for** $k = 150$**, and its envelope** $\pm 1/\pi |x|$**.**

Example 4. Another oscillatory example. Contrary to intuition kernels weakly approximating Dirac delta need not be convergent to zero outside the origin. Indeed, consider the sequence

$$f_k(x) = \frac{1}{\pi} \frac{\sin(kx)}{x} . \tag{15}$$

Unlike Gaussian $\{\gamma_{1/\kappa}(x)\}$, and Cauchy sequences, it does not converge, as $k \to \infty$, to zero for any $x \neq 0$. Instead, it "fills in" the area between the four branches of the hyperbola $\pm 1/\pi|x|$ (see Fig. A.1.5). Nevertheless, sequence $\{f_k(x)\}$ (15) weakly converges to the Dirac delta, as $k \to \infty$. ∎

As a concluding remark we should say again that although some of the above functional sequences approximating Dirac delta look rather strange and counterintuitive at the first glance, they actually arise in a quite natural fashion in real physical situations.

A.1.3.   Derivatives of distributions. Infinite differentiability of the chosen set $\mathcal{D}$ of test functions $\phi(x)$ makes it possible to define, for any distribution T, the derivative of arbitrary order, thus freeing us for a constant worry about differentiability within the class of regular functions. It is one of the main advantages theory of distributions has over the classical calculus of regular functions. Before providing a general definition, let us observe that the familiar integration-by-parts formula of the integral calculus applied to a differentiable function $f(x)$, and any test function $\phi(x)$ in $\mathcal{D}$, reduces to

$$\int f'(x)\phi(x)dx = -\int f(x)\phi'(x)dx \ , \tag{19}$$

since the boundary term

$$f(x)\phi(x)|_{-\infty}^{\infty} = 0 \ ,$$

because the test function $\phi(x)$ is zero outside of a certain bounded interval on the x-axis. If we think about a regular function $f(x)$ as representing a distribution T, then equation (19) can be rewritten as

$$T'[\phi(x)] = -T[\phi'(x)] \ . \tag{20}$$

Noticing that, if $\phi(x) \in \mathcal{D}$, then, due to its infinite differentiability, $\phi'(x) \in \mathcal{D}$ as well. Thus, one can take the equality (20) as a definition of the distributional derivative T'. Namely:

If T is a distribution then its derivative T' is defined as a distribution which is equal to the right-hand side of equality (20).

Analogously, the nth derivative of a distribution T is determined by the right-hand side of the equality

$$T^{(n)}[\phi(x)] = (-1)^n T[\phi^{(n)}(x)] \ . \tag{21}$$

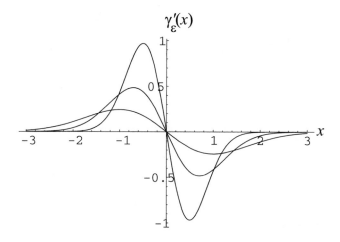

FIGURE A.1.6.
Plots of function $\gamma'_\varepsilon(x)$ (22), weakly converging to $\delta'(x)$ as $\varepsilon \to 0$, for $\varepsilon = 1; 2; 4$.

Example 1. Derivative of Dirac delta. Consider shifted Dirac delta distribution $\delta(x - a)$, which is defined by the probing property (4). Then, in view of (21), its nth derivative $\delta^{(n)}(x - a)$ is defined by the equality

$$\int \delta^{(n)}(x - a)\phi(x)dx = (-1)^n\phi^{(n)}(a) .$$

In particular,

$$\int \delta'(x - a)\phi(x)dx = -\phi'(a) .$$

Notice that one can find regular functions, weakly converging to the first derivative of the Dirac delta $\delta'(x)$, by differentiating, with respect to x, functions, weakly converging to the Dirac delta $\delta(x)$. For instance, derivative of the Gaussian functions (8),

$$\gamma'_\varepsilon(x) = -\frac{x}{\varepsilon\sqrt{2\pi\varepsilon}} \exp\left(-\frac{x^2}{2\epsilon}\right) \tag{22}$$

weakly converges, as $\varepsilon \to 0$, to the derivative of the Dirac delta, $\delta'(x)$. Plot of function $\gamma'_\varepsilon(x)$ for some ε values are depicted in Fig. A.1.6. ∎

Obviously, Heaviside function $\chi(x)$ (14) does not have the classical derivative at the jump point, $x = 0$. But the derivative exists, if we interpret the Heaviside function in the distributional sense. Indeed, the corresponding functional is equal to

$$\chi[\phi(x)] = \int \chi(x)\phi(x)dx = \int_0^\infty \phi(x)dx ,$$

and

$$\chi'[\phi(x)] = -\int_0^\infty \phi'(x)dx = -\phi(x)\Big|_0^\infty = \phi(0) \ .$$

It means that

$$\chi'(x) = \delta(x) \ . \tag{32}$$

Example 2. Differentiation of truncated exponential functions. As an exercise let us find the nth derivative of function

$$y = e^{\lambda x}\chi(x) \ .$$

As the first step we calculate the first derivative. Applying the rule of differentiation of a product of functions we get

$$y' = (e^{\lambda x})'\chi(x) + e^{\lambda x}\chi'(x) \ .$$

In view of (32), and the multiplier probing property of Dirac delta, $y' = \lambda e^{\lambda x} + \delta(x)$. Continuing this line of reasonings, we eventually arrive at the formula,

$$y^{(n)} = \lambda^n e^{\lambda x}\chi(x) + \sum_{k=1}^n \lambda^{k-1}\delta^{(n-k)}(x),$$

with $\delta^{(0)}(x) = \delta(x)$.   ∎

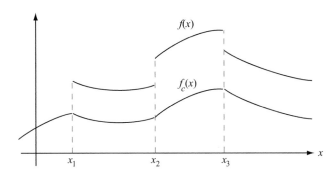

FIGURE A.1.7.
A schematic illustration of decomposition (33). Shown are function $f(x)$ with jumps and the corresponding continuous function $f_c(x)$ which has been obtained from $f(x)$ by the removal of its jumps.

Having found the derivative of the Heaviside function, one can compute easily the distributional derivative of any piecewise-smooth function $f(x)$, which has jump discontinuities at points x_k, $k = 1, 2, \ldots, n$. Such a function can be always represented as a sum of its everywhere continuous piecewise-smooth part, $f_c(x)$, without jumps, and pure jump part:

$$f(x) = f_c(x) + \sum_{k=1}^{n} \lfloor f_k \rfloor \chi(x - x_k) \ . \tag{33}$$

Here $\lfloor f_k \rfloor = f(x_k + 0) - f(x_k - 0)$ denotes the size of the jump at $x = x_k$. A schematic illustration of decomposition (33) is shown in Fig. A.1.7

Since derivation is a linear operation, we immediately see that

$$f'(x) = \{f'_c\}(x) + \sum_{k=1}^{n} \lfloor f_k \rfloor \delta(x - x_k) \ . \tag{34}$$

Here $\{f'\}(x)$ denoted function, which is equal to derivative $f'(x)$ where this derivative exists in the classical sense, and complemented by arbitrary bounded values, where it does not.

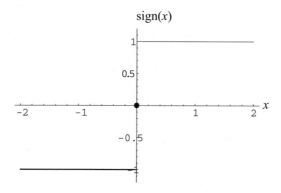

FIGURE A.1.8.
A plot of the pure jump function sign(x).

It can happen that a function has the first $n - 1$ classical derivatives, and only the derivative of order $n - 1$ displays some discontinuities. In that case derivative of the latter function has to be taken in the distributional sense. Before providing an example, let us define the function sign(x), which is another of those special discontinuous functions that we shall encounter often in what follows. By definition,

$$\text{sign}(x) = \begin{cases} +1 \ , & x > 0 \ , \\ 0 \ , & x = 0 \ , \\ -1 \ , & x < 0 \ . \end{cases} \tag{35}$$

Its graph is shown in Fig. A.1.8. Its distributional derivative $\text{sign}'(x) = 2\delta(x)$.

Example 3. Consider the function $f(x) = x^2 \text{sign}(x)$. It is differentiable in the classical sense, and $f'(x) = 2|x|$ for any point x. The derivative, however, is not differentiable at $x = 0$, but in the distributional sense, $f''(x) = 2\,\text{sign}(x)$, so that $f^{(3)}(x) = 4\delta(x)$.   ∎

A.1.4.   Distributions of several variables. By analogy with distributions on $\mathbf{R}$, distributions on $\mathbf{R}^n$ are defined as linear continuous functionals on the space $\mathcal{D}(\mathbf{R}^n)$ of infinitely differentiable test functions $\phi(\boldsymbol{x})$ with compact support in $\mathbf{R}^n$. Again, if a function $f(\boldsymbol{x})$ of an n-dimensional variable $\boldsymbol{x} = (x_1, x_2, \ldots, x_n)$ is locally integrable, then it defines a distribution on $\mathbf{R}^n$ via the multiple integral,

$$T[\phi(\boldsymbol{x})] = \int \cdots \int f(\boldsymbol{x})\phi(\boldsymbol{x})d^n x \;,$$

where the integral is an n-tuple integral with respect to the differential $d^n \boldsymbol{x} = dx_1 \ldots dx_n$. In the future, to avoid unwieldy formulas, we shall denote the multiple integral by a single integral sign $\int$ without any risk of misunderstanding. The dimension will be clear from what appears under the integral sign.

All of the conclusions about distributions in $\mathcal{D} = \mathcal{D}(\mathbf{R})$ may be extended, with obvious adjustments, to distributions on multidimensional spaces. In particular, one can define a multidimensional Dirac delta distribution, $\delta(\boldsymbol{x}-\boldsymbol{a})$, by the formula

$$\delta(\boldsymbol{x} - \boldsymbol{a})[\phi(\boldsymbol{x})] = \phi(\boldsymbol{a}) \;. \tag{43}$$

As in the 1D space, the 3D Dirac delta can be obtained as weak limit of distributions, represented by regular functions $f_k(\boldsymbol{x}), k = 1, 2, \ldots$, on $\mathbf{R}^3$. For example, it is convenient to put

$$f_k(\boldsymbol{x}) = g_k(x_1)g_k(x_2)g_k(x_3) \;,$$

where $g_k(x_i), i = 1, 2, 3$, are regular functions of one variable, approximating the one-dimensional Dirac delta $\delta_1(x_i)^2$. In this context, 3D Dirac delta can be intuitively viewed as a direct product of 1D Dirac deltas

$$\delta(\boldsymbol{x}) = \delta_1(x_1)\delta_1(x_2)\delta_1(x_3) \;. \tag{44}$$

However, the above construction hides the important property of *isotropy* of the 3D Dirac delta, that is, its invariance with respect to the group of rotations of $\mathbf{R}^3$. This property becomes more transparent if we take the Gaussian function $g_\varepsilon(x) = \gamma_\varepsilon(x)$ (8), with $\varepsilon = 1/k$, as the approximating 1D regular function of $\delta_1(x)$. Its coordinatewise product,

$$f_\varepsilon(\boldsymbol{x}) = \left(\frac{1}{\sqrt{2\pi\varepsilon}}\right)^3 \exp\left(-\frac{r^2}{2\varepsilon^2}\right) \;, \tag{45}$$

---

^{2}In what follows we shall express 3D and 2D Dirac deltas via their one-dimensional counterparts. So, to avoid confusion, we shall denote 1D Dirac deltas appearing in multidimensional formulas by $\delta_1(x)$, where the subscript explicitly indicates its one-dimensional origin.

which weakly converges to 3D Dirac delta, as $\varepsilon \to 0$, depends only on the magnitude $r = |\boldsymbol{x}| = \sqrt{x_1^2 + x_2^2 + x_3^2}$ of vector $\boldsymbol{x}$ and not on its orientation. Fig. A.1.11 plots a 2D version of function $f_\varepsilon(\boldsymbol{x})$ (45), illustrating its isotropy property.

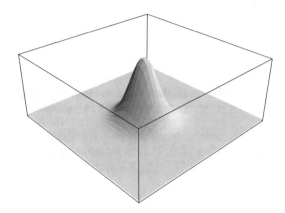

FIGURE A.1.11.
A 2D illustration of the isotropy property of the multidimensional Dirac delta.

Computations involving 3D Dirac delta of a composite argument follow the lines of similar operations in the 1D case, with the obvious adjustment for the more involved 3D Jacobian machinery in the change-of-variables formula. So, assume that $\delta(\boldsymbol{\alpha}(\boldsymbol{x}) - \boldsymbol{a})$, where $\boldsymbol{a}$ is a vector constant, while $\boldsymbol{\alpha}(\boldsymbol{x})$ is a vector function of a vector argument $\boldsymbol{x}$, which is defined by its three components $y_k = \alpha_k(\boldsymbol{x})$, $k = 1, 2, 3$. If we assume that function $\boldsymbol{y} = \boldsymbol{\alpha}(\boldsymbol{x})$ maps $\mathbf{R}^3$ onto $\mathbf{R}^3$ and satisfies the required differentiability conditions, then the equality

$$\delta(\boldsymbol{\alpha}(\boldsymbol{x}) - \boldsymbol{a}) = \frac{\delta(\boldsymbol{x} - \boldsymbol{\beta}(\boldsymbol{a}))}{|J(\boldsymbol{x})|} \tag{46}$$

is valid, with $\boldsymbol{x} = \boldsymbol{\beta}(\boldsymbol{y})$ representing the inverse mapping to $\boldsymbol{\alpha}(\boldsymbol{x})$, and J standing for the Jacobian

$$J(\boldsymbol{x}) = \left| \frac{\partial \alpha_k(\boldsymbol{x})}{\partial x_m} \right| . \tag{47}$$

To demonstrate validity of (46) consider the classical change-of-variables formula for 3D integrals,

$$\int f(\boldsymbol{\alpha}(\boldsymbol{x}) - \boldsymbol{a})\phi(\boldsymbol{x})d^3x = \int f(\boldsymbol{y} - \boldsymbol{a})\phi(\boldsymbol{\beta}(\boldsymbol{y}))|I(\boldsymbol{y})|d^3y , \tag{48}$$

which holds true if the Jacobian of the inverse mapping,

$$I(\boldsymbol{y}) = \left| \frac{\partial \beta_k(\boldsymbol{y})}{\partial y_m} \right| \tag{49}$$

is nonzero everywhere. Replacing $f(\boldsymbol{x})$ in (48) by $\delta(\boldsymbol{x})$, and using the probing property (43) of the 3D Dirac delta, we obtain

$$\int \delta(\boldsymbol{\alpha}(\boldsymbol{x}) - \boldsymbol{a})\phi(\boldsymbol{x})d^3x = \phi(\boldsymbol{\beta}(\boldsymbol{a}))|I(\boldsymbol{a})|, \tag{50}$$

or, in symbolic notation,

$$\delta(\boldsymbol{\alpha}(\boldsymbol{x}) - \boldsymbol{a}) = |I(\boldsymbol{a})|\ \delta(\boldsymbol{x} - \boldsymbol{\beta}(\boldsymbol{a}))\ . \tag{51}$$

Since $I(\boldsymbol{a})J(\boldsymbol{\beta}(\boldsymbol{a})) \equiv 1$, the multiplier probing property for multidimensional Dirac deltas permits us to transform (51) into the sought formula (46).

Example 1. Mapping polar into Cartesian coordinates. Sometimes it is convenient to apply formulas (46), (51) in coordinate notation, analogous to the right-hand side of formula (44). In 2D case, formula (51) then takes the form,

$$\delta(\alpha_1 - a_1)\delta(\alpha_2 - a_2) = |I(a_1, a_2)|\delta(x_1 - \beta_1(a_1, a_2))\delta(x_2 - \beta_2(a_1, a_2))\ . \tag{52}$$

For instance, if function $\boldsymbol{\beta}$ represents the mapping from polar to Cartesian coordinates,

$$\boldsymbol{\beta}: \qquad x_1 = \rho\cos\theta\ , \qquad x_2 = \rho\sin\theta\ ,$$

where $\boldsymbol{a}$ is a given point with polar coordinates equal to (ρ_0, θ_0), then

$$I(\boldsymbol{a}) = \begin{vmatrix} \dfrac{\partial x_1}{\partial \rho} & \dfrac{\partial x_2}{\partial \rho} \\ \dfrac{\partial x_1}{\partial \theta} & \dfrac{\partial x_2}{\partial \theta} \end{vmatrix} = \rho_0,$$

and equality (52) takes the form

$$\delta(\rho - \rho_0)\delta(\theta - \theta_0) = \rho_0\delta(x_1 - \rho_0\cos\theta_0)\delta(x_2 - \rho_0\sin\theta_0)\ .$$

In view of the multiplier probing property one can rewrite the above equality as follows:

$$\delta(x_1 - x_1^0)\delta(x_2 - x_2^0) = \frac{1}{\rho}\delta(\rho - \rho_0)\delta(\theta - \theta_0)\ ,$$

where $x_1^0 = \rho_0\cos\theta_0$ and $x_2^0 = \rho_0\sin\theta_0$, or, equivalently,

$$\delta(\rho - \rho_0)\delta(\theta - \theta_0) = \sqrt{x_1^2 + x_2^2}\ \delta(x_1 - x_1^0)\delta(x_2 - x_2^0)\ . \qquad \blacksquare$$

Of course, in the 3D case, the Dirac delta distribution with a one-point support is but the simplest of the singular distributions. Indeed, the structure of zero-volume support sets in 3D can be geometrically very complex and include objects such as lines and surfaces.

To get a taste of this physically extremely important variety of Dirac deltas in 3D, consider a surface σ in $\mathbf{R}^3$. Then the *surface Dirac delta* δ_σ is defined by the probing property,

$$\delta_\sigma[\phi(\boldsymbol{x})] = \int_\sigma \phi(\boldsymbol{x})d\sigma \; ,$$

where $\phi(\boldsymbol{x}) \in \mathcal{D}(\mathbf{R}^3)$, and the integral on the right-hand side is the surface integral of the first kind[3]. In the same fashion, one defines a *line Dirac delta* for a curve ℓ in $\mathbf{R}^3$ by the equality

$$\delta_\ell[\phi(\boldsymbol{x})] = \int_\ell \phi(\boldsymbol{x})d\ell$$

with the line integral of the first kind on the right-hand side.

In some particular cases, the notations introduced above for surface and line Dirac deltas are not used since one can express them via regular 1D Dirac deltas. Thus, a surface Dirac delta corresponding to the plane $x_1 = 0$ can be more readily interpreted as 1D $\delta_1(x_1)$, while the line Dirac delta concentrated on the x_3-axis can be written in the form of a product of 1D Dirac deltas $\delta_1(x_1)\delta_1(x_2)$. In the same manner, the scalar field of a spherical wave, radiated by an instantaneous point source, and propagating away from the origin with velocity c, can be expressed with the help of the one-dimensional Dirac delta as follows

$$U(\boldsymbol{x}, t) = \frac{A}{|\boldsymbol{x}|}\delta_1(|\boldsymbol{x}| - ct) \; ,$$

where amplitude A depends on the source energy.

Also, there are other, more general, formulas expressing some surface and line integrals via 1D Dirac deltas. Some of them are quoted below without proofs.

(a) If σ is the level surface

$$\sigma : \qquad g(\boldsymbol{x}) = a \in \mathbf{R} \; , \qquad \boldsymbol{x} \in \mathbf{R}^3 \; , \tag{53}$$

where $g(\boldsymbol{x})$ is a smooth scalar function of the 3D argument, then, using arguments similar to those that gave us formula (51), one can show that

$$\delta_\sigma = |\nabla g(\boldsymbol{x})|\delta_1(g(\boldsymbol{x}) - a) \; . \tag{54}$$

Here $\nabla g(\boldsymbol{x})$ denotes the *gradient* of scalar field $g(\boldsymbol{x})$, which is equal, in Cartesian coordinate system $\boldsymbol{x} = (x_1, x_2, x_3)$, to

$$\nabla g(\boldsymbol{x}) = \boldsymbol{j}_1\frac{\partial g(\boldsymbol{x})}{\partial x_1} + \boldsymbol{j}_2\frac{\partial g(\boldsymbol{x})}{\partial x_2} + \boldsymbol{j}_3\frac{\partial g(\boldsymbol{x})}{\partial x_3} \; ; \tag{55}$$

---

[3]Of course one can define a surface Dirac delta of the second kind via the surface integral of the second kind of vector test functions.

$(\boldsymbol{j}_1, \boldsymbol{j}_2, \boldsymbol{j}_3)$ are Cartesian unit vectors.

One can justify relation (54) relying on its obvious geometrical meaning. Indeed, as in (41), factor $|\nabla g|$ on the right-hand side of the multidimensional relation (54), takes into account the compression (if $|\nabla g| > 1$) phenomenon, influencing the volume underneath the surface Dirac delta $\delta_1(g(\boldsymbol{x}) - a)$.

(b) Analogously, one can justify the expression,

$$\delta_\ell = |[\nabla g_1(\boldsymbol{x}) \times \nabla g_2(\boldsymbol{x})]| \, \delta_1(g_1(\boldsymbol{x}) - a_1)\delta(g_2(\boldsymbol{x} - a_2) \qquad (56)$$

for the line Dirac delta δ_ℓ, concentrated on the intersection of two level surfaces

$$\ell : \qquad g_1(\boldsymbol{x}) = a_1 , \qquad g_2(\boldsymbol{x}) = a_2 . \qquad (57)$$

Recall that $[\boldsymbol{c} \times \boldsymbol{d}]$ denotes the cross-product of vectors $\boldsymbol{c}$ and $\boldsymbol{d}$.

Example 2. Flux of a vector field. Until now, we have discussed *scalar distributions of a vector argument*, as linear functionals acting on *scalar test functions* $\phi(\boldsymbol{x})$ of vector argument $\boldsymbol{x}$. However, in physical applications one has sometimes need to deal with a vector-valued test function $\boldsymbol{\phi}(\boldsymbol{x})$. A typical example of such vector distributions is

$$\boldsymbol{P} = \delta(\psi(\boldsymbol{x}) - c)\, \nabla \psi(\boldsymbol{x}) , \qquad (58)$$

which acts on an arbitrary vector-valued test function $\boldsymbol{\phi}(\boldsymbol{x})$ as follows:

$$\boldsymbol{P}[\boldsymbol{\phi}(\boldsymbol{x})] = \int \delta(\psi(\boldsymbol{x}) - c)\,(\nabla \psi(\boldsymbol{x}) \cdot \boldsymbol{\phi}(\boldsymbol{x}))\, d^3 x , \qquad (59)$$

where $(\boldsymbol{a} \cdot \boldsymbol{b})$ denotes the *dot product* of vectors $\boldsymbol{a}$ and $\boldsymbol{b}$.

The geometrical interpretation of the above distribution becomes clear if one notices that

$$(\nabla \psi(\boldsymbol{x}) \cdot \boldsymbol{\phi}(\boldsymbol{x})) = |\nabla \psi|\,(\boldsymbol{n} \cdot \boldsymbol{\phi}) ,$$

where

$$\boldsymbol{n}(\boldsymbol{x}) = \frac{\nabla \psi(\boldsymbol{x})}{|\nabla \psi(\boldsymbol{x})|}$$

is the unit vector, normal to the level surface of function $\psi(\boldsymbol{x})$. Thus, in view of relation (54),

$$\boldsymbol{P}[\boldsymbol{\phi}(\boldsymbol{x})] = \int_\sigma (\boldsymbol{n} \cdot \boldsymbol{\phi})d\sigma ,$$

where σ is the level surface, $\psi(\boldsymbol{x}) = c$. So, from the physical point of view, $\boldsymbol{P}[\boldsymbol{\phi}(\boldsymbol{x})]$ is equal to the flux of the vector field $\boldsymbol{\phi}(\boldsymbol{x})$ into the area, where $\psi(\boldsymbol{x}) > c$. ∎

A.2   A few elementary applications

A.2.2.   Fractional-order integration.   *Fractional-order integrals* are linear operators generalizing the usual n-tuple (nth-order) integrals,

$$I^n g(t) = \int_0^t dt_1 \int_0^{t_1} dt_2 \ldots \int_0^{t_{n-1}} dt_n g(t_n) \tag{1}$$

via an extension of the classical Cauchy formula expressing (1) by the single integral

$$I^n g(t) = \frac{1}{(n-1)!} \int_0^t (t-s)^{n-1} g(s) ds . \tag{2}$$

It is instructive here to retrace the proof of (2). For this purpose, let us introduce an auxiliary function

$$x(t) = \int_0^t dt_1 \int_0^{t_1} dt_2 \ldots \int_0^{t_{n-1}} dt_n\, g(t_n) , \tag{3}$$

which coincides with the right-hand side of equality (1). Differentiating both sides of the above equality n times with respect to t, we see that $x(t)$ is a forced solution of the differential equation

$$\frac{d^n x(t)}{dt^n} = g(t) , \tag{4}$$

satisfying the causality principle. Moreover,

$$g(t) \equiv 0, \qquad \text{for all} \qquad t < 0 . \tag{5}$$

Recall that the sought solution is equal to convolution integral (9), which, in view of (4), takes the form

$$x(t) = \int_0^t k_n(s) g(t-s) ds , \tag{6}$$

where $k_n(t)$ is a solution of the corresponding homogeneous differential equation

$$\frac{d^n k_n(t)}{dt^n} = 0 ,$$

satisfying the initial conditions

$$k_n(0) = \dot{k}_n(0) = \cdots = k_n^{(n-2)}(0) = 0 , \qquad k_n^{(n-1)}(0) = 1 .$$

Obviously, the solution of the above initial-value problem is equal to

$$k_n(t) = \frac{1}{(n-1)!} t^{n-1} \chi(t) . \tag{7}$$

We deliberately inserted here the Heaviside function $\chi(t)$, which explicitly takes into account the causality principle. Now we can rewrite solution (6) in a more general form utilizing the convolution integral

$$x(t) = k_n(t) * g(t) . \tag{8}$$

It is easy to show, that expressions (18), and (20), with $k_n(t)$ of the form (7), are equal to the right-hand side of Cauchy formula (2), which completes the required proof.

At this point we are in a good position to introduce the concept of integration of fractional order α. Replacing integer n in the kernel $k_n(t)$ (7) by an arbitrary nonnegative real number α, and the factorial $(n-1)!$ by the gamma function $\Gamma(\alpha)$, which is the natural generalization of the factorial function to noninteger arguments, we arrive at the generalized kernel,

$$k_\alpha(t) = \frac{1}{\Gamma(\alpha)} \, t^{\alpha-1} \chi(t) . \tag{9}$$

So, it is natural to call the convolution integral,

$$I^\alpha g(t) = k_\alpha(t) * g(t) \tag{10}$$

integral of fractional order α of function $g(t)$. Fractional integration is a linear operator. Moreover, observe that in view of the causality principle and condition (5), one can rewrite definition (10) of integration of fractional order in the more explicit form,

$$I^\alpha g(t) = \frac{1}{\Gamma(\alpha)} \int_0^t (t-s)^{\alpha-1} g(s) ds \qquad (\alpha > 0) . \tag{23}$$

Integration of fractional order preserves the usual iteration or *semigroup* property of regular multiple integrals, i.e.,

$$I^\alpha I^\beta \equiv I^{\alpha+\beta} , \tag{11}$$

which is equivalent to the equality

$$k_\alpha * k_\beta = k_{\alpha+\beta} . \tag{12}$$

The left-hand side

$$k_\alpha * k_\beta = \frac{1}{\Gamma(\alpha)\Gamma(\beta)} \int_0^t s^{\alpha-1}(t-s)^{\beta-1} ds ,$$

so that, passing to the new dimensionless integration variable $\tau = s/t$, we get

$$k_\alpha * k_\beta = \frac{t^{\alpha+\beta-1}}{\Gamma(\alpha)\Gamma(\beta)} B(\alpha, \beta) , \tag{13}$$

where
$$B(\alpha,\beta) = \int_0^1 \tau^{\alpha-1}(1-\tau)^{\beta-1}d\tau = \frac{\Gamma(\alpha)\Gamma(\beta)}{\Gamma((\alpha+\beta)}.$$

is the well-known *beta function*. Substituting it into (13) we obtain equality (12).

Remark 1. Note that in view of relation (11) the operation of fractional integration of order $\alpha > 1$ can always be decomposed as follows: $I^\alpha = T^\gamma I^n$, where $n = \lfloor\alpha\rfloor$ is the "floor" of α, i.e., the largest integer $\leq \alpha$, while γ is the fractional part of α: $\gamma = \alpha - n$, $0 \leq \gamma < 1$. So, essentially, studying properties of fractional integrals I^α one can always restrict one's attention to the case $0 < \alpha < 1$.

Remark 2. As $\gamma \to 0+$. the operator I^γ has the following consistency property: $I^\gamma \to$ Id, where Id is the identity operator. Clearly, it suffices to check the weak convergence of kernel $k_\gamma(t)$ (9), as $\gamma \to 0+$, to Dirac delta $\delta(t+)$, so that, for any continuous function $g(t)$, $\lim_{\gamma\to0+} I^\gamma g(t) = g(t)$. To prove it, notice that, in view of recurrence formula $\Gamma(\gamma+1) = \gamma\Gamma(\gamma)$, and the fact that $\Gamma(1) = 1$, we have the asymptotics $\Gamma(\gamma) \sim 1/\gamma$ ($\gamma \to 0+$). Hence $k_\gamma(t) \sim \gamma\, t^{\gamma-1}\chi(t) \to \delta(t+)$.

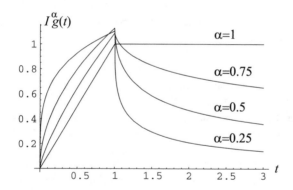

FIGURE A.2.3.
Plots of fractional-order integrals of rectangular function $g(t) = \chi(t) - \chi(t-1)$.

Example 1. Fractional-order integrals of a rectangular function. Figure A.2.3 shows results of applying fractional-order integration of different orders to rectangular function $g(t) = \chi(t) - \chi(t-1)$. ∎

A.3   Fourier transform

A.3.1.   Definition and basic properties. In this section we shall discuss the *Fourier transform* and explore its properties for functions $f(t)$, depending on a single variable t, which will be interpreted as time. The Fourier transform (or *Fourier image*) $\tilde{f}(\omega)$ of an (original) function $f(t)$ is defined by the formula,

$$\tilde{f}(\omega) = \frac{1}{2\pi} \int f(t)e^{-i\omega t} dt \; , \tag{1}$$

whenever the integral on the right-hand side exists. Notice that if, for instance, $f(t)$ is absolutely integrable on the whole axis t, the above integral is well defined.

Formula (1) describes a mapping $f(t) \mapsto \tilde{f}(\omega)$, which transforms a function f of the time variable t into a function $\tilde{f}$ of another variable ω. In this context ω will be called *angular velocity* or *angular frequency*. Directly from (1), we obtain that

$$f(t + \tau) \quad \mapsto \quad \tilde{f}(\omega)e^{i\omega\tau} \; , \tag{2a}$$

$$f(t)e^{i\Omega t} \quad \mapsto \quad \tilde{f}(\omega - \Omega) \; , \tag{2b}$$

$$f(-t) \quad \mapsto \quad \tilde{f}(-\omega) \; , \tag{2c}$$

$$f^*(t) \quad \mapsto \quad \tilde{f}^*(-\omega) \; , \tag{2d}$$

where the asterisk denotes the complex conjugate. In particular, it follows from the two last relations, that if $f(t)$ is real valued, then its Fourier image satisfies the symmetry condition

$$\tilde{f}(-\omega) = \tilde{f}^*(\omega) \; . \tag{3}$$

So for scientists, who often deal with real processes $f(t)$, it is sufficient to operate only with the nonnegative angular frequency ($\omega \geqslant 0$).

Many useful properties of the Fourier transform depend on the kernel of Fourier integral $g(t) = e^{i\omega t}$ satisfying the remarkable functional equation

$$g(t + \tau) = g(t)g(\tau) \; . \tag{4}$$

It reflects the invariance of the kernel $g(t)$ under the time shifts: If the time variable t is shifted by a fixed τ, then the kernel g does not change, except for a constant factor $g(\tau) = e^{i\omega\tau}$. Moreover, if $\tau = kT$, where $T = 2\pi/\omega$, and k is any integer, then $e^{i\omega\tau} = 1$, so that $g(t + \tau) = g(t)$, and the kernel does not change at all. In other words, function $g(t) = e^{i\omega t}$ is periodic with period T.

Remark 1. Fourier transform in the frequency domain. In many applications it is more convenient to use the frequency $\nu = \omega/(2\pi)$ (measured in cycles per second, i.e., Hertz) as the Fourier variable, rather than the angular frequency. Then

$$\hat{f}(\nu) = \int f(t)e^{-2\pi i \nu t} dt \; . \tag{5}$$

In what follows, however, we will stick with the Fourier variable ω, and for the sake of convenience will also call it "frequency."

A.3.2.   Smoothness, inverse transform, and convolution. We begin with the relationship between the smoothness of function $f(t)$ and the asymptotic behavior of its Fourier image, $\tilde{f}(\omega)$, as $\omega \to \infty$. Assume that $f(t)$ is n-times continuously differentiable on $\mathbf{R}$ and absolutely integrable together with its first n derivatives. Multiply (1) by $(-i\omega)^n$ and note that $(-i\omega)^n e^{-i\omega t}$ is nth derivative of function $e^{-i\omega t}$ with respect to t. Hence, integrating n times by part, we arrive at the equality

$$(i\omega)^n \tilde{f}(\omega) = \frac{1}{2\pi} \int f^{(n)}(t) e^{-i\omega t} dt , \qquad (6)$$

which expresses one of the most useful for applications property of the Fourier transform:

Differentiation of the original function $f(t)$ corresponds to multiplication of its Fourier image by $i\omega$, i.e.,

$$\dot{f}(t) \quad \mapsto \quad i\omega \tilde{f}(\omega) . \qquad (7)$$

Also, if $g(t)$ is the derivative of an absolutely integrable function $f(t)$ $(g(t) = f'(t))$, then inverse to (7) relation is valid:

$$f(t) = \int_{-\infty}^{t} g(t') dt' \quad \mapsto \quad \frac{\tilde{g}(\omega)}{i\omega} . \qquad (8)$$

Absolute integrability of the integrand in (6) implies that the expression on the left-hand side is bounded, so that we obtain the following important principle:

Sufficiently smooth functions with absolutely integrable nth derivative have Fourier images that decay at infinity not slower than $|\omega|^{-n}$.

Due to the symmetry between the Fourier transform and its inverse transform (to be established below), the following statement is also true:

If the Fourier image $\tilde{f}(\omega)$ is a smooth function with an absolutely integrable nth derivative (with respect to ω), then the original function $f(t)$ decays at infinity not slower than $|t|^{-n}$.

The *inverse transform* formula permits recovery of an original function from its Fourier transform. Let us assume, for simplicity, that $f(t)$ is absolutely integrable and sufficiently smooth function, so that $\tilde{f}(\omega)$ is absolutely integrable as well. Multiplying (1) by $\tilde{\phi}(\omega) e^{i\omega\tau}$, where $\tilde{\phi}(\omega)$ is an absolutely integrable function, and integrating both sides of the equality with respect to ω, we obtain

$$\int \tilde{\phi}(\omega) \tilde{f}(\omega) e^{i\omega\tau} d\omega = \frac{1}{2\pi} \int f(t) \left[\int \tilde{\phi}(\omega) e^{i\omega(\tau-t)} d\omega \right] dt . \qquad (9)$$

Note that we have changed the order of integration on the right-hand side. This is justified by the absolute integrability of the integrands. Now, let us take

$$\tilde{\phi}(\omega) = \exp\left(-\frac{\varepsilon^2 \omega^2}{2} \right) ,$$

and evaluate the inner integral on the right-hand side using the well-known formula

$$\int e^{-bx^2+iyx}dx = \sqrt{\frac{\pi}{b}}\exp\left(-\frac{y^2}{4b}\right) , \tag{10}$$

valid for any Re $b \geqslant 0$, $b \neq 0$. As a result, equality (9) is transformed into

$$\int \tilde{f}(\omega)\exp\left(-\frac{\varepsilon^2\omega^2}{2}+i\omega\tau\right)d\omega = \int f(t)\frac{1}{\sqrt{2\pi}\varepsilon}\exp\left(-\frac{(t-\tau)^2}{2\varepsilon^2}\right)dt .$$

For $\varepsilon \to 0$, the Gaussian kernel in the right-hand side integral weakly converges to a shifted Dirac delta $\delta(t-\tau)$, and its probing property recovers the value $f(\tau)$. On the left-hand side, in view of assumed absolute integrability of $\tilde{f}(\omega)$, one can just set $\varepsilon = 0$. Finally, replacing τ by t, we arrive at the inverse Fourier integral

$$f(t) = \int \tilde{f}(\omega)e^{i\omega t}d\omega , \tag{11}$$

which expresses the original function $f(t)$ through its Fourier image.

Remark 1. Variations in the definition of the Fourier transform. Our definition of Fourier transform (1) included factor $1/2\pi$ in front of the integral, and the sign minus in the exponent of the kernel. Such a choice of constants results in the formula (11) for the inverse transform which has an immediate physical interpretation: original function $f(t)$ is representable as a continuum superposition of harmonic oscillations $e^{i\omega t}$ with $\tilde{f}(\omega)$ being the complex amplitude corresponding to the angular frequency ω.

But if one defines the Fourier transform using a different constant, say, via the formula

$$\tilde{f}(\omega) = \frac{1}{\sqrt{2\pi}}\int f(t)e^{-i\omega t}dt ,$$

then the corresponding inverse Fourier transform takes on a pleasingly symmetric form,

$$f(t) = \frac{1}{\sqrt{2\pi}}\int \tilde{f}(\omega)e^{i\omega t}dt ,$$

which can be an advantage in certain situations.

Engineers, who often prefer to work with frequencies ν, rather than with angular frequencies ω, and use (5) as definition of the Fourier transform, also can enjoy a symmetric formula for the inverse transform: to (A.4.1.5) transform looks like

$$\hat{f}(t) = \int f(\nu)e^{-2\pi i\nu t}d\nu . \tag{12}$$

∎

The convolution of two functions is often a complicated operation to perform. However, in the frequency domain it is replaced by a simple multiplication of the Fourier transforms:

The Fourier image of a convolution $f * \phi$ is, up to the factor 2π, a product of their Fourier images:

$$f(t) * \phi(t) \quad \longmapsto \quad 2\pi \tilde{f}(\omega)\tilde{\phi}(\omega) . \tag{13}$$

Indeed, in view of (11), the integral inside the square brackets in (9) can be replaced by $\phi(\tau - t)$, and we obtain that

$$\int \tilde{f}(\omega)\tilde{\phi}(\omega)e^{i\omega\tau}d\omega = \frac{1}{2\pi}\int f(t)\phi(\tau - t)dt . \tag{14}$$

The integral on the right-hand side is the convolution of functions f and ϕ. Comparing (14) and (11) we obtain the desired conclusion.

Multiplying (11) by $\phi(t)e^{-i\Omega t}$, integrating it with respect to t, and recalling the Fourier transform definition (1) to evaluate the right-hand side integral, we arrive at the relation,

$$\int f(t)\phi(t)e^{-i\Omega t}dt = 2\pi \int \tilde{f}(\omega)\tilde{\phi}(\Omega - \omega)d\omega ,$$

dual to the equality (14). The above formula implies, in particular, that the Fourier transform of a product of two functions is the convolution of their Fourier images:

$$f(t)\phi(t) \quad \longmapsto \quad \tilde{f}(\omega) * \tilde{\phi}(\omega) . \tag{15}$$

We shall call relation (14) *Parseval equality*, although that name is often reserved for the following special case obtained from (14) by setting $\tau = 0$, and substituting $f^*(-t)$ for $\phi(t)$, which gives

$$\int |f(t)|^2 dt = 2\pi \int |\tilde{f}(\omega)|^2 d\omega . \tag{16}$$

In engineering applications, $P(t) = |f(t)|^2$ often represents the signal's *power function*, so that the integral on the left-hand side of (16) is the signal's energy. Thus, in view of the Parseval equality (16), function $\tilde{P}(\omega) = |\tilde{f}(\omega)|^2$ shows how energy is distributed over the frequencies.

A.3.3.   Generalized Fourier transform. Tools of the theory of distributions permit an extension of the Fourier transform's domain beyond the class of integrable, in classical sense, Fourier integral (1). Recall that distribution was defined as a linear continuous functional, which assigned a number to each function ϕ from a certain set of test functions. In other words, you can tell a distribution by its action on test functions. To extend the domain of the Fourier transform equality (14) offers a similar opportunity.

Suppose that $\tilde{F}$ is a distribution on some set of test functions $\mathcal{S}$ (to be defined later) each of which, say, $\tilde{\phi}(\omega)$, is the Fourier transform of an absolutely integrable, original function $\phi(t)$. We shall call $\tilde{F}$ the *generalized Fourier transform* of function $f(t)$, if the integral on the right-hand side of (14) is equal to the value of the linear continuous functional $\tilde{F}$ on the test function $\tilde{\phi}(\omega)e^{i\omega t}$, that is,

$$\tilde{F}[\tilde{\phi}(\omega)e^{i\omega\tau}] = \frac{1}{2\pi} \int f(t)\phi(\tau - t)dt . \tag{17}$$

Although we have not defined the appropriate to the circumstances set $\mathcal{S}$ of test functions $\tilde{\phi}(\omega)$, we already able to calculate some generalized Fourier images. For instance, if $f(t) \equiv 1$, then (1),

$$\frac{1}{2\pi} \int \phi(\tau - t)dt = \frac{1}{2\pi} \int \phi(t)dt = \tilde{\phi}(\omega = 0) .$$

On the left-hand side of relation (17), the same result is secured by taking the distributional Fourier image to be equal to Dirac delta $\delta(\omega)$. Thus, we have found the first concrete formula for the generalized Fourier transform:

$$1 \quad \mapsto \quad \delta(\omega) . \tag{18}$$

Equality (17) is asymmetric in the sense that whereas its left-hand side contains a distribution $\tilde{F}$, its right-hand side is still a regular integral. To remove such an asymmetry we can replace the integral on the right-hand side of (17) by action of a distribution T on the test function $\phi(\tau - t)$:

$$\tilde{F}[\tilde{\phi}(\omega)e^{i\omega\tau}] = \frac{1}{2\pi}T[\phi(\tau - t)] . \tag{19}$$

This gives the equality, defining distributional Fourier transforms $\tilde{F}$ of original distributions T, and vice verse. Taking, for example, as original distribution $\delta(t)$, we obtain

$$\tilde{F}[\tilde{\phi}(\omega)e^{i\omega\tau}] = \frac{1}{2\pi}\phi(\tau) ,$$

which, in view of (11), gives

$$\tilde{F}[\tilde{\phi}(\omega)e^{i\omega\tau}] = \frac{1}{2\pi} \int \tilde{\phi}(\omega)e^{i\omega\tau}d\omega .$$

In other words, we have established the inverse to (18) formula for the generalized Fourier transform:

$$\delta(t) \quad \mapsto \quad \frac{1}{2\pi} . \tag{20}$$

Symbolically, this can be written in the form

$$\delta(t) = \frac{1}{2\pi} \int e^{i\omega t} d\omega \ . \tag{21}$$

To complete the above definition we still need define the appropriate space of test functions $\mathcal{S}$. Our original space $\phi(t) \in \mathcal{D}$ will not do in this context as the Fourier image of a function with compact support need not have compact support. The solution is to expand the space $\mathcal{D}$ of test functions $\{\phi(t)\}$ by demanding that the sets of $\phi(t)$'s, and of $\tilde{\phi}(\omega)$'s, be identical. It turns out that the right space $\mathcal{S}$ here is the set of all infinitely differentiable functions $\phi(t)$, which decrease at infinity, together with all their derivatives, faster than any power function $|t|^{-n}$. Such functions are often called *rapidly decreasing*. In other words, $\phi(t) \in \mathcal{S}$, if for any positive integers $n, m \geqslant 0$, one can find constants $K_{nm} < \infty$ such that, for arbitrary t,

$$|t^n \phi^{(m)}(t)| < K_{nm} \ . \tag{22}$$

It is easy to show that if function $\phi(t) \in \mathcal{S}$ then its Fourier image is also infinitely differentiable and rapidly decreasing. Indeed, according to (6), the infinite differentiability of $\phi(t)$ implies that its Fourier image decays faster than an arbitrary power $|\omega|^{-n}$. Vice versa, due to the symmetry of direct and inverse Fourier transforms, from the fact that $\phi(t)$ is rapidly decreasing, it follows that $\tilde{\phi}(\omega)$ is infinitely differentiable.

Note that the expansion of the test function space from $\mathcal{D}$ to $\mathcal{S}$ results in the set $\mathcal{S}'$ of corresponding distributions being smaller than the set of $\mathcal{D}'$ described in Section A.1, i.e., $\mathcal{S}' \subset \mathcal{D}'$. Traditionally, the set $\mathcal{S}'$ of continuous linear functionals on $\mathcal{S}$ with convergence related to the conditions (22) is called the space of *tempered distributions*.

Importantly, although the set $\mathcal{S}'$ is smaller than $\mathcal{D}'$ it still contains all the distributions with compact support. Thus, in particular, the Dirac delta and its derivatives are tempered distributions. Also, any function $f(t)$ which grows slower than a certain power of t defines a continuous functional on $\mathcal{S}$ via the regular integral

$$\int f(t)\phi(t)dt < \infty \ , \qquad \phi(t) \in \mathcal{S} \ .$$

However, function e^t represent distribution in the space $\mathcal{D}'$ but not in $\mathcal{S}'$. Thus, generalized Fourier transform for the exponential e^t is not well defined but it is for any power functions t^n.

All the basic operations applicable to ordinary Fourier transform remain valid for their generalized cousins. So, in view of (2a), the shifted Dirac delta $\delta(t - \tau)$ has Fourier image $e^{-i\omega\tau}/2\pi$, while (6) implies that the Fourier transform of $\delta^{(n)}(t)$ is equal to $(i\omega)^n/2\pi$. Two additional formulas involving generalized Fourier transform are useful:

$$t^n e^{i\Omega t} \quad \longmapsto \quad i^n \delta^{(n)}(\omega - \Omega) \,,$$
$$\delta^{(n)}(t - \tau) \quad \longmapsto \quad \frac{(i\omega)^n}{2\pi} e^{-i\omega\tau} \,. \tag{23}$$

Example 1. Weak limits of Fourier images. As was the case for ordinary distributions, the generalized Fourier images could be defined as weak limits of regular Fourier images. Here, let us consider an example of the functions

$$\tilde{f}(\omega, \lambda) = \frac{1}{\pi} \frac{\sin(\omega\lambda)}{\omega} \,,$$

dependent on parameter λ, weakly converging to $\delta(\omega)$ as $\lambda \to \infty$. The last statement can be verified by considering the functional

$$F(\lambda) = \int \tilde{\phi}(\omega) \frac{\sin(\omega\lambda)}{\omega} \, d\omega \,,$$

which is called *Dirichlet integral*. Differentiating the above equality with respect to λ, one obtains

$$F'(\lambda) = \int \tilde{\phi}(\omega) \cos(\omega\lambda) \, d\omega = \frac{1}{2} [\phi(+\lambda) + \phi(-\lambda)] \,.$$

Now, taking definite integrals of the left- and right-hand sides over the interval $(0, \lambda)$, and noticing that $F(0) = 0$, we arrive at the formula,

$$F(\lambda) = \frac{1}{2} \int_{-\lambda}^{\lambda} \phi(t) dt \,.$$

If the Fourier image $\tilde{\phi}(\omega)$ is sufficiently smooth, then the original function $\phi(t)$ is absolutely integrable, and the above integral converges, as $\lambda \to \infty$, to

$$F(\infty) = \frac{1}{2} \int \phi(t) dt = \pi \tilde{\phi}(0) \,,$$

which is the value of the test function at 0, multiplied by π. This proves that, weakly,

$$\frac{1}{\pi} \frac{\sin(\omega\lambda)}{\omega} \quad \longmapsto \quad \delta(\omega) \operatorname{sign}(\lambda) \quad (\lambda \to \pm\infty) \,,$$

although pointwise, the same function does not converge to zero for $\omega \neq 0$, as it "fills out" the area between the branches of the hyperbola $\pm 1/\pi\omega$ (see Fig. A.1.5).   ∎

Example 2. A function that is not absolutely integrable. Function

$$f(t) = \frac{t^2}{\gamma^2 + t^2}$$

is not absolutely integrable and does not have the classical Fourier image. Nevertheless its generalized Fourier transform is well defined. To find it let us rewrite $f(t)$ in the form

$$f(t) = 1 - \frac{\gamma^2}{\gamma^2 + t^2} \ .$$

In view of (2), Fourier transform of the first term on the right-hand side is equal to Dirac delta. Fourier image of the second last term exists in the classical sense so that

$$\tilde{f}(\omega) = \delta(\omega) - \frac{\gamma}{2} e^{-\gamma|\omega|} \ . \qquad\blacksquare$$

Sometimes, knowledge of the generalized Fourier transform facilitates calculations of classical Fourier integrals. The following examples, which are exploiting equalities related to (8), provide an illustration of this remarkable fact.

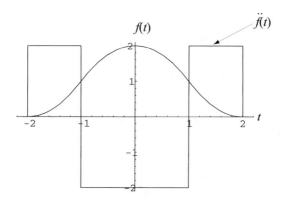

FIGURE A.3.1.
Plots of original function $f(t)$ (24) and its piecewise constant second derivative $\ddot{f}(t)$, illustrating the idea of regular Fourier image calculations via generalized Fourier transform.

Example 3. Classical Fourier transform via generalized Fourier transform.
Function

$$f(t) = \begin{cases} 2 - t^2 \ , & \text{for } |t| < 1 \ ; \\ (2 - |t|)^2 \ , & \text{for } 1 < |t| < 2 \ ; \\ 0 \ , & \text{for } 2 < |t| \ . \end{cases} \qquad (24)$$

Its Fourier image can be found directly but the calculation is rather tedious. So, let find it not by integration, but by differentiation. The third derivative of the above function

$$\dddot{f}(t) = 2\delta(t + 2) - 4\delta(t + 1) + 4\delta(t - 1) - 2\delta(t - 2) \ .$$

Taking Fourier images of left- and right-hand sides of this equality gives

$$-i\omega^3 \tilde{f}(\omega) = \frac{1}{\pi}\left(e^{2i\omega} - 2e^{i\omega} + 2e^{-i\omega} - e^{-2i\omega}\right) \, .$$

Thus we finally obtain

$$\tilde{f}(\omega) = \frac{4}{\pi\omega^3}\sin\omega \cdot (1 - \cos\omega) \, .$$

Plot of original function $f(t)$ (24) and its second derivative, illustrating the idea of the above calculations, is shown Fig. A.3.1. ∎

Example 4. Linear interpolation of discretely sampled function. The values f_n of function $f(t)$ are known at equidistant instants $t_n = sn$, $n = \ldots, -1, 0, 1, 2 \ldots$. We will find the Fourier image of its linear interpolation

$$f^l(t) = f_n + (f_{n+1} - f_n)\frac{t - ns}{s} \, , \qquad ns < t < (n+1)s \, . \tag{25}$$

Here $f_n = f(sn)$. Notice that the second derivative of the above function

$$\ddot{f}^l(t) = \sum_{n=-\infty}^{\infty} \Delta_2 f_n \delta(t - ns) \, ,$$

where we have used the standard notation, $\Delta_2 f_n = f_{n+1} - 2f_n + f_{n-1}$, for the second-order difference of function f. The corresponding Fourier image

$$\tilde{f}^l(\omega) = -\frac{1}{2\pi s\omega^2}\sum_{n=-\infty}^{\infty} \Delta_2 f_n e^{-i\omega sn} \, .$$

Regrouping terms of the above series we obtain

$$\tilde{f}^l(\omega) = \frac{2\sin^2(\Omega/2)}{\pi\Omega^2}\, s \sum_{n=-\infty}^{\infty} f_n e^{-i\Omega n} \, , \tag{26}$$

where $\Omega = s\omega$ is the dimensionless frequency.

Example 5. Convolution of triangular impulses. Let

$$f(t) = \int h(\tau)h(\tau + t)d\tau \, , \tag{27}$$

where

$$h(t) = \begin{cases} t \, , & for \quad 0 < t < \theta \, ; \\ 0 \, , & otherwise \, ; \end{cases}$$

be a triangular impulse. To find its Fourier image first observe that the sought Fourier image is related to the Fourier image of impulse $h(t)$ via the formula

$$\tilde{f}(\omega) = 2\pi|\tilde{h}(\omega)|^2 \ . \tag{28}$$

The Fourier image of impulse $h(t)$ itself can be calculated by remembering that its second derivative is $\ddot{h}(t) = \delta(t) - \delta(t-\theta) - \theta\dot{\delta}(t-\theta)$. The corresponding Fourier image of impulse $h(t)$ is $\tilde{h}(\omega) = \frac{\theta^2}{2\pi\Omega^2}\left[e^{-i\Omega}(1+i\Omega) - 1\right]$, where $\Omega = \omega\theta$ is the new dimensionless argument. Thus, finally,

$$\tilde{f}(\omega) = \frac{\theta^4}{2\pi\Omega^4}\left[\Omega^2 + 2(1 - \cos\Omega - \Omega\sin\Omega)\right] \ . \qquad \blacksquare$$

Example 6. Fourier series. The generalized Fourier transform theory gives specific interpretation to interplay between Fourier transform and Fourier series. First, recall some basic results of Fourier series theory. If $f(t)$ is a continuous periodic function, such that $f(t) \equiv f(t + 2\pi)$, then it may be represented in the form of the Fourier series

$$f(t) = \sum_{m=-\infty}^{\infty} \tilde{f}_m e^{imt} \ , \tag{29}$$

where

$$\tilde{f}_m = \frac{1}{2\pi}\int_{-\pi}^{\pi} f(t)e^{-imt}dt \ . \tag{30}$$

Multiplying (29) by $\phi(\tau - t)/2\pi$ and integrating both sides of resulting equality over the whole t-axis, we obtain the identity

$$\frac{1}{2\pi}\int f(t)\phi(\tau - t)dt = \sum_{m=-\infty}^{\infty} \tilde{f}_m\tilde{\phi}(\omega)e^{im\tau} \ .$$

Comparing this equality and (14), we get

$$\sum_{m=-\infty}^{\infty} \tilde{f}_m\tilde{\phi}(\omega)e^{im\tau} = \int \tilde{f}(\omega)\tilde{\phi}(\omega)e^{i\omega\tau}d\omega \ ,$$

where $\tilde{f}(\omega)$ is the generalized Fourier image of periodic function $f(t)$. It is clear that last equality holds true if

$$\tilde{f}(\omega) = \sum_{m=-\infty}^{\infty} \tilde{f}_m\delta(\omega - m) \ . \tag{31}$$

Now, let $f(t)$ be a periodic Dirac delta:

$$f(t) = \sum_{n=-\infty}^{\infty} \delta(t - 2\pi n) \ .$$

Then all coefficients (30) are identical and are equal to $1/2\pi$. Thus, equality (29) can be rewritten in the form,

$$\sum_{n=-\infty}^{\infty} \delta(t - 2\pi n) = \frac{1}{2\pi} \sum_{m=-\infty}^{\infty} e^{imt} \ . \tag{32}$$

An obvious extension of this formula for the case of an arbitrary period $2\pi/\Omega$ has the form

$$\sum_{n=-\infty}^{\infty} \delta\left(t - \frac{2\pi n}{\Omega}\right) = \frac{\Omega}{2\pi} \sum_{m=-\infty}^{\infty} e^{im\Omega t} \ . \tag{33}$$

Bibliography

Chapter 16

[1] Feller, W. (1978). *An Introduction to Probability Theory and its Applications* (Vols. 1 (528 pp), 3rd Ed. and 2 (704 pp), 2nd Ed.). New York: Wiley.

[2] Loéve, M. (1978). *Probability Theory* (Vols. I (415 pp), and II (395 pp), 4th Ed.). Berlin: Springer.

[3] Billingsley, P. (2012). *Probability and Measure* (Anniversary Edition) New York: Wiley.

[4] Kallenberg, O. (2002). *Foundations of Modern Probability* (638 pp). Berlin: Springer.

[5] Denker, M., & Woyczyński, W. A. (1998). *Introductory Statistics and Random Phenomena: Uncertainty, Complexity and Chaotic Behavior in Engineering and Science.* Birkhauser: Boston.

[6] Keller, J. (1986). The probability of heads. *American Mathematical Monthly, 93,* 191–196.

[7] Guttorp, P. (1995). *Stochastic Modeling of Scientific Data* (372 pp). London: Chapman and Hall.

[8] Watson, G. N. (1995). *A Treatise on the Theory of Bessel Functions* (2nd Ed.). Cambridge: Cambridge University Press.

[9] Marcinkiewicz, J. (1939). Sur une propriété de la loi de Gauss. *Mathematische Zeitschrift, 44,* 612–618.

[10] Granas, A., & Dugundji, J. (2003). *Fixed Point Theory.* New York: Springer.

Chapter 17

[1] Itô, K. (1953). Stationary random distributions. *Memoirs of the College of Science, University of Kyoto, 26,* 209–223.

[2] Gelfand, I. M., & Vilenkin, N. Ya. (1964). *Generalized Functions, Volume 4: Applications of Harmonic Analysis.* New York and London: Academic Press.

© Springer International Publishing AG, part of Springer Nature 2018

A. I. Saichev and W. A. Woyczynski, *Distributions in the Physical and Engineering Sciences, Volume 3,* Applied and Numerical Harmonic Analysis, https://doi.org/10.1007/978-3-319-92586-8

[3] Billingsley, P. (1986). *Probability and Measure*. New York: Wiley.

[4] Dvoretzky, A., Erdös., & Kakutani, S. (1961). Nonincrease everywhere of the brownian motion process. In *Proceedings of the 4th Berkeley Symposium, II* (pp. 103–116).

[5] Schwartz, L. (1966). *Théorie des Distributions*, Hermann, Paris.

[6] Woyczyński, W. A. (2013). *A First Course in Statistics for Signal Analysis*. New York: Birkhauser-Springer.

[7] Mikusiński., & Sikorski, R. (1957). *Elementaryl Theory of Distributions*, Part I (1957), Part II (1961). Warsaw: Polish Scientific Publishers.

[8] Urbanik, K. (1958). Generalized stochastic processes. *Studia Mathematica, 16*(3), 268–334.

[9] Urbanik, K. (1958). Local characteristics of generalized stochastic processes. *Studia Mathematica, 17*(3), 199–266.

[10] Urbanik, K. (1958). The conditional expectations and the ergodic theorem for strictly stationary generalized stochastic processes. *Studia Mathematica, 16*(3), 267–283.

[11] Zieleźny, Z. (1955). Sur la définition de Łojasiewicz de la valeur d'une distribution dans un point. *Bull. Acad. Polon. Sci. 3*, 519–520.

[12] Doob, J. (1953). *Stochastic Processes*. New York: Wiley Inc.

Chapter 18

[1] Klyatskin, V. I., & Woyczyński, W. A. (1997). Dynamical and statistical characteristics of geophysical fields and waves and related boundary-value problems. In *Stochastic Models in Geosystems* (pp. 171–208). Berlin: Springer.

[2] Klyatskin, V. I., & Saichev, A. I. (1992). Statistical and dynamical localization of plane waves in randomly layered media. *Soviet Physics Usp., 35*(3), 231–247.

[3] Klyatskin, V. I. (1985). *Ondes et Équations Stochastiques dans les milieus Aléatoirement non Homogènes*, Eddition de Physique. Besançon-Cedex.

[4] Klyatskin, V. I. (1994). The imbedding method in statistical boundary-value wave problems. In E. Wolf (Ed.), *Progress in optics* (Vol. XXXIII). North-Holland, Amsterdam.

[5] Chandrasekhar, S. (1943). Stochastic problems in physics and astronomy. In *Reviews of modern physics* (Vol. 15, pp. 1–89).

[6] Saichev, A. I. (1993). Chaotic motion of particles flows. *Dynamics of Systems, 1*(1), 1–31.

[7] Guzev, M. A., Klyatskin, V. I., & Popov, G. V. (1992). Phase fluctuations and localization length in layered randomly inhomogeneous media. *Waves in Random Media, 2*(2), 117–123.

[8] Guzev, M. A., & Klyatskin, V. I. (1993). Influence of boundary conditions on statistical characteristics of wavefield in layered randomly inhomogeneous medium. *Waves in Random Media, 3*(4), 307–315.

[9] Kulkarny, V. A., & White, B. S. (1982). Focusing of rays in a turbulent inhomogeneous medium. *Physics of Fluids, 25*(10), 1770–1784.

[10] White, B. S. (1983). The stochastic caustic. *SIAM Journal on Applied Mathematics, 44*(1), 127–149.

[11] Gardiner, C. W. (1985). *Handbook of Stochastic Methods for Physics, Chemistry and the Natural Sciences.* Berlin: Springer.

[12] Furutsu, K. (1963). On the statistical theory of electromagnetic waves in a fluctuating medium. *Journal of Research of the NBS, D-67*, 303.

[13] Novikov, E. A. (1964). Functionals and the random-force method in turbulence theory. *Soviet Physics JETP, 20*(5), 1290–1294.

[14] Klyatskin, V. I. (1991). Approximations by delta-correlated random processes and diffusive approximation in stochastic problems. In W. Kohler & B. S. White (Eds.), *Mathematics of Random Media.* Lectures in applied mathematics (Vol. 27, pp. 447–476).

Chapter 19

[1] Saichev, A. I., & Woyczyński, W. A. (1996). Density fields in Burgers and KdV-Burgers turbulence, *SIAM Journal on Applied Mathematics, 56*, 1008–1038.

[2] Saichev, A. I., & Woyczyński, W. A. (1996). Model description of passive tracer density fields in the framework of Burgers' and other related model equations. In *Nonlinear Stochastic PDE's: Burgers Turbulence and Hydrodynamic Limit*, IMA volumes (pp. 167–192). Berlin: Springer.

[3] Saichev, A. I., & Woyczyński, W. A. (1996). Density fields in Burgers' and KdV-Burgers' turbulence. *SIAM Journal on Applied Mathematics, 56*, 1008–1038.

[4] Saichev, A. I., & Woyczyński, W. A. (1997). Evolution of Burgers' turbulence in presence of external forces. *Journal of Fluid Mechanics, 331*, 313–343.

[5] Saichev, A. I., & Woyczyński, W. A. (1997). Advection of passive and reactive tracers in multidimensional Burgers' velocity field. *Physica D, Nonlinear Phenomena, 100*, 119–141.

[6] Woyczynski, W. A. (1998). *Burgers-KPZ Turbulence: Göttingen Lectures.* Lecture notes in mathematics (Vol. 1700, 318 pp). Berlin: Springer.

[7] Sinai, Ya. (1991). Two results concerning asymptotic behavior of solution of the Burgers equation with force. *Journal of Statistical Physics 64*, 1–12.

[8] Sinai, Ya. (1996). *Burgers system driven by a periodic stochastic flow* (9 pp), Preprint. Princeton: Princeton University.

[9] Da Prato, G., & Zabczyk, J. (1992). *Stochastic Equations in Infinite Dimension.* Cambridge: Cambridge University Press.

[10] Funaki, T., Surgailis, D., & Woyczyński, W. A. (1995). Gibbs-Cox random fields and Burgers' turbulence. *The Annals of Applied Probability, 5,* 701–736.

[11] Molchanov, S. A., Surgailis, D., & Woyczyński, W. A. (1995). Hyperbolic asymptotics in Burgers' turbulence and extremal processes. *Communications in Mathematical Physics 168,* 209–226.

[12] Molchanov, S. A., Surgailis, D., & Woyczyński, W. A. (1997). The large-scale structure of the Universe and quasi-Voronoi tessellation of shock fronts in forced inviscid Burgers turbulence in $\mathbf{R}^d$. *Annals of Applied Probability 7,* 200–228.

[13] Hu, Y., Woyczyński, W. A. (1994). An extremal rearrangement property of statistical solutions of the Burgers' equation. *The Annals of Applied Probability, 4,* 838–858.

[14] Hu, Y., Woyczyński, W. A. (1995). Shock density in Burgers' turbulence. In *Nonlinear Stochastic PDE's: Burgers Turbulence and Hydrodynamic Limit,* IMA volumes (pp. 211–226). Berlin: Springer.

[15] Burgers, J. M. (1974). *The Nonlinear Diffusion Equation.* Dordrecht: Reidel.

[16] Shandarin, S. F., & Zel'dovich, Ya. B. (1989). Turbulence, intermittency, structures in a self-gravitating medium: The large scale structure of the Universe. *Reviews of Modern Physics, 61,* 189.

[17] Oleinik, O. (1957). Discontinuous solutions of nonlinear differential equations. In *Russian Mathematical Surveys* (Vol. 26, pp. 95–172). American Mathematical Society Translations: Series 2.

[18] Lions P. A. (1982). *Generalized Solutions of Hamilton–Jacobi Equation,* Pitman.

[19] Gurbatov, S., Malakhov, A., & Saichev, A. (1991). *Nonlinear Random Waves and Turbulence in Nondispersive Media: Waves, Rays and Particles.* Cambridge: Manchester U Press.

[20] Weinberg, D. H., & Gunn, J. E. (1990). Large scale structure and the adhesion approximation. *Monthly Notices of the Royal Astronomical Society, 247,* 260–286 .

[21] Vergassola, M., Dubrulles, B., Frisch, U., & Noullez, A., (1994). Burgers' equation, devil's staircases and the mass distribution for large–scale structures. *Astronomy and Astrophysics, 189,* 325–356.

[22] Woyczyński, W. A. (1993). Stochastic Burgers' flows. In N. Fitzmaurice et al. (Eds.), *Nonlinear Waves and Weak Turbulence* (pp. 279–311). Boston: Birkhäuser.

[23] Janicki, A., Surgailis, D., & Woyczyński, W. A. (1995). Statistics and geometric-thermodynamics of passive tracer densities in forced 2-D Burgers turbulence, CWRU preprint.

Chapter 20

[1] Saichev A. I., & Woyczyński, W. A. (1997). Probability distributions of passive tracer in randomly moving media. In S. A. Molchanov & W. A. Woyczynski (Eds.), *Stochastic models in geosystems*, IMA Volumes (pp. 359–399). Berlin: Springer.

[2] Avellaneda, M., & Majda, A. J. (1992). Approximate and exact renormalization theories for a model for turbulent transport. *Physics of Fluids, A4*, 41–57.

[3] Majda, A. J. (1993). Explicit inertial range renormalization theory in a model for turbulent diffusion. *Journal of Statistical Physics, 73*, 515–542.

[4] Batchelor, G. K. (1959). Small-scale variation of convected quantities like temperature in turbulent fluid. 1. General discussion and the case of small conductivity. *Journal of Fluid Mechanics, 5*, 113.

[5] Kraichnan, R. H. (1968–70). Small scale structure of scalar field convected by turbulence. *Physics of Fluids, 11*, 945; Diffusion by a random velocity field. *Physics of Fluids 13*, 22.

[6] Csanady, G. T. (1980). *Turbulent Diffusion in the Environment.* Boston: Reidel.

[7] Davis, R. E. (1982). On relating Eulerian and Lagrangian velocity statistics: Single particles in homogeneous flow. *Journal of Fluid Mechanics, 74*, 1–26.

[8] Lipscomb, T. C., Frenkel, A. L., & ter Haar, D. (1970). On the convection of a passive scalar by a turbulent Gaussian velocity field. *Journal of Statistical Physics, 63*, 305–313.

[9] Careta, A., Sagues, F., Ramirez-Piscina, L., & Sancho, J. M. (1993). Effective diffusion in a stochastic velocity field. *Journal of Statistical Physics, 71*, 235–313.

[10] Day, T., Hickey, W., Parakkal, B., & Woyczyński, W. A. (2016). Deepwater horizon: Locating submerged oil in ocean water using the k-means clustering technique. *Chance, 29*(2), 46–52.

[11] Papanicolaou, G. C. (1971). Wave propagation in one-dimensional random medium. *SIAM Journal on Applied Mathematics, 21*, 13–18.

[12] Kesten, H., & Papanicolaou, G. C. (1979). A limit theorem for turbulent diffusion. *Communications in Mathematical Physics, 65*, 97–128.

Chapter 21

[1] Applebaum, D. (2004). *Lévy Processes and Stochastic Calculus.* Cambridge: Cambridge University Press.

[2] Bertoin, J. (1996). *Lévy Processes.* Cambridge: Cambridge University Press.

[3] Samorodnitsky, G., & Taqqu, M. S. (1994). *Stable Non-Gaussian Random Processes: Stochastic Models with Infinite Variance.* London: Chapman and Hall.

[4] Sato, K. -I. (1999). *Lévy Processe and Infinite Divisibility.* Cambridge: Cambridge University Press.

[5] Zolotarev, V. M. (1986). *One-Dimensional Stable Distributions.* Providence: American Mathematical Society.

[6] Rosiński, J. (2007). Tempering stable distributions. *Stochastic Processes and their Applications, 117*(6), 677–707.

[7] Rolski, T., & Woyczyński, W. A. (2017). In memoriam: Czeslaw Ryll-Nardzewski's contributions to probability theory. *Probability Theory and Mathematical Statistics, 37*, 1–20.

[8] Stieltjes, T. J. (1894). Recherches sur les fractions continues. *Annales de la Faculté des Sciences de Toulouse, 8*(1894), 1–122.

[9] Chambers, J. M., Mallows, C. L., & Stuck, B. W. (1976). A method for simulating stable random variables. *Journal of the American Statistical Association, 71*, 340–344.

[10] Woyczyński, W. A. (2001). Levy processes in the physical sciences. In T. Mikosch, O. Barndorff-Nielsen & S. Resnick (Eds.), *Lévy Processes - Theory and Applications* (pp. 241–266). Boston: Birkhäuser.

[11] Holtsmark, J. (1919). Über die Verbreiterung von Spektrallinien. *Annalen der Physik, 58*, 577–630.

[12] Pittel, B., Mann, J. A., & Woyczyński, W. A. (1987). From Gaussian subcritical to Holtsmark (3/2-Levy stable) supercritical asymptotic behavior in "rings-forbidden" Flory-Stockmayer model of polymerization". In *Conference on Graph Theory and Topology in Chemistry* (pp. 362–370). Amsterdam: Elsevier.

[13] Pittel, B., Mann, J. A., & Woyczyński, W. A. (1990). Random tree-type partitions as a model for acyclic polymerization: Holtsmark 3/2 stable distribution of the supercritical gel. *Annals of Probability, 18*, 319–341.

[14] Pittel, B., Mann, J. A., & Woyczyński, W. A. (1990). Random tree-type partitions as a model for acyclic polymerization: Gaussian behavior of the subcritical sol phase. In *Random Graphs '87* (pp. 223–273). New York: Wiley Ltd.

[15] Pittel, B., Mann, J. A., & Woyczyński, W. A. (1992). Correction to "Random tree-type partitions as a model for acyclic polymerization: Holtsmark 3/2 stable distribution of the supercritical gel". *Annals of Probability, 20*, 1105–1106.

[16] Weeks, E. R., Solomon, T. H., Urbach, J. S. & Swinney, H. L. (1995) Observation of anomalous diffusion and Lévy flights. In M. F. Shlesinger et al. (Eds.), *Lévy Flights and Related Topics in Physics* (pp. 51–71). Berlin: Springer.

Chapter 22

[1] Montroll, E. W., & Weiss, G. H. (1965). Random walks on lattices II. *Journal of Mathematical Physics*, *6*, 167–181.

[2] Shlesinger, M. F. (1988). Fractal time in condensed matter. *Annual Review of Physical Chemistry*, *39*, 269.

[3] Saichev, A. I., & Zaslavsky, G. M. (1997). Fractional kinetic equations: Solutions and applications. *Chaos*, *7*, 753.

[4] Metzler, R., & Klafter, J. (2000). The random walk's guide to anomalous diffusion: A fractional dynamics approach. *Physics Reports*, *339*, 1–77.

[5] Scalas, E., Gorenflo, R., & Mainardi, F. (2000). Fractional calculus and continuous-time finance. *Physica A 284*, 376–384.

[6] Mainardi, F., Raberto, M., Gorenflo, R., & Scalas, E. (2000). Fractional calculus and continuous-time finance II: The waiting-time distribution. *Physica A, 287*, 468–481.

[7] Kwapien, S., & Woyczynski, W. A. (1992). *Random Series and Stochastic Integrals: Single and Multiple*. Boston: Birkhäuser.

[8] Bertoin, J. (1996). *Lévy Processes*. Cambridge: Cambridge University Press.

[9] Sato, H. (1999). *Lévy processes and Infinitely Divisible Distributions*. Cambridge: Cambridge University Press.

[10] Piryatinska, A., Saichev, A. I., & Woyczyński, W. A. (2005). Models of anomalous diffusion: the subdiffusive case. *Physica A: Statistical Mechanics and Applications*, *349*, 375–420.

[11] Terdik, Gy., Woyczyński, W. A., & Piryatinska, A. (2006). Fractional- and Integer-Order Moments, and Multiscaling for Smoothly Truncated Lévy Flights. *Physics Letters A 348*, 94–109.

[12] Samko, S. G., Kilbas, A. A., & Marichev, O. I. (1993). *Fractional integrals and derivatives. Theory and applications*. New York: Gordon and Breach Sci. Publishers.

[13] Gorenflo, R., & Mainardi, F. (1997). Fractional calculus: Integral and differential equations of fractional order. In: A. Carpinteri & F. Mainardi (Eds.), *Fractals and fractional calculus in continuum mechanics* (pp. 223–276). Wien: Springer.

[14] Podlubny, I. (1999). *Fractional Differential Equations*. San Diego: Academic Press.

[15] West, B. J., Bologna, M., & Grigolini, P. (2003). *Physics of Fractal Operators*. New York: Springer.

[16] Seshadri, V. (1999). *The Inverse Gaussian Distribution: Statistical Theory and Applications*. Berlin: Springer.

[17] Bochner, S. (1955). *Harmonic Analysis and The Theory of Probability*. Berkeley: University of California Press.

[18] Meerschaert, M. M., Benson, D. A., Scheffler, H. P., & Becker-Kern, P. (2002). Governing equations and solutions of anomalous random walk limits. *Physical Review E, 66*, 060102-1/4 (2002).

[19] Rosiński, J. (2007). Tempering stable processes. *Stochastic Processes and Their Applications, 117*, 677–707.

[20] Terdik, Gy., & Woyczyński, W. A. (2006). Rosiński measures for tempered stable and related Ornstein-Uhlenbeck processes. *Probability and Mathematical Statistics, 26*, 300–327.

[21] Meerschaert, M. M., & Sikorskii, A. (2012). *Stochastic Models for Fractional Calculus*. Berlin: De Gruyter.

[22] Meerschaert, M. M., Nane, E., & Vellaisamy, P. (2011). The fractional Poisson process and the inverse stable subordinator. *Electronic Journal of Probability, 16*(59), 1600–1620.

[23] Leonenko, N. N., Meerschaert, M. M., & Sikorskii, A. (2013). Fractional Pearson diffusions. *Journal of Mathematical Analysis and Applications, 403*, 532–546.

[24] Leonenko, N. N., Meerschaert, M. M., Schilling, R. L., & Sikorskii, A. (2014) Correlation structure of time-changed Lvy processes. *Communications in Applied and Industrial Mathematics, 6*(1), e-483 (22 pp). (Special Issue in Honor of Francesco Mainardi).

[25] Leonenko, N. N., & Merzbach, E. (2015). Fractional Poisson fields. *Methodology and Computing in Applied Probability, 17*, 155–168

Chapter 23

[1] Mann, J. A., & Woyczyński, W. A. (2001). Growing fractal interfaces in the presence of self-similar hopping surface diffusion. *Physica A. Statistical Mechanics and Its Applications, 291*, 159–183.

[2] Biler, P., Funaki, T., & Woyczyński, W. A. (1998). Fractal Burgers equations. *Journal of Differential Equations, 148*, 9–46.

[3] Biler, P., Funaki, T., & Woyczyński, (1999) Interacting particle approximation for nonlocal quadratic evolution problems. *Probability and Mathematical Statistics, 19*, 267–286.

[4] Biler, P., Karch, G., & Woyczyński, W. A. (2001). Critical nonlinearity exponent and self-similar asymptotics for Lévy conservation laws. *Annales d'Institute H. Poincaré- Analyse Nonlineaire (Paris), 18*, 613–637.

[5] Biler, P., Karch, G., & Woyczyński, W. A. (2001). Asymptotics for conservation laws involving Lévy diffusion operators. *Studia Mathematica, 148*, 171–192

[6] Karch, G., & Woyczyński, W. A. (2008). Fractal Hamilton–Jacobi–KPZ equations. *Transactions of the American Mathematical Society, 360*, 2423–2442.

[7] Górska, K., & Woyczyński, W. A. (2015). Explicit representations for multiscale Lévy processes, and asymptotics of multifractal conservation laws. *Journal of Mathematical Physics, 56*, 083511, 1–19.

[8] Kardar, M., Parisi, G., Zhang, Y. -C. (1986). Dynamic scaling of growing interfaces. *Physical Review Letters, 56*(9), 889–892.

[9] Fritz, P., & Hairer, M. (2014). *A Course on Rough Paths. With an Introduction to Regularity Structures.* Berlin: Springer.

[10] Woyczyński, W. A. (1998). *Burgers-KPZ Turbulence: Göttingen Lectures.* Berlin: Springer.

[11] Barabási, A. -L., & Stanley, H. E. (1995). *Fractal Concepts in Surface Growth.* Oxford: Oxford University Press.

[12] Molchanov, S. A., Surgailis D., & Woyczyński W. A. (1995). Hyperbolic asymptotics in Burgers turbulence. *Communications in Mathematical Physics, 168*, 209–226.

[13] Senft, D. C., & Ehrlich, G. (1995). Long jumps in surface diffusion: One-dimensional migration of isolated adatoms. *Physical Review Letters, 74*(2), 294–297.

[14] Linderoth T.R., Horch S., Laegsgaard E., Stensgaard I., & Besenbacher F. (1997). Surface diffusion of Pt on Pt(110): Arrhenius behavior of long jumps. *Physical Review Letters, 78*(26), 4978–4981.

[15] Tully, J. C., Gilmer, G. H., & Shugard, M. (1979). Molecular dynamics of surface diffusion. I. The motion of adatoms and clusters. *Journal of Chemical Physics, 71*, 1630.

[16] Mruzik, M. R., & Pound, G. M. (1981). A molecular dynamics study of surface self-diffusion. *Journal of Physics F, 11*, 1403.

[17] DeLoranzi, G., & Jacucci, G. (1985). The migration of point defects on BCC surfaces using a metallic pair potential. *Surface Science, 164*, 526.

[18] Sanders, D. E., & DePristo, A. E. (1992). A non-unique relationship between potential energy surface barrier and dynamical diffusion barrier: fcc(111) metal surface. *Surface Science, 264*, L169.

[19] Beenakker, J. J. M., & Krylov, S. Tu. (1998). Jump length distribution in molecule-on-substrate diffusion. *Surface Science, 411*, L816–L821.

[20] Biler, P., Karch, G., & Woyczyński, W. A. (1999). Asymptotics of multifractal conservation laws. *Studia Mathematica, 135*, 231–252.

[21] Gunaratnam, B., & Woyczyński, W. A. (2015). Multiscale conservation laws driven by Lévy stable and Linnik diffusions: Asymptotics, shock creation, preservation and dissolution. *Journal of Statistical Physics, 56*, 1–33.

[22] Richie, K., Shan, X-Y., Kondo, J., Iwasawa, K., Fujiwara, T., & Kusumi, A. (2005). Detection of non-Brownian diffusion in the cell membrane in single molecule tracking. *Biophysical Journal 88*, 2266–2277.

[23] Kusumi, A., Sako, Y., & Yamamoto, M. (1993). Confined lateral diffusion of membrane receptors as studied by single particle tracking (nanovid microscopy). Effects of calcium-induced differentiation in cultured epithelial cells. *Biophysics Journal, 65,* 2021–2040.

[24] Smith, P. R., Morrison, I. E. G., Wilson, K. M., Fernandez, N., & Cherry, R. J. (1999). Anomalous diffusion of major histocompatibility complex class I molecules on HeLa cells determined by single particle tracking. *Biophysics Journal, 76,* 3331–3344.

[25] Fujiwara, T., Ritchie, K., Murakoshi, H., Jacobson, K., & Kusumi, A. (2002). Phospholipids undergo hop diffusion in compartmentalized cell membrane. *Journal of Cell Biology, 157,* 1071–1081.

[26] Golding, I., & Cox, E. C. (2006). Physical nature of bacterial cytoplasm. *Physical Review Letters, 96,* 98–102.

[27] Zeldovich, Y. B., Einasto, J., & Shandarin, S. F. (1982). Giant voids in the Universe. *Nature, 300,* 407–413.

[28] Peebles, P. J. E. (1980). The large scale structure of the universe. Princeton: Princeton University Press.

[29] Kofman, L., & Raga, A. C. (1992). Modeling structures of knots in jet flows with the Burgers equation. *Astrophysical Journal,* 390, 359–364 (1992).

[30] Molchanov, S. A., Surgailis, D., & Woyczyński, W. A. (1997). The large-scale structure of the universe and quasi-Voronoi tesselation of shock fronts in forces Burgers turbulence in $\mathbf{R}^d$. *Annals of Applied Probability, 7,* 200–228.

[31] Slobodrian, R. J. (2005). Fractal cosmogony: Similarity of the early universe to microscopic fractal aggregates. *Chaos, Solitons and Fractals, 23,* 727–729.

[32] Joyce, M., Gabrielli, A., & Labini, F. S. (2005). Basic properties of galaxy clustering in the light of recent results from the Sloan digital sky survey. *Astronomy and Astrophysics: A European Journal, 443,* 11–16.

[33] Baryshev, Y., & Teerikorpi, P. (2005). Fractal approach to Large Scale Galaxy Distribution. arXiv:astro-ph/0505185v1.

[34] Shlesinger, M. F., Zaslavsky, G. F., & Frisch, U. (Eds.), (1995). *Lévy Flights and Related Topics in Physics.* Berlin: Springer.

[35] Bertoin, J. (1996). *Lévy Processes.* Cambridge: Cambridge University Press.

[36] Funaki, T., & Woyczyński, W. A. (1998). Interacting particle approximation for fractal Burgers equation. In I. Karatzas, B. S. Rajput & M. S. Taqqu (Eds.), *Stochastic Processes and Related Topics: In Memory of Stamatis Cambanis 1943–1995* (pp. 141–166). Boston: Birkhäuser.

[37] Biler, P., Funaki, T., & Woyczyński, W. A. (1999). Interacting particle approximation for nonlocal quadratic evolution problems. *Probability and Mathematical Statistics, 19,* 267–286.

[38] Prudnikov, A. P., Brychkov, Yu. A., & Marichev, O. I. (1992). *Integrals and Series: More Special Functions* (Vol. 3). Amsterdam: Gordon and Breach.

[39] Lukacs, E. (1970). *Characteristic Functions.* Griffith, London.

[40] Feller, W. (1970). *An Introduction to Probability Theory and its Applications* (Vol. 2). New York: Wiley.

[41] Górska, K., & Penson, K. A. (2011). Lévy stable two-sided distributions: Exact and explicit densities for asymmetric case. *Physical Review E, 83*, 061125.

Index

© Springer International Publishing AG, part of Springer Nature 2018
A. I. Saichev and W. A. Woyczynski, *Distributions in the Physical and Engineering Sciences, Volume 3*, Applied and Numerical Harmonic Analysis, https://doi.org/10.1007/978-3-319-92586-8

Printed in the United States
By Bookmasters